Corrosion control in the aerospace industry

Related titles

Aircraft system safety: military and civil aeronautical applications
(ISBN 978-1-84569-136-3)
Demonstrating safety for the application of ever more complex technologies is a formidable task. System engineers often do not have the appropriate training, are unfamiliar with the range of safety approaches, tools and techniques, and their managers do not know when and how these may be applied and appropriately resourced. *Aircraft system safety* provides a basic skill set for designers, safety practitioners, and their managers by exploring the relationship between safety, legal liability and regulatory requirements. Different approaches to measuring safety are discussed, along with the appropriate safety criteria used in judging acceptability. Engineering students, designers, safety assessors (and their managers), regulatory authorities (especially military), customers and projects teams should find *Aircraft system safety* provides an invaluable guide in appreciating the context, value and limitations of the various safety approaches used in cost-effectively accomplishing safety objectives.

Techniques for corrosion monitoring
(ISBN 978-1-84569-187-5)
Corrosion monitoring technologies are a vital step in efforts to combat corrosion, which can have major economic and safety implications. The first part of the book reviews electrochemical techniques for corrosion monitoring. A second group of chapters analyses other physical or chemical methods of corrosion monitoring. Part III examines corrosion monitoring in special environments and conditions. Section IV covers the selection of monitoring techniques and probes and a final group of chapters discusses applications and case studies.

The air transport system
(ISBN 978-1-84569-325-1)
Major operational elements of the world air transport system are examined in this important book, which provides an invaluable single information source to managers in all air transport sectors. Initial chapters consider route structure options in terms of operational impacts and describe the context and boundaries of the industry – the natural, regulatory and operational environments. These introduce 'systems' perspectives that integrate the content of further chapters, examining aircraft, airlines, airports and airspace issues. Finally, the challenge of ensuring symbiosis of all elements of the changing scene and the scope for developing strategies that will suit a balance of all stakeholder requirements are considered in depth.

Details of these and other Woodhead Publishing books can be obtained by:

- visiting our web site at www.woodheadpublishing.com
- contacting Customer Services (e-mail: sales@woodheadpublishing.com; fax: +44 (0) 1223 893694; tel.: +44 (0) 1223 891358 ext. 130; address: Woodhead Publishing Limited, Abington Hall, Granta Park, Great Abington, Cambridge CB21 6AH, England)

If you would like to receive information on forthcoming titles, please send your address details to: Francis Dodds (address, tel. and fax as above; e-mail: francis.dodds@woodheadpublishing.com). Please confirm which subject areas you are interested in.

Corrosion control in the aerospace industry

Edited by
Samuel Benavides

CRC Press
Boca Raton Boston New York Washington, DC

WOODHEAD PUBLISHING LIMITED
Cambridge New Delhi

Published by Woodhead Publishing Limited, Abington Hall, Granta Park, Great Abington, Cambridge CB21 6AH, England
www.woodheadpublishing.com

Woodhead Publishing India Pvt Ltd, G-2, Vardaan House, 7/28 Ansari Road, Daryaganj, New Delhi – 110002, India

Published in North America by CRC Press LLC, 6000 Broken Sound Parkway, NW, Suite 300, Boca Raton, FL 33487, USA

First published 2009, Woodhead Publishing Limited and CRC Press LLC

British Library Cataloguing in Publication Data
A catalogue record for this book is available from the British Library.

Library of Congress Cataloging in Publication Data
A catalog record for this book is available from the Library of Congress.

Woodhead Publishing ISBN 978-1-84569-345-9 (book)
Woodhead Publishing ISBN 978-1-84569-553-8 (e-book)
CRC Press ISBN 978-1-4200-7965-4
CRC Press order number: WP7965

The publishers' policy is to use permanent paper from mills that operate a sustainable forestry policy, and which has been manufactured from pulp which is processed using acid-free and elemental chlorine-free practices. Furthermore, the publishers ensure that the text paper and cover board used have met acceptable environmental accreditation standards.

Typeset by SNP Best-set Typesetter Ltd., Hong Kong
Printed by TJ International Limited, Padstow, Cornwall, England

Contents

	Contributor contact details	*x*
1	Corrosion in the aerospace industry S. Benavides, US Coast Guard Aging Aircraft Branch, USA	1
1.1	Introduction	1
1.2	Aerospace corrosion	2
1.3	Impact of corrosion	4
1.4	Corrosion prediction	7
1.5	Conclusion	13
1.6	Sources of further information and advice	13
1.7	References	13
Part I	**Corrosion fundamentals and cost of corrosion**	**15**
2	Assessing the cost of corrosion to the aerospace industry E. Herzberg, LMI Research Fellow, USA	17
2.1	Introduction	17
2.2	Corrosion impacts	18
2.3	Corrosion cost elements and characterizations	19
2.4	Corrosion cost measurement methodology	23
2.5	General case studies	31
2.6	Conclusion	33
2.7	References	33
2.8	Appendix A: Cost element definitions	33
2.9	Appendix B: Typical corrosion activities	34

3 Corrosion and the threat to aircraft structural integrity 35
T. Mills, S. Prost-Domasky, K. Honeycutt and C. Brooks, Analytical Processes / Engineered Solutions, Inc. (APES), USA
3.1 Introduction 35
3.2 Types of corrosion and their impact on aircraft structure 38
3.3 Structurally significant and unique corrosion forms 39
3.4 Other forms of corrosion attack 42
3.5 Insidious synergisms 43
3.6 Design paradigms and corrosion structural effects 47
3.7 Damage tolerance 51
3.8 Holistic structural integrity 55
3.9 Conclusions 62
3.10 References 63

4 Effect of corrosion on the mechanical behaviour of aircraft aluminum alloys 67
S. G. Pantelakis, University of Patras, Greece; and A. T. Kermanidis, University of Thessaly, Greece
4.1 Introduction 67
4.2 Corrosion behaviour of aircraft aluminum alloys 68
4.3 Effect of corrosion on the mechanical behaviour of aircraft aluminum alloys 70
4.4 Modeling the corrosion effect on damage tolerance characteristics 97
4.5 Conclusions 105
4.6 References 105

Part II Corrosion monitoring, evaluation and prediction 109

5 Nondestructive testing of corrosion in the aerospace industry 111
D. S. Forsyth, TRI/Austin, USA
5.1 Introduction to nondestructive testing 111
5.2 Data fusion for nondestructive testing 115
5.3 Reliability of nondestructive testing for corrosion 117
5.4 Typical applications of nondestructive testing to corrosion in aerospace systems 119
5.5 Summary and conclusions 126
5.6 References 128

6	**Corrosion prediction in the aerospace industry**	**131**
	J. Ullett, S&K Technologies, USA	
6.1	Introduction	131
6.2	Material and environmental influences	133
6.3	Data generation and correlation	135
6.4	Model development and implementation	145
6.5	Summary	148
6.6	Future trends	148
6.7	Sources of further information and advice	149
6.8	References	149
7	**Integrated health and corrosion monitoring systems in the aerospace industry**	**151**
	A.-D. Nguyen, Los Gatos Research, USA; and V. Godinez, Physical Acoustic Corporation, USA	
7.1	Introduction	151
7.2	Acoustic emission	153
7.3	Ultrasonic guided waves (or Lamb waves)	156
7.4	Strain monitoring	159
7.5	Active corrosion detection	161
7.6	Integrated strain, temperature, and stress wave monitoring sensor	162
7.7	Crack detection	164
7.8	Corrosion monitoring	168
7.9	Conclusions	169
7.10	Acknowledgements	170
7.11	References	170
8	**Corrosion and fatigue modeling of aircraft structures**	**172**
	N. C. Bellinger and M. Liao, Institute for Aerospace Research, Canada	
8.1	Introduction	172
8.2	Corrosion pitting and fatigue modeling	175
8.3	Exfoliation corrosion and modeling	180
8.4	Residual fatigue life analysis	184
8.5	Risk assessment of corrosion maintenance actions	184
8.6	Conclusions	189
8.7	References	190

Part III Corrosion protection and prevention 193

9 Corrosion control in space launch vehicles 195
L. M. Calle, NASA, Kennedy Space Center, USA
9.1 Introduction 195
9.2 Space launch vehicles environment 196
9.3 Materials selection 200
9.4 Corrosion control 203
9.5 Corrosion control in the Space Shuttle Orbiter 204
9.6 Corrosion control in the launch pad and ground support equipment 212
9.7 Corrosion control and treatment program 218
9.8 Space Shuttle Orbiter corrosion history 219
9.9 Summary 220
9.10 Future trends 220
9.11 Sources of further information 221
9.12 References 222

10 Coating removal techniques in the aerospace industry 225
D. L. Monette, Canada
10.1 Introduction 225
10.2 Background 226
10.3 When to remove the coating 228
10.4 A brief history of coating removal 228
10.5 How to remove coatings 229
10.6 Chemical strippers 231
10.7 Thermal coating removal methods 235
10.8 High-pressure water 236
10.9 Dry stripping 237
10.10 Dry stripping media types 238
10.11 Mechanical effects of coating removal by impact methods 242
10.12 Engineered bio-based media 244
10.13 Future trends 246
10.14 Conclusions 247
10.15 References 247

11 Novel corrosion schemes for the aerospace industry 248
F. Gui, CC Technologies – a DNV Company, USA
11.1 Introduction 248
11.2 Corrosion prevention compounds 249
11.3 Novel inhibitors and coatings 252
11.4 Novel surface treatment 259
11.5 References 263

12 Greases and their role in corrosion control in the aerospace industry 266
K. D. Akin, Nye Lubricants, Inc., USA
12.1 Introduction 266
12.2 Grease composition 267
12.3 Corrosion-inhibiting additives 274
12.4 Lubricant selection and the design cycle 274
12.5 Lubricant testing 276
12.6 Grease manufacturing 282
12.7 Future trends 284
12.8 Sources of further information and advice 285
12.9 References 286

13 Business strategies for corrosion control in fleet maintenance 288
M. W. Schleider, Mercer University, USA
13.1 Introduction 288
13.2 Acquisition requirements 289
13.3 Sustainment requirements 290
13.4 Training 291
13.5 Corrosion prevention and control 292
13.6 Corrosion tracking 293
13.7 Costs tracking 295
13.8 Communications 296
13.9 Sources of further information and advice 297
13.10 References 297

Index *299*

Contributor contact details

(* = main contact)

Editor, Chapter 1

Samuel Benavides
Senior Materials Engineer
US Coast Guard
Aircraft Repair and Supply Center (ARSC)
Aging Aircraft Branch
BLDG 78
Weeksville Road
Elizabeth City, NC 27909
USA
E-mail: Samuel.Benavides@uscg.mil

Chapter 2

Eric Herzberg
LMI Research Fellow
8109 Shoal Creek Drive
Laurel
Maryland 20724
USA
E-mail: eherzberg@lmi.org

Chapter 3

Dr T Mills*, Dr S Prost-Domasky, K Honeycutt and C Brooks
Analytical Processes / Engineered Solutions, Inc. (APES)
6669 Fyler Avenue
Saint Louis
MO 63139
USA
E-mail: tmills@apesolutions.com

Chapter 4

Prof Spiros G Pantelakis*
Laboratory of Technology and Strength of Materials (LTSM)
Department of Mechanical Engineering and Aeronautics
University of Patras
Panepistimioupolis Rion 26500
Patras
Greece
E-mail: pantelak@mech.upatras.gr

Dr Alexis T Kermanidis
Laboratory of Mechanics and Strength of Materials
Department of Mechanical and Industrial Engineering
University of Thessaly
Leoforos Athinon
Pedion Areos 38334
Volos
Greece
E-mail: akermanidis@mie.uth.gr

Chapter 5

David S Forsyth
NDE Division Manager
TRI/Austin
9063 Bee Caves Road
Austin
TX 78733
USA
E-mail: dforsyth@tri-austin.com

Chapter 6

Dr Jill Ullett
S&K Technologies
3139 Research Blvd, Suite 101
Dayton, OH 45420
USA
E-mail: jullett@sktdayton.com

Chapter 7

Dr An-Dien Nguyen*
Los Gatos Research
Mountain View
CA 94041
USA
E-mail: a.d.nguyen@lgrinc.com

Dr Valery Godinez
Physical Acoustic Corporation
NJ 08550
USA

Chapter 8

N C Bellinger* and M Liao
Institute for Aerospace Research
National Research Council Canada
1200 Montreal Road
Ottawa
Ontario K1A 0R6
Canada
E-mail: Nick.Bellinger@nrc-cnrc.gc.ca; Min.Liao@nrc-cnrc.gc.ca

Chapter 9

Dr Luz Marina Calle
NASA, Mail Code KT-E-3
Kennedy Space Center
FL 32899
USA
E-mail: Luz.M.Calle@nasa.gov

Chapter 10

Mr Denis L Monette
Consultant
C/O 995 Mill Street
Montreal
QC H3C 1Y5
Canada
E-mail: denislmonette@sympatico.ca

Chapter 11

Dr Feng Gui*
CC Technologies – a DNV company
Dublin
OH 43017
USA
E-mail: Feng.Gui@dnv.com

Chapter 12

Kevin D Akin
Nye Lubricants, Inc.
12 Howland Road
Fairhaven
MA 02719
USA
E-mail: kakin@nyelubricants.com

Chapter 13

M W Schleider, PE
Mercer Engineering Research Center
Mercer University
135 Osigian Blvd
Warner Robins
GA 31088
USA
E-mail: MSchleider@merc-mercer.org

1 Corrosion in the aerospace industry

S. BENAVIDES, US Coast Guard Aging
Aircraft Branch, USA

Abstract: This chapter serves as an introduction to the rest of the book, which is a conduit for moving beyond the basics of standard corrosion principles to a real-world perspective of aerospace corrosion, prediction and prevention. A holistic approach to corrosion sustainment processes is presented and alternative methods to modeling corrosion growth are discussed. Experts provide insight into cutting-edge knowledge of the complex variables that combine to influence the corrosion that poses such a threat to the aging fleets of aircraft. The importance of accurate corrosion prediction models, health monitoring systems, novel inspection methods for aircraft maintenance, and cradle-to-grave corrosion management in order to continue to fly safely is stressed.

Key words: corrosion modeling, aerospace industry, aircraft maintenance.

1.1 Introduction

Commercial and military aircraft fleets are aging. Aircraft designers and manufacturers never suspected many of their air vehicles would be flying beyond their designed life (Fig. 1.1). As aircraft become older, they are subject to the insidious time-dependent effects of corrosion. Forgotten, ignored and often deferred, corrosion eventually threatens the integrity of aircraft structure. Traditional approaches to understanding the effects of aircraft corrosion have relied on the classic textbook models of general corrosion. However, the multifarious variables involved in aircraft corrosion are convoluted, complex and interacting. This chapter shows how this book is a conduit for moving beyond the basics of standard corrosion principles to a real-world perspective of aerospace corrosion, prediction and prevention.

An understanding of aerospace corrosion should account for the disparate confounding variables that influence corrosion. These variables include the interacting material properties that work in conjunction with one another such as fatigue strength and load carrying capability, both of which are compromised in the presence of corrosion. Coating protection systems designed to prevent corrosion can also vary. A fleet of aircraft may have an array of surface conversion coatings, primers, topcoats and corrosion

1.1 A Dassault Falcon jet undergoing corrosion inspection and repair. Manufactured in 1983 and designed as a corporate business jet, this aircraft flies low-altitude maritime search-and-rescue missions in a corrosive marine environment. The aircraft is scheduled to retire in 2021.

preventative compounds, some of which may be traditional chromate-based while others are environmentally friendly. Of course, as long as the protection system functions, corrosion will not occur. However, an exceptionally unknown variable is the time at which corrosion begins. Determining the onset of corrosion has implications on the corrosion growth-rate, structural viability, inspection intervals, repair schedules and the remaining life of the structure. Finally, aircraft move from one location to another changing the environmental severity that effects corrosion. Variables such as humidity, sheltering, chloride levels, pollutants and distance from water also influence aircraft corrosion (Fig. 1.2).

1.2 Aerospace corrosion

Corrosion compromises the properties of aircraft structure carrying people and cargo. In the event of a failure due to corrosion, the consequences can be catastrophic. This was brought to light by Aloha Airline's Flight 243 that suffered an explosive decompression in 1988 as a result of metal fatigue exacerbated by corrosion. This watershed event, in which an airline attendant was swept from the cabin at 24 000 feet, established US civil and military programs of aging aircraft of which corrosion was a principal area of focus.[1] Although corrosion inspections on the Aloha Airlines class of 737s were proposed by a Boeing alert service bulletin, no one predicted the apocalyptic damage from accelerated corrosion that would result from operating in the severe Hawaiian environment.[2] Since this mishap, the avia-

1.2 Corroded aircraft skin resulting from the failure of the coating protection system. The onset of corrosion occurred when the primer and surface conversion coating no longer prevented electrolytes from reacting on the metallic substrate.

tion community has struggled to accurately predict the onset of corrosion or the extent of structural damage it induces. Despite advances in corrosion algorithms, computation material research, simulation, reliability and maintenance data analysis, a reliable model for predicting corrosion on aging aircraft has yet to be produced. This book presents holistic approaches to corrosion sustainment processes and discusses alternative methods to modeling corrosion growth.

1.2.1 Corrosion and structural integrity

In the preliminary design of an aircraft, requirements such as weight, aerodynamics, fuel efficiency and other specifications are determined. Unfortunately, not all the requirements are attainable and compromises to the design and production of an aircraft result. Historically, corrosion prevention has not been appreciably designed into an aircraft. Dissimilar metals, lack of drain paths and poor corrosion resistant lavatory spill containment designs have been the source of corrosion on many aircraft.

Corrosion can compromise the designed factor-of-safety built into the ultimate load carrying capabilities of the structure. Variations in the physical properties of materials, aircraft fabrication, flight profiles and emergency conditions necessitate a reserve factor-of-safety for an aircraft. Nonetheless, corrosion can affect the load-carrying capability of structure in several ways. First, it reduces the cross-sectional area of a structure, thereby creating a scenario where less material must carry more load per area. Second, corrosion – especially pits – can become a site for cracks. This is especially disconcerting if multiple small corrosion initiated cracks join to create a long

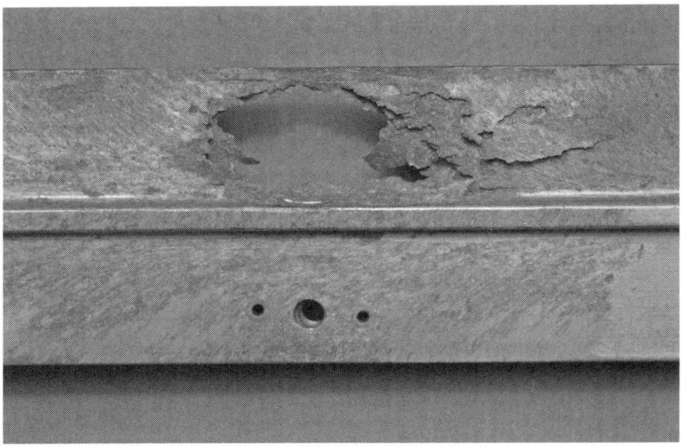

1.3 Corroded aircraft structure. The ability for the stringer to carry and transfer its load has been compromised. Corrosion started at the dissimilar interface between the aluminum structure and the steel rivets used to attach the nutplate. A bolt hole and smaller rivet holes can be seen on the lower section of the stringer.

crack. Finally, repair of a corroded structure usually entails removal of the corroded material and some of the surrounding structure. If the corrosion is light or moderate, enough of the structure remains to safely carry its load. Eventually, continued repairing and removal of corroded material will exceed allowable repair limits and the structure will need to be replaced (Fig. 1.3).

1.3 Impact of corrosion

Corrosion impacts safety, cost and aircraft availability. Although fatalities and injuries that result from corrosion-related mishaps are well documented, reliable cost-of-corrosion studies are only now emerging. To quantify corrosion improvements and establish sound corrosion prevention strategies requires effective fiscal metrics to measure the cost of corrosion.

1.3.1 Corrosion of military aircraft

Corrosion, a $20 billion annual expense for the US Department of Defense (DoD), affects the military service's 15 000 aircraft and helicopters. Corrosion involves 700 000 military and DoD civilian personnel in addition to several thousand commercial firms.[3] Corrosion impacts aircraft readiness, availability, armament, hangar facilities and, most importantly, the safety of the aircraft.

The US Government Accounting Office (GOA) cited the DoD as lacking reliable corrosion data to develop an effective prevention and mitigation strategy.[4] While the GOA recognizes that the Department of Defense operates in high-salt and wet environments that accelerate corrosion, the DoD has nevertheless lacked reliable data, mechanisms or methodologies to accurately quantify corrosion. The databases that exist are often incomplete as not all corrosion-related damage is reported.[5] An examination of the effectiveness of DoD corrosion programs finally became law when the US Congress enacted the National Defense Authorization Act of Fiscal Year 2006. In chapter two, a methodology is introduced to capture corrosion costs in the DoD. It defines corrosion costs and characterizes this expense using a novel 'top-down/bottom-up' approach that analyzes maintenance budgets from a high organizational perspective and correlates it to corrosion maintenance labor, parts and material supplies from existing data.

Understanding the impact of corrosion is a primary concern to the US Air Force where the average age of its fleet is 23.6 years. Many of these aircraft suffer from corrosion and other age-related problems to such a degree that policy decisions regarding repair or replacement of a fleet are often debated. The corrosion on engine struts is one example of the military debate that led to the grounding of twenty-nine KC-135 refueling-tankers. The decision to repair or replace these air assets still continues in the Air Force and US Congress. Another example is the nearly 50-year-old B-52 bombers that are scheduled to fly until 2040. Finally, there is the cost of corrosion. In 1991, the Air Force estimated corrosion to cost $700 million; in 2001, the cost had increased to over $1 billion.[6]

In support of Air Force studies to further understand the effects of corrosion, Abbott and Kinzie of the Battelle Institute determined the effects of local corrosion rates on a number of airframes including C-141, C-130, F-15, F-16 and KC-135.[7] From this study, an algorithm was developed to predict local geographic corrosion rates:

$$c = Af_1(m)f_2(T)f_3(C) + Bf_4(r)f_2(T)f_5(C) \qquad [1.1]$$

The resulting corrosion growth rates provided material thinning data to AFGROW, the Air Force's structural analysis tool. AFGROW is a crack-analysis software tool that predicts the crack growth on aircraft structure and uses corrosion data to determine the compounding affects of corrosion on the crack.[8]

Other Air Force corrosion models include that developed by Ullett, who correlated corrosion rate data from accelerated laboratory testing, outdoor exposures and analysis of real aircraft repair data.[9] In chapter six, Dr Ullett summarizes factors that influence corrosion rates and reviews the legacy of Air Force corrosion repair data in developing a statistical approach for variation in parameters.

The US Army maintains the largest fleet of aircraft in the US Department of Defense. The average age of these aviation assets continues to increase along with increasing corrosion discrepancies. Major General James Pillsbury, Commander of the United States Army Aviation Command, recently expressed his concern of the challenges to his Army's aging fleet. In a keynote speech, the General shared his concern on the adverse effects of corrosion on aging aircraft and the effect of the Army's ability to fulfill its missions. One solution the General offered was the use of Condition Based Maintenance in which data is used to optimize, predict and schedule maintenance resources. A key enabler for the success of Condition Based Maintenance is the ability to 'predict remaining component life'.[10] This will necessitate an Army predictive model that incorporates real maintenance data to determine the remaining life of corroded aircraft structure.

The US Navy's 3880 aircraft, averaging 18 years in age, are currently the oldest aviation fleet in its history. Like the other armed services, the Navy faces corrosion and age-related challenges in extending the life span of many of its aircraft into the middle of the 21st century. An example is the in-flight refueling tankers and the maritime surveillance aircraft that share an average age of 29 years.[11] Keeping these older aircraft flying is becoming increasingly more costly. Recently, the Navy spent $408.6 million to upgrade rotor and engine drive trains on CH-46 helicopters. To address these issues, the Navy is focusing on collecting and analyzing data, including corrosion data, to forecast incipient component failures.

The US Coast Guard's fleet of aging aircraft operates in extremely corrosive environments. Many of these aircraft are flying beyond their designed life where corrosion has become not only a structural integrity challenge but a major cost of operation. In order to continue meeting the challenges placed upon aviation assets, it has become imperative that the Coast Guard effectively manage corrosion of its aviation assets to enhance operational availability, extend service life, reduce cost and ensure safety.

1.3.2 Corrosion of civilian transport aircraft

The civilian commercial airline industry is not immune to the effects of corrosion where the average cost of corrosion can be 10% of total aircraft maintenance costs. In 1996, the US annual cost of corrosion was estimated to be $2.225 billion.[12] The commercial industry's databases of age information suggest that maintenance costs increase with age. Boeing maintains a maturity curve to describe corrosion cost as a function of age.[13] The following empirical equation is used by Boeing to calculate annual corrosion maintenance cost:

$$\text{Corrosion maintenance cost} = R + NR + P\&C \qquad [1.2]$$

Table 1.1 Airline annual cost of corrosion

Corrosion category	Cost, $ billions
Corrosion maintenance	1.7
Downtime due to corrosion	0.3
Design and manufacturing for corrosion	0.225
Total	2.225

where R is routine maintenance, NR is non-routine maintenance and $P\&C$ are parts and consumable costs. Given a technician's hourly rate as HR, the values for a 747-B airplane are: routine maintenance (4500 h × HR); non-routine maintenance (3000 h × HR); and parts and consumables $4500.[12] The annual cost of corrosion is given in Table 1.1.[12]

Corrosion greatly affects the first generation of jet transports that were designed to a fail-safe strength criterion with little or no attention incorporated into corrosion protection. These aircraft include early models such as the 747-B, 737-B, DC-10 and DC-8, to name a few. In the 1970s and 1980s, damage tolerance was incorporated into the design of the second generation of commercial passenger aircraft. By then, it was realized that corrosion in aircraft was becoming an economic burden and could possibly become detrimental to the structural integrity of the airplane. Accordingly, the US Federal Aviation Administration issued an Airworthiness Directive related to corrosion control in design and maintenance. In their design, third generation civilian transport aircraft incorporate significant improvements in corrosion prevention.

1.4 Corrosion prediction

The ability to accurately predict corrosion will become more critical as aircraft continue to age and strategic life-cycle management of air assets is incorporated. Advance analytical methods for predicting when and where corrosion will initiate and how quickly corrosion will manifest are needed for structural life assessment and condition-based maintenance. For accurate prediction models that approach realism, empirical and deterministic approaches must work with algorithms derived from real-world data.

1.4.1 Traditional modeling

Traditional corrosion modeling of aviation structures has been based on electrochemical mechanisms that drive materials to a lower energy state – that is, the material corrodes. This classical representation is strictly physical. Countless laboratory and field experiments have strived to produce empirical functions to account for the complex physical phenomena that occurs as a result of corrosion. Equation 1.3 shows the General Corrosion Equation. Many models include interacting variables to capture elements of realism, such as that of Miyata *et al.*, whose extended function (equation 1.4) includes corrosion as a result of dew, rain, relative humidity, temperature and sea salt particles.[14]

$$c = At^{B} \quad [1.3]$$

$$c = Af_1(m)f_2(T)f_3(C) + Bf_4(r)f_2(T)f_5(C) \quad [1.4]$$

Models for various types of corrosion growth mechanisms in aluminum aircraft structure can be found. They include models for pitting, crevice corrosion, intergranular corrosion and filiform corrosion. Despite advances in corrosion algorithms, computation material research, simulation, reliability and maintenance data analysis, a reliable model for predicting the onset and damage of corrosion on aging aircraft has yet to be produced. Often, these models are based on simulated laboratory tests or a specific type of corrosion in a particular geographic location. Generally, these models cannot predict the level of corrosion to be found on actual aircraft.

1.4.2 Survival analysis

A novel application of an advance probability tool that allows for interacting variables and the use of censored data is the use of survival analysis for corrosion modeling. The US Coast Guard is exploring survival analysis as a statistical method to analyze the probability of failure as a result of structural corrosion. This technique accounts for the disparate confounding variables that influence corrosion and develops time-to-failure corrosion rate functions for specific structural components. Using historic corrosion data, which is censored and truncated, survival analysis provides failure probability plots for aircraft assigned to varying environmentally severe geographical locations.

Survival analysis is a biostatistical branch of reliability that involves modeling data until an event occurs. It is a well-grounded statistical technique used primarily in biomedical research. Survival analysis models time-to-event data with the event usually associated with death, the onset of a disease, relapse or recovery. For engineering analysis an 'event' can be defined to be a failure, such as an individual component or system failure. For example, survival analysis has been used to model electrical overhead distribution systems and municipal water infrastructure water pipe failures. The Coast Guard is extending the application of survival analysis to the failure of aircraft structure due to corrosion (Table 1.2).

Table 1.2 Application of survival analysis to corrosion

Event	Time
The onset of corrosion on aircraft	Time to onset, months
Failure due to corrosion	Time to failure, months
Aircraft life	Time until retirement due to corrosion, years
Repaired structure	Time until re-occurrence, weeks
Avionic failures due to corrosion	Time to fail, hours

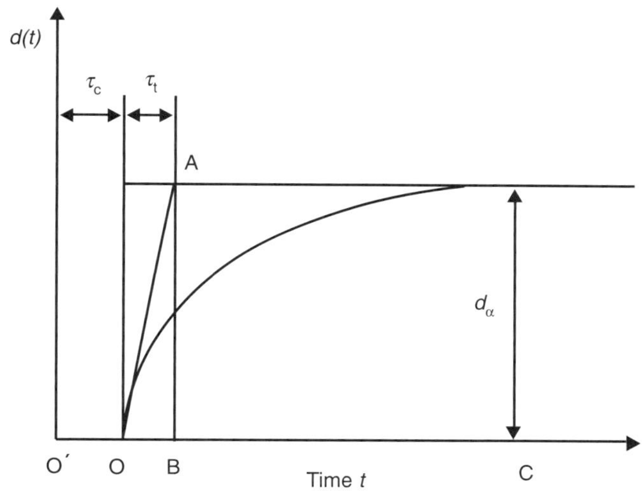

1.4 Guedes Soares, Garbatov 3-phase corrosion model. While corrosion protection systems are functioning, no corrosion occurs in phase τ_c.

Censored corrosion data

Guedes *et al.* proposed a three-phase model to describe the various stages of corrosion.[15] In the application of this model to aircraft structure, the first phase, τ_c, is defined by the endurance of the protection systems to prevent the initiation of corrosion. In other words, for a period of time, there is no corrosion in the aircraft (Fig. 1.4).[16] In the second phase, these protective systems eventually wear and the corrosion of airframe structure begins. Its discrete time is unknown during the second phase. Corrosion initiates and proceeds at a non-linear rate, eventually reaching a steady asymptotic rate in phase three. In aircraft, the existence of corrosion is likely to be discovered by maintenance personnel only after a significant time has passed since

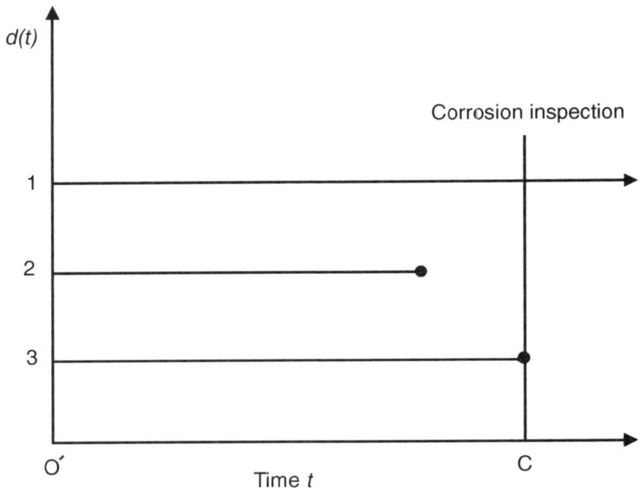

1.5 Censored Data. Time $t = C$ could represent the time at which an aircraft undergoes a corrosion inspection. Line 1 represents right censored data where failure due to corrosion has not yet occurred. Line 2 represents structural failure at some time prior to $t = C$, but the exact time is unknown. Line 3 represents corrosion failure at exactly $t = C$.

the corrosion initiated. As in any reliability study, the exact time the failure occurs is desired, but in survival analysis it is not necessary. Survival analysis handles these unknown occurrences through censoring. Unlike other reliability models, the ability to work with censored data is one of the strong benefits of survival analysis.

Right-censoring occurs when the component has not experienced the event at the end of the observation period (Fig. 1.5). For example, it is noted during aircraft depot maintenance that a specific structural component has not failed due to corrosion. Given enough time, T, the structure would eventually succumb to corrosion, but at the time of maintenance, t, the condition observed is that the component has survived – that is, not failed. When the true failure time T is greater then the observed time t, $T > t$, the data point is classified as a right-censored data point. A left-censored data point would occur when the true survival time is less than or equal to the time at which the observation is made. Other censoring such as left and right truncation can also be handled by survival analysis.

The survival function

The cornerstone of survival analysis is the time-to-event (failure) phenomena described by the survivor function, $S(t)$. The survivor function describes the probability of surviving beyond a specified time, t. It analyzes not only the known failure times but takes into consideration those components that may not have failed at the observed time. This is significant because these un-failed censored data points provide valuable information about the survivability of components up to the observed time, t. The survivor function benefits corrosion engineering in that the exact time that corrosion initiated is not necessary. The survivor function is described as:

$$S(t) = Pr\ (T > t) \qquad [1.5]$$

At time $t = 0$, the probability of surviving is 100%. As t approaches infinity, failure will eventually occur and the probability of surviving becomes 0%.

The survivor function is the complement of the cumulative distribution function:

$$S(t) = 1 - F(t) \qquad [1.6]$$

where

$$F(t) = Pr(T < t) \qquad [1.7]$$

By definition, the survival function then becomes the integral of the probability density function:[17]

$$S(t) = Pr(T > t) = \int_0^\infty f(t)\mathrm{d}t \qquad [1.8]$$

Censored corrosion maintenance data

The Coast Guard maintains a robust depot corrosion mapping program at its aviation depot center. This facility performs depot maintenance on C-130, HU-25, HH-60 and HH-65 aircraft. During the overhaul process, rotary wing corrosion is documented and charted on a Coast Guard corrosion mapping program (Table 1.3). The mapping data is used to identify corrosion 'hot spots', track corrosion mitigation efforts and provide structural

Table 1.3 US Coast Guard HH-65 Corrosion Mapping Program: instances of corrosion

Zone	Light	Moderate	Severe	Total
Cockpit	2162	579	489	3230
Cabin	2066	228	414	2708
Transition	1824	120	494	2438
Belly	2078	1640	2725	6443
Tailcone/Fenstron	99	203	369	671
Total	8229	2770	4491	15490

reports of the aircraft (Table 1.4). The historic data residing in the mapping program offers an opportunity to analyze and predict the failure probability owing to corrosion using survival analysis. Explanatory variables such as geographic severity, time-of-wetness, sheltering or diurnal data can be accounted in the probability plots.

Survival analysis probability plots

An example of a survival analysis probability plot is shown in Fig. 1.6. In this analysis, aircraft from six Coast Guard air stations were selected to develop the two survival curves. The corrosion severities of the air stations were categorized as mild or severe depending on their boldly exposed weight loss corrosion rates. The data derives from airframe structure below

Table 1.4 US Coast Guard HH-65 Depot Corrosion Mapping Program: man hours to repair corrosion

Zone	Light	Moderate	Severe	Total
Cockpit	4390	1761	1856	8007
Cabin	2472	952	2300	5724
Transition	4019	488	2478	6985
Belly	5521	8585	19206	33312
Tailcone/Fenstron	2283	912	1604	4799
Total	18695	12698	27444	58827

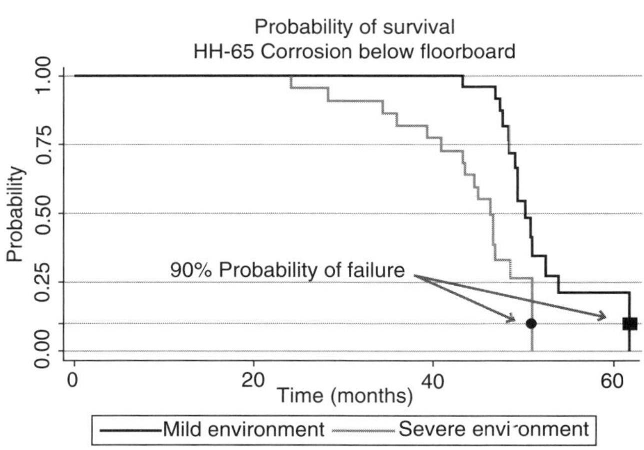

1.6 Kaplan-Meier survival analysis probability plot.

the HH-65 helicopter floorboard, a highly corrosive area susceptible to corrosion failure. From the Kaplan–Meier survival plot, it is possible to determine the probability of failure as a function of time. In 50 months, the probability of failure would be 90% for aircraft assigned to a severe environment while aircraft operating in a mild corrosion environment could expect to fail in about 62 months. This type of analysis, which can account for influencing variables and censored data, provides a statistical framework for structural life assessment. Survival analysis can be used as a management tool to minimize detrimental aircraft corrosion damage as a result of geographic and environmental locations.

1.5 Conclusion

We soar through the sky in great flying machines. For the moment, we triumph over nature by disembarking from Earth in a colossal system of carefully assembled metallic structure. Nature, however, has a patient way of prevailing. Given an opportunity, she will reclaim the alloys that comprise an aircraft by reducing the metals to their stable state in the process known as corrosion. When the structure is compromised, the safety of the aircraft is jeopardized.

The authors that contributed to this book are passionate about corrosion. They understand the special threat that corrosion poses to the aerospace industry. Their chapters offer a framework for incorporating expert insight into cutting-edge corrosion knowledge. To solve the menace of corrosion requires a holistic approach of the complex variables that combine to influence corrosion. As aerospace fleets age, they will rely on advances in corrosion mitigation technologies, accurate prediction models, health monitoring systems, novel inspection methods and cradle-to-grave corrosion management in order to continue to fly safely.

1.6 Sources of further information and advice

KLEIN, JOHN P. and MELVIN L. Moeschberger. *Survival Analysis, Techniques for Censored and Truncated Data*. New York: Springer, 2003.

KLEINBAUM, DAVID G. and MITCHEL KLEIN. *Survival Analysis, a Self-Learning Text*. New York: Springer, 2005.

THERNEAU, TERRY M. and PATRICIA M. GRAMBSCH. *Modeling Survival Data, Extending the Cox Model*. New York: Springer, 2001.

1.7 References

1 CHRIS SEHER, 'Managing the Aging Aircraft Problem,' in *the AVT Symposium on Ageing Mechanisms and Control Specialists' Meeting on Life Management Techniques for Ageing Air Vehicles* (Manchester, UK, 2001).

2 NTSB, 'Aloha Airlines, Flight 243, Boeing 737-200, N737311, near Maui, Hawaii', ed. National Transportation Safety Board (1988).
3 UNITED STATES GOVERNMENT ACCOUNTING OFFICE, 'Opportunites to Reduce Costs and Increase Readiness', (2003).
4 UNITED STATES GOVERNMENT ACCOUNTING OFFICE, 'Report to Congressional Committees, High-Level Leadership Commitment and Actions Are Needed to Address Corrosion Issues', (Washington, DC: 2007).
5 UNITED STATES GOVERNMENT ACCOUNTING OFFICE, 'Status of Department of Defense (DoD) Corrosion Prevention and Mitigation Efforts (Preliminary Observations)', (2002).
6 UNITED STATES GOVERNMENT ACCOUNTING OFFICE, Report to Congressional Committees, Defense Management 'Opportunites to Reduce Costs and Increase Readiness', (2003).
7 WILLIAM H. ABBOTT and RICHARD KINZIE, 'Corrosion Monitoring on Operational Aircraft Status of Recent Work' (paper presented at the Joint NASA/FAA/DoD Conference on Aging Aircraft Conference, New Orleans, LA, 2003).
8 J. HARTER, *Afgrow Users Guide and Technical Manual* (AFRL-VA-WP-TR-2004-XXX).
9 J. S. ULLETT, 'Prediction of Corrosion Growth Rates in Legacy Alloys' (paper presented at the US Air Force Corrosion Conference, Macon, GA, 6–8 March, 2007).
10 MAJOR GENERAL JAMES PILLSBURY, 'Aging Aircraft: An Army Perspective' (paper presented at the 10th Joint DoD/NASA/FAA Conference on Aging Aircraft, Palms Spring, CA, 2007).
11 JOHN MILLIMAN, 'The War on Aging Aircraft: One Battle Down, Many to Go', *Naval Aviation News*, July–August, 2002.
12 GERHARDUS H. KOCH, MICHIEL P. H. BRONGERS and NEIL G. THOMPSON, 'Corrosion Cost and Preventive Strategies in the United States', (Dublin, OH: CC Technologies Laboratories, Inc. and NACE International, 2001).
13 MATTHEW C. DIXON, 'The Costs of Aging Aircraft: Insights from Commercial Aviation' (Dissertation, Pardee RAND Graduate School, 2006).
14 R. E. MELCHERS, 'Transition from Marine Immersion to Coastal Atmospheric Corrosion for Structural Steels', *Corrosion* **63**, 6, 500–514 (2007).
15 GUEDES SOARES, C. and GARBATOV, Y. 'Reliability of Corrosion Protected and Maintained Ship Hulls Subjected to Corrosion and Fatigue', *Journal of Ship Research* **43**, 2 (1999).
16 GUEDES SOARES, C. and GARBATOV, Y. 'Reliability of Maintained, Corrosion Protected Plates Subject to Non-linear Corrosion and Compressive Loads', *Marine Structures*, p 425–445, (1999).
17 MOESCHBERGER, MELVIN L. *Survival Analysis*, Secaucus, NJ, USA: Springer-Verlag New York, Incorporated, 1997. p 21.

Part I

Corrosion fundamentals and cost of corrosion

2
Assessing the cost of corrosion to the aerospace industry

E. HERZBERG, LMI Research Fellow, USA

Abstract: Corrosion has significant cost, readiness and safety impacts on weapon systems, facilities and infrastructure in the aerospace industry. The Department of Defense (DoD) in the United States has developed a methodology to measure the financial impact of corrosion, which involves measuring actual expenditures through a combined top-down and bottom-up approach. The top-down portion makes use of summary-level cost and budget documentation to establish spending ceilings for depot, intermediate, and operational-level maintenance for both organic and commercial maintenance activities. This establishes a maximum cost of corrosion in each area of activity. The bottom-up portion uses detailed work order records to aggregate actual occurrences of corrosion maintenance and activity. This establishes a minimum level of corrosion costs in each activity area. The cost elements measured include labor, materials/parts, corrosion facilities, training, premature replacement and research and development. Costs are further classified as preventive or corrective, and applying to either the structure (non-replaceable) or parts (replaceable) of a weapon system, facility or infrastructure. The methodology is explained and some general case studies are presented in which the results from the corrosion cost studies within the US DoD have provided value to decision makers.

Key words: corrosion cost, aerospace industry, weapon systems, combined top-down and bottom-up, aircraft structure.

2.1 Introduction

In this chapter, the following areas are covered:

- The impact of corrosion is discussed and reasons why measuring the direct financial cost of corrosion provides the most value to decision makers are explored.
- The various elements and characterizations that contribute to corrosion costs are discussed.
- A methodology is outlined for determining corrosion costs for all weapon systems, facilities and infrastructure for the US Department of Defense (DoD).

- Some general case studies are presented to demonstrate that the results from corrosion cost studies have provided value to decision makers.

2.2 Corrosion impacts

Corrosion has three major negative impacts – financial cost, aircraft availability and safety.

- It affects financial cost mainly in terms of labor hours for maintenance and materials needed to mitigate corrosion.
- It can cause an aircraft to be deemed unavailable to perform its mission or to have a degraded capability. (The DoD uses the term 'readiness' to measure weapon system non-availability and/or degraded capability.)
- Corrosion has also been the cause of aircraft failures in flight that have resulted in injury and death.

Past corrosion studies have had difficulty isolating corrosion costs from non-corrosion costs. The clearest course of action is to treat these three areas of negative impact separately and not try to determine the cost implications of corrosion-induced readiness issues or safety concerns. Cost information is extremely useful for facilitating decision making. Decision makers cannot use readiness and safety information to judge the cost–benefit trade-offs on a project-by-project basis; nor can they use this information to measure the scope of the corrosion problem or judge the overall effectiveness of a chosen corrosion mitigation strategy. Focusing on cost information also eliminates the difficult task of turning non-cost measurements into costs. For example, imagine the difficulty in trying to put a value on the loss of life or a lost training opportunity. Trying to quantify the cost of loss of readiness due to corrosion is similarly elusive.

2.2.1 What is a corrosion cost?

The task of defining a corrosion cost is still a challenge, even when its effects on readiness and safety are excluded. To illustrate, we use a generic example of an obviously corroded aircraft as shown in Fig. 2.1. Is there a corrosion cost if the aircraft has all of its capabilities, and merely looks unpleasing? If the aircraft were inspected for corrosion and an accurate estimate of corrosion treatment costs were determined, would these become corrosion costs, even if the maintenance was deferred on the aircraft owing to a lack of currently available funds? If we design a more expensive aircraft with a metal alloy that corrodes at a slower rate but also is lighter (which results in fuel savings), how much of the increased cost of the metal alloy is a corrosion cost?

2.1 Corroded helicopter.

We address these types of questions by defining corrosion costs as historical costs incurred because of corrosion correction or prevention after the aircraft is placed into operation. This is known as the operating, support, or sustainment phase of a weapon system's life cycle.

2.3 Corrosion cost elements and characterizations

The following specific cost elements of corrosion are measured:

- labor hours (e.g., for inspection, repair, and treatment);
- materials and parts usage;
- premature replacement of the aircraft or its major components;
- corrosion facilities;
- training;
- research, development, testing, and evaluation (RDT&E).

The RDT&E costs are included even though they may occur before the weapon system is placed into operation because it is possible to separate expenditures specifically for corrosion from other RDT&E spending. The definitions of each of these cost elements are presented in Appendix A.

2.3.1 Identifying corrosion cost elements

Maintenance required as a result of corrosion is rarely identified as such in reporting systems. Therefore, it is necessary to develop a list of typical maintenance activities that counter the effects of corrosion. Corrosion costs are found by looking for the costs associated with these activities. Typical corrosion activities include cleaning, sand blasting, and painting. The complete list of the anti-corrosion activities, which serve as surrogates for corrosion costs, is provided in Appendix B.

2.3.2 Use of corrosion cost information

Decision makers can use cost information to pick which 'battles' to fight first, choose the level of resources to dedicate, and predict or monitor the effect of chosen solutions on overall cost. Such information is 'tactically useful'. Cost as a tactical indicator is a useful measure of the effect of changes to potential root causes of corrosion. For example, the impact of a new aircraft corrosion treatment compound can be measured by its effect on the rate of aircraft degradation due to corrosion. This change in degradation rate is eventually reflected in higher or lower maintenance costs.

However, not all costs are useful for these tactical decisions. Only costs that vary according to changes in root-cause corrosion conditions should be used. Because some costs are more useful in this type of tactical decision making than others, they have more value and we consider them a higher priority to acquire. Table 2.1 indicates which cost elements are the most tactically useful and their acquisition priority for corrosion cost studies.

Training and RDT&E are not tactically useful because, although they represent real expenditures, their costs and potential benefits are generally not attributable to a specific source of corrosion. While there are occasional exceptions (such as a training class that deals with a specific type of corrosion on a specific weapon system), the cost and benefits of training and RDT&E are spread over many different sources of corrosion and weapon systems. Knowledge of these expenditures is necessary to determine the overall cost of corrosion.

Facilities costs can be tactically useful if their potential benefits can be closely tied to a single or a few weapon systems or root causes of corrosion. For example, the cost of a new maintenance facility has little tactical cost-

Table 2.1 Prioritization of corrosion cost elements

Cost element	Tactically useful?	Priority to acquire
Labor hours	Yes	1
Materials	Yes	1
Premature replacement	Yes	1
Corrosion facilities	Potentially	2
Training	No	3
RDT&E	No	3

of-corrosion benefit because it can be used by several types of weapon systems and it has many uses other than corrosion mitigation. The cost of a wash and corrosion treatment facility for helicopters, on the other hand, may be tactically useful because the costs and benefits associated with this facility can be tied directly to a type of aviation platform, and the main purpose of the facility is to prevent corrosion.

For the remainder of this chapter, we refer to corrosion facilities, training, and RDT&E costs as 'outside normal reporting' costs because they are not normally reported in maintenance databases.

2.3.3 Characterization of corrosion costs

Corrosion costs are characterized into categories that provide some additional insight into the nature of these costs. Two of the most useful characterizations are discussed below.

Corrective and preventive costs

We classify all corrosion costs as either corrective or preventive.

- Corrective costs are incurred when removing an existing nonconformity or defect. Corrective actions address actual problems.
- Preventive costs involve steps taken to remove the cause of potential nonconformities or defects. Preventive actions address future problems.

From a management standpoint, it is useful to determine the ratio between corrective costs and preventive costs. Over time, it is usually more expensive to fix a problem than it is to prevent a problem, but it is also possible to overspend on preventive measures. As shown in Fig. 2.2, classifying the cost elements into categories helps decision makers find the proper balance between preventive and corrective expenses to minimize the overall cost of corrosion.

The task of classifying each cost element as either preventive or corrective could become an enormously challenging undertaking, one that involves thousands of people trying to classify millions of activities and billions of dollars of cost in a standard method. The real value of characterizing costs into preventive and corrective categories is to determine the ratio between the nature of these costs; the classification does not require precision. To simplify, we categorize the preventive and corrective cost elements as depicted in Table 2.2.

The classification of labor hours and the associated materials as corrective or preventive must be determined on a case-by-case basis. To ensure

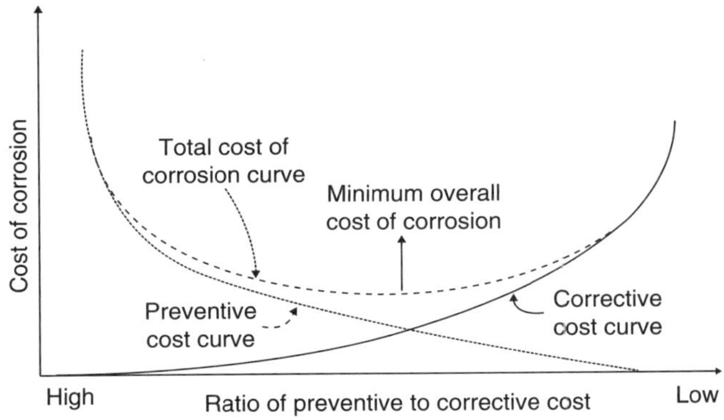

2.2 Preventive and corrective corrosion cost curves.

Table 2.2 Classification of corrosion cost elements into preventive or corrective natures

Cost element	Classification
Labor hours	Corrective or preventive
Materials	Corrective or preventive
Premature replacement	Corrective
Corrosion facilities	Preventive
Training	Preventive
RDT&E	Preventive

consistency, direct labor hours and the associated material costs were classified based on the following convention:

- Hours and materials spent repairing and treating corrosion damage, including surface preparation and sandblasting, are classified as corrective costs.
- Hours and materials spent gaining access to equipment that has corrosion damage so that it can be treated are classified as corrective costs.
- Hours spent on maintenance requests and planning for the treatment of corrosion damage are classified as corrective costs.
- Hours and materials spent cleaning, inspecting, painting, and applying corrosion prevention compounds or other coatings are classified as preventive costs.
- Hours spent at a facility built for the purpose of corrosion mitigation (such as a wash facility) are classified as preventive costs.

Structure and parts costs

Corrosion costs are also characterized as either structure or parts costs. All direct materials and direct labor costs fall into one of these two categories. Direct costs can be attributed to a specific system or end item.

Structure and parts are defined as follows:

- Structure is the body frame of the system or end item. It is not normally removable or detachable.
- Parts are items that can be removed from the system or end item, and can be ordered separately through government or commercial supply channels.

Segregating direct corrosion costs into structure and parts categories helps decision makers to give the design community more precise feedback about the source of corrosion problems.

The DoD has a major concern about the effects and costs of aging of weapon systems. The age of a typical weapon system is calculated starting with the year of manufacture of the individual piece of equipment – essentially, the age measures the structural age of the weapon system. The age of a removable part is not tracked, with the exception of major, more expensive components like engines. Separating the corrosion costs related to the structure of the weapon system (which has an age measurement) from the corrosion costs related to removable parts (which do not have an age measurement) may give further insight into the relationship between structural costs and the effects of aging on weapon systems.

2.4 Corrosion cost measurement methodology

A methodology called the 'top-down/bottom-up' approach is used to quantify the cost of corrosion. It is useful to illustrate this method using an analogy. Suppose we need to know how much of our monthly household budget we spend on meat. Now, it is clear that this is a strange piece of information to need to know but suppose it were vitally important. Normally, one would not be able to determine this answer by looking at the information at hand – specifically check book logs and credit card records. Even if we were diligent at logging our expenses, it is highly unlikely that 'meat' expenditures would be recorded in their own separate category. This is also the reason that corrosion costs are not easily found – they simply are not identified in maintenance databases in their own category. So how would we answer the meat question? We start with the 'top-down' portion of the analysis.

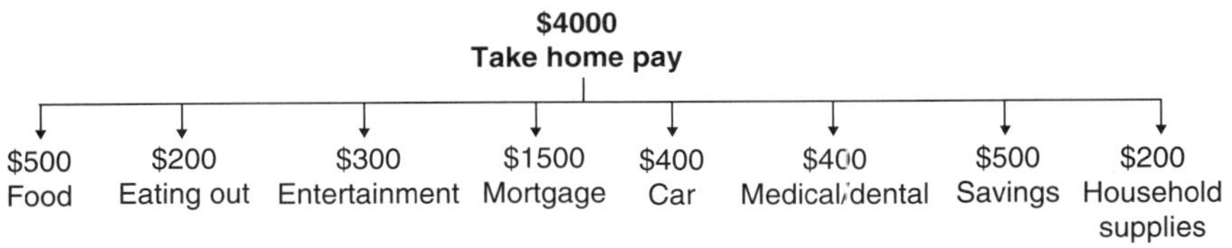

2.3 Top-down analysis using household budget example.

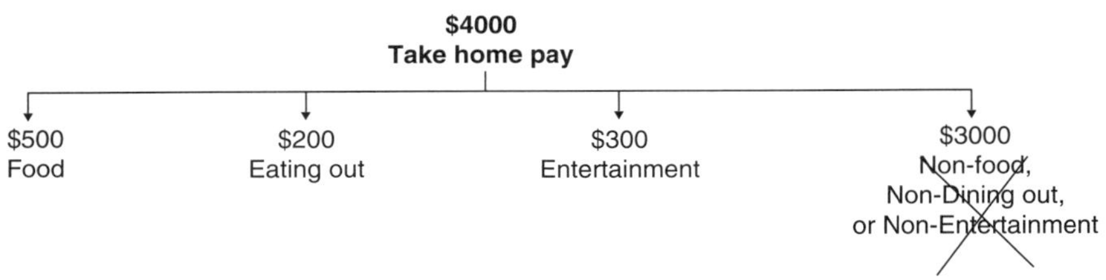

2.4 Consolidation of budget categories showing those that potentially contain meat.

2.4.1 Top-down analysis

In this analogy, the top-down portion of the analysis begins with identifying the combined net household monthly income. For illustration purposes, let us say it is $4000 per month. The next step is to separate this income amount into the major categories of spending that are visible in the check book logs, credit card records and other normal expense recording done in the household. A typical example might be similar to that contained in Fig. 2.3.

This diagram is called a 'cost-tree'. Note that the typical categories of spending in the second level of the cost tree accounts for the entire $4000 monthly household income. This will always be the case in the top-down portion of the analysis – each level of the cost tree has to account for the entire spending amount of the level above it.

Now that we have some breakdown of the expenses, we eliminate those categories that could not possibly contain any spending for meat and focus on the three categories remaining – groceries, dining out and entertainment. We still do not have an answer to the amount of spending on meat but we now know the answer cannot be more than $1000, as shown in Fig. 2.4. The spending within these three categories is further examined and characterized into more detail, as shown in Fig. 2.5. Note how each level of the cost tree below accounts for all the spending in the level above. This is as far as we can go with the top-down analysis. We still do not have a definitive answer to the question about meat spending but have narrowed the scope

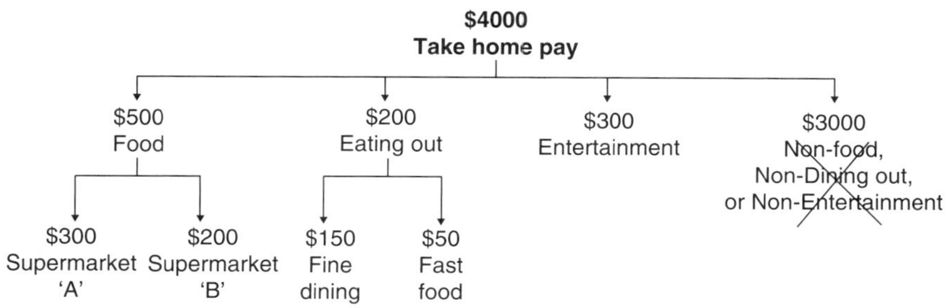

2.5 Expansion of potential meat spending into more detail.

of the problem dramatically. We now complete the methodology with the 'bottom-up' portion of the analysis.

2.4.2 Bottom-up analysis

We know that all our spending on meat is contained in the five categories at the lowest level of the cost tree in Fig. 2.5. These categories are:

- Supermarket 'A'
- Supermarket 'B'
- Fine dining
- Fast food
- Entertainment

The bottom-up portion of the analysis requires us to obtain as many detailed receipts for the spending in each of these five categories as possible. It requires us to secure grocery receipts for supermarkets 'A' and 'B', restaurant receipts for fine dining and fast food spending and receipts for all the entertainment expenses. The best method to secure these receipts is to ask the respective organization for them. For example, supermarket 'A' may have a record of our spending there because they have issued us a customer card and track our expenses to a member number. The methodology does not require every receipt to be obtained but the more of the spending that can be accounted for with the receipts, the more accurate the spending estimate becomes.

2.4.3 Combined top-down/bottom-up analysis

Once we have obtained all the receipts for the month in question, it is now possible to determine the answer about how much was spent on meat by examining every entry for every grocery receipt and extracting the spending

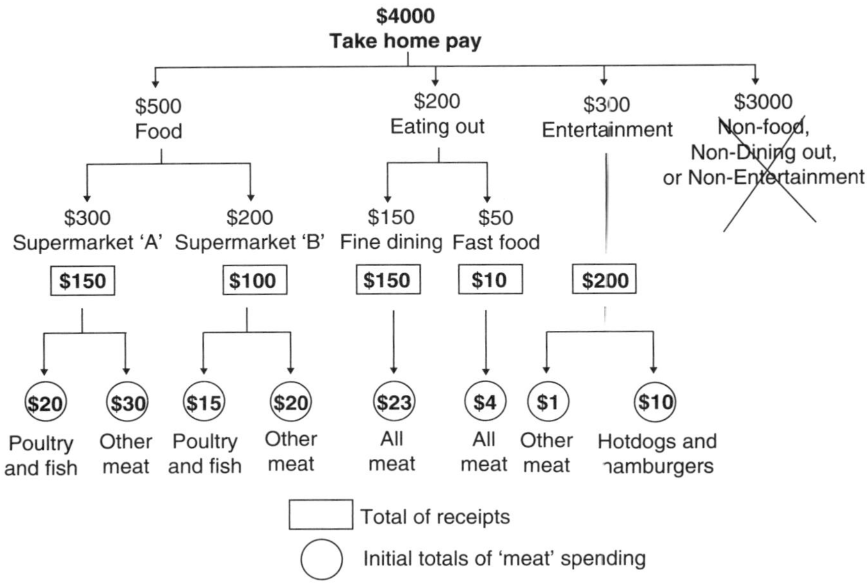

2.6 Initial calculation of meat spending.

on meat. It is also possible at this step to categorize the meat spending by type. For example, categories of meat could include pork, chicken, beef, etc. Figure 2.6 shows the analysis to this point.

It is important not only to quantify the amount of spending on meat, but also to calculate the total amount of the receipts for each of the five categories of spending. By comparing the top-down amount for each category (the 'should have' amount) with the total of the receipts for each category (the 'did have' amount), it is possible to identify gaps in the bottom-up data collection, or to re-examine some of the top-down assumptions should the two totals not converge.

Once we are comfortable that all data that can be acquired has been obtained and the top-down calculations are correct, it is time for the final step in the analysis, which is illustrated in Fig. 2.7. In the simplest terms, the meat totals from each category of spending are multiplied by the ratio of the top-down to bottom-up numbers. For example, the top-down total for Supermarket 'A' is $300. From the grocery receipts obtained from Supermarket 'A', the total amount of spending that was accounted for was $150. Of this amount, $20 was initially determined to have been spent on 'poultry and fish' and another $30 on 'other meat'. Therefore, the initial spending on meat from Supermarket 'A' is $50. The ratio of obtained receipts ($150) to the top-down spending at Supermarket 'A' of $300 is determined to be ½ or 50%. To compensate for the fact that we have obtained receipts that

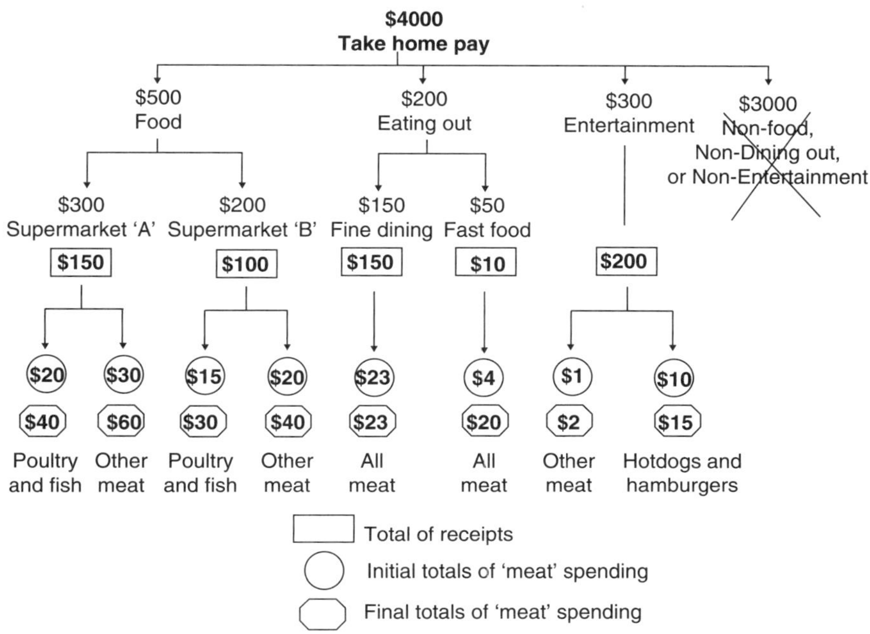

2.7 Final calculation of meat spending.

account for only 50% of the top-down spending amount, we multiply the 'poultry and fish' and 'other meat' totals by two. Our total meat expenditure for poultry and fish from Supermarket 'A' is therefore $40. For other meat, the total becomes $60.

This calculation can then be replicated for each of the spending categories in Fig. 2.7 to give the total monthly spending on meat, which is $230.

2.4.4 What does 'meat' have to do with corrosion?

Clearly, the example above is a simplified explanation of the cost of corrosion methodology. For the actual corrosion methodology, corrosion takes the place of spending on meat while the different types of maintenance expenditures are the categories of spending on food. Figure 2.8 shows a cost tree from one of the recently completed US DoD corrosion studies. This is an aviation example outlining spending for a major level of maintenance called depot.[1] Notice the logic and appearance is similar to the meat example with each lower level of the cost tree summing to the level above it. The real challenge is to conduct the bottom-up analysis – in essence, extracting the meat from the grocery and restaurant receipts.

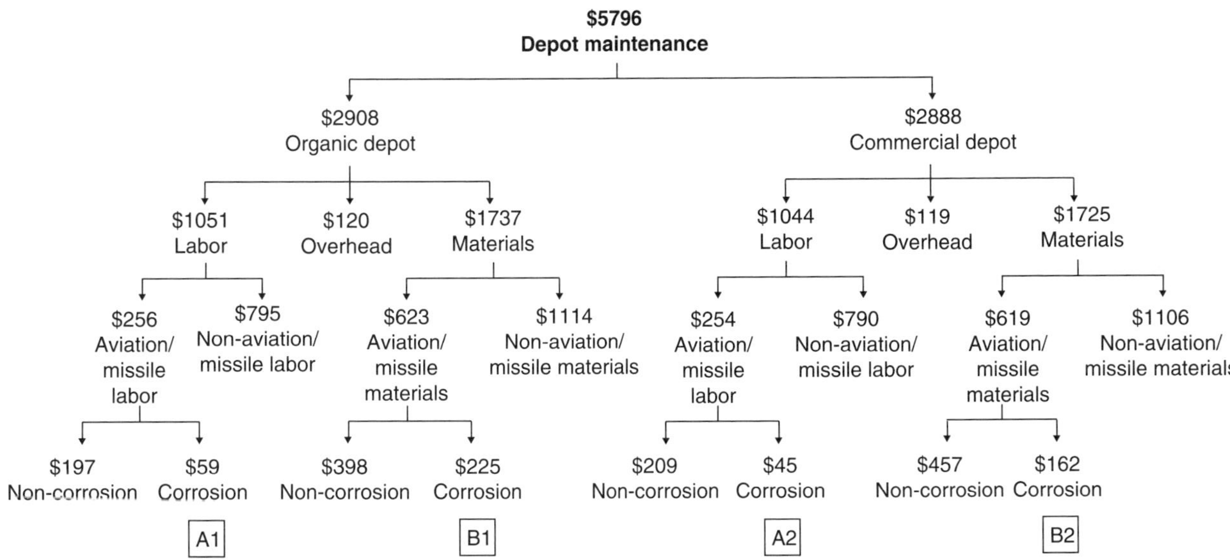

2.8 Cost tree from a recently completed United States Department of Defense corrosion study.

Conducting bottom-up analysis to extract corrosion costs

Instead of a few grocery receipts, this analysis involves millions of maintenance labor, parts and other material supply records. A computerized search algorithm is built based on the corrosion activities listed in Appendix B, subject matter expert input, applicable coding of the maintenance records for corrosion, work center information and any other detail contained in the records that would help identify corrosion-related work. Once all the bottom-up data is acquired, it is placed into a standard format and the search algorithm is run. Work records which involve corrosion-related work are flagged. For some types of corrosion related activities, we then apply a percentage based on discussions with the maintenance technicians who perform the work to determine the final amount of corrosion-related work. The flagged records with their labor and materials totals are added to determine the initial totals. Just like the meat example, the final step is to apply the top-down to bottom-up ratio to account for data that was not obtained.

Possible flaws

Both the top-down and bottom-up methods by themselves have their flaws. Determining the total cost of an enterprise can be a challenge in itself (in our meat example above, this was the total household income). Starting with an incorrect 'all there is' estimate will almost guarantee an incorrect top-down outcome. The results of a well implemented top-down analysis can yield a good estimate of overall costs, but that estimate can lack the detail necessary to pinpoint major cost drivers within the enterprise.

The bottom-up method can produce very accurate, auditable information so long as maintenance data collection systems accurately capture all relevant labor and materials costs, identify corrosion-related events, and are used with discipline. If any of these three boundary conditions are missing, corrosion costs are likely to be determined incorrectly. In most cases, they will be understated.

By combining both the top-down and bottom-up methods and determining if the results are approaching each other, we can validate our overall method and assumptions. Theoretically, the top-down method could produce the same estimate as the bottom-up. If the values produced using both approaches simultaneously converge, it is confirmation that the corrosion data collection methods and analysis assumptions are acceptable, and the data is adequate. If the two results initially do not converge showing a large top-down to bottom-up gap, we correct our approach to prevent erroneous cost information, assumptions, or incomplete data from corrupting the final outcome.

Implications and some surprising advantages

Using the combined top-down/bottom-up cost estimation method yields some significant advantages over previous methods.

- We can not only understand total corrosion cost but can also determine the corrosion cost by type of weapon system and subcomponent.
- We can understand the costs by level of maintenance and by work center – who is doing the work.
- The method allows us to characterize cost by their preventive and corrective natures as well as by parts and structure. This characterization applies not only to corrosion flagged records but to each of the millions of maintenance records in our bottom-up data.
- The methodology allows subject matter experts to help build the recipe (extract the meat from the receipts). This leads to a high level of ownership of the data once it is finalized.
- We can understand not only corrosion but total maintenance costs, by type and by weapon system. This has been surprisingly useful for maintenance managers at all levels because there is no central system that compiles complete maintenance cost information by weapon system.

Corrosion database and data structure

To accommodate the anticipated variety of decision makers and data users, we designed a corrosion cost data structure that maximizes analysis flexibility. This data structure is shown in Fig. 2.9.

Using this data structure, we were able to analyze the data against the following:

- equipment type;
- age of equipment type;
- corrective versus preventive costs;
- depot, field-level, or outside normal reporting costs;
- structure versus parts cost;
- material costs;
- labor costs;
- work breakdown structure (WBS).[2]

Any of these data elements can be grouped with another (with the exception of outside normal reporting costs) to create a new analysis category. For example, a data analyst can isolate corrective corrosion costs for field level maintenance materials if desired.

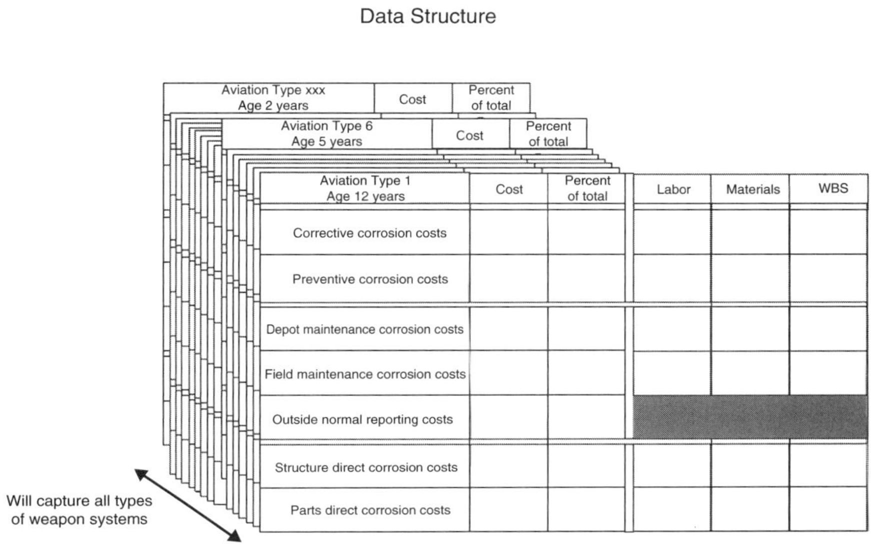

2.9 Corrosion cost data structure and methods of analysis.

2.5 General case studies

The cost-of-corrosion studies for DoD are only two years old, but despite these studies being in their infancy, there have been some interesting uses of the cost data.

2.5.1 New Navy ship example

Within six months of the release of the study of the cost of corrosion for Navy ships,[3] the United States Department of the Navy acquisition personnel entered a stage of contract negotiations with the commercial shipbuilder concerning the type of coating to apply to the interior liquid storage tanks on a new Navy ship. Because the corrosion study results showed liquid storage 'tanks and voids' as the highest contributor of corrosion costs to Navy ships at over $200 million per year (shown in Table 2.3), Navy acquisition personnel were able to convince the shipbuilder to use a more expensive high epoxy solid coating material for these storage tanks. The high epoxy solid provides more protection against corrosion and will significantly reduce the corrosion cost in the tanks and voids of this proposed new ship in the long run. Had the acquisition personnel not known the cost of corrosion information, it is likely the coating material used would have been the same as the coating used on previous ships.

Table 2.3 Highest 20 contributors to Navy ships corrosion cost by ESWBS*

Rank	ESWBS*	ESWBS description	Corrosion cost ($)	Maintenance cost ($)	Corrosion (%)
1	123	Tanks and voids	204	211	96.7
2	992	Bilge cleaning and gas freeing	182	330	55.1
3	631	Painting	166	167	99.3
4	863	Dry-docking and undocking	149	471	31.6
5	634	Deck covering	103	107	96.6
6	993	Crane and rigging services/preservation	60	61	98.8
7	251	Combustion air system	57	116	48.7
8	130	Hull decks	55	123	44.9
9	176	Masts, kingposts and service platforms	39	42	92.1
10	593	Environmental pollution control systems	34	100	34.1
11	864	Care and preservation	24	24	99.4
12	233	Propulsion internal combustion	21	106	19.6
13	505	General piping requirements	20	32	64.8
14	551	Compressed air systems	19	218	8.5
15	514	Air conditioning system	17	82	20.2
16	261	Fuel service system	17	38	43.2
17	150	Deck house structure	15	25	61.4
18	713	Ammunition stowage	15	18	82.2
19	131	Main decks	15	21	69.2
20	980	Contractual and production support service	14	80	17.0

* Expanded Ships Work Breakdown Structure – a five digit code that specifies the subsystem of the ship being worked on.

2.5.2 Navy ships work team

The Navy formed a working team to address the highest corrosion cost drivers as determined by the cost-of-corrosion study. After assessing the data and consulting maintenance experts in the field, they made nine recommendations, including re-examining Navy policy concerning touch-up painting, cosmetic appearance standards, relative humidity requirements of the coating application process, and tank inspection periodicity. The Navy is currently in the process of reviewing these policies.

2.5.3 Government Accountability Office (GAO) recommendations

In the US, the GAO's work includes oversight of federal programs and providing insight into ways to make government more efficient, effective, ethical and equitable. They are known as the 'investigative arm' of the US Congress. The GAO recently audited the workings of the DoD corrosion program. One of their main recommendations was for the Office of the Secretary of Defense (OSD) to develop an action plan to exploit the data from the cost of corrosion studies. The GAO concludes that the data provides the military Services with an opportunity to achieve long-term cost savings were it to be properly exploited.

2.6 Conclusion

Corrosion is an issue with significant cost, readiness and safety impacts. The cost-of-corrosion study methodology has been widely accepted among the DoD community, as well as within the GAO. As teams like the Navy's corrosion working team start to exploit the possibilities uncovered by this data, we expect to see many more success stories in the future.

2.7 References

1 DEPARTMENT OF DEFENSE DIRECTIVE 4151.18, *Maintenance of Military Materiel*, 12 August 1992, Enclosure 2.

2 *DoD Financial Management Regulation*, Volume 6, Chapter 14, Addendum 4, January 1998.

3 THE ANNUAL COST OF CORROSION FOR ARMY GROUND VEHICLES AND NAVY SHIPS, LMI, April 2006.

2.8 Appendix A: Cost element definitions

- Labor hours – Any time spent in corrosion prevention or correction that can be attributed directly to a specific system or end item. The labor can be military, civilian, or contract.
- Materials and parts usage – The cost of any materials used for corrosion prevention or correction. This includes both consumables (paints, sealants, rags, etc.) and reparables (engines, avionics systems, etc.).
- Premature replacement – The cost of removing and discarding any end item, subcomponent, or material primarily because of corrosion, or its use in preventing or correction corrosion, less the salvage value recouped from the end item, subcomponent, or material. The scrap cost included a percentage of the cost of replacing the end item, subcomponent, or material if it was disposed of before the end of its useful life.

- Corrosion facilities – The acquisition and installation costs of an asset constructed primarily or partially for corrosion prevention or correction. The labor spent to acquire and install the facility will be counted in this cost category. The labor to operate a facility that is used for corrosion correction or prevention will be counted in the labor-hours cost category if the labor can be attributed to a specific weapon system or family of systems.
- Training – The cost of training related to corrosion. This cost will include all labor, materials, educational aids, and travel. It includes the cost of training development as well as the actual training itself.
- Research, development, testing and evaluation – The cost of creating a new product, process, or application that may be used for corrosion correction or prevention. All labor costs spent in research and development will be collected in this cost category rather than in the labor-hours category.

2.9 Appendix B: Typical corrosion activities

The following list of corrosion activities was used to develop keyword searches and other methods to extract corrosion costs from maintenance reporting databases:

1. Cleaning to remove surface contaminants;
2. Stripping of protective coatings;
3. Inspection to detect corrosion or corrosion related damage;
4. Repair or treatment of corrosion damage:
    a. Corrosion removal,
    b. Sheet metal or machinist work,
    c. Replacement of part;
5. Application of surface treatment (alodine, other surface, etc.);
6. Application of protective coatings, regardless of reason;
7. Maintaining facilities for performing corrosion maintenance;
8. Time spent gaining access to and closure from parts requiring any of activities 1–6;
9. Preparation and clean up activities associated with activities 1–7;
10. Documentation of inspection results;
11. Maintenance requests and planning for corrosion correction;
12. Replacing cathodic protection systems (for example, zinc);
13. Maintaining environmental control facilities (e.g., dehumidification tents).

3
Corrosion and the threat to aircraft structural integrity

T. MILLS, S. PROST-DOMASKY,
K. HONEYCUTT and C. BROOKS
Analytical Processes / Engineered Solutions, Inc. (APES), USA

Abstract: The threat of corrosion specific to the structural integrity of aircraft is examined. A synopsis of the economic and safety issues associated with corrosion is given and the principal types of corrosion affecting aircraft structure are described. The tenuous relationship between corrosion and aircraft structural integrity methodologies, particularly safe-life and damage tolerance, is explored in depth. An evolved, physically based, holistic structural integrity process is presented that provides methods for properly accounting for the structural effects of corrosion in critical aircraft structure. The need for this new process is illustrated by providing examples of failure scenarios that pose an increasing threat to perceived integrity when viewed under existing structural management philosophies.

Key words: structural integrity, corrosion, metal fatigue, aircraft safety.

3.1 Introduction

3.1.1 Living with corrosion

Corrosion runs rampant almost everywhere we look and often in places we do not look. The fenders of cars provide some of the best examples, but rusted and cracked bridge structures illustrate the crumbling infrastructure we as commuters must tolerate daily during our drive to work and back. Many people use aircraft to commute, and in a sense, the aircraft they travel in show many parallels to a slowly deteriorating freeway overpass. We rely on corrosion protection schemes, so corrosion tolerance is not a design feature. We count on nondestructive testing (NDT) to tell us when corrosion is present, unless the area is inaccessible. We do not necessarily know the best time to look, unless we have managed to learn from the fleet over time; alas, there is always something new cropping up.

Corrosion is alarmingly proficient at chewing away the common aluminum and steel structures of aircraft. Occasionally, we read about the results in the newspaper (loss of life, health, property, aircraft), and the outcomes often tarnish aircraft manufacturers' and operators' reputations. The real causes of these accidents and incidents is often lost in translation. Evidence uncovered by Hoeppner *et al.* (1995) suggests that the technical community, let alone the general public, does not have the entire story related to the involvement of corrosion in aircraft accidents. For instance, in cases where corrosion pits led to 'premature' fatigue nucleation and subsequent failure (although right on schedule from the corrosion pits' perspective) the fact that corrosion was the root cause can often take a back seat, to the point of becoming irrelevant or even non-existent, to the fatigue crack which ultimately fractured. Evidence of this was illustrated clearly in Hoeppner's paper.

Fortunately, enough evidence exists to allow us to condemn corrosion (and its partner, fretting) and to recognize it as an expensive and often violent offender. In a 1984 worldwide accident survey involving fatigue failures in both fixed-wing and rotary-wing varieties of aircraft, Campell and Lahey (1984) concluded that two of the five most common fatigue crack nucleation sites were corrosion and fretting damage. Of the 574 accidents investigated where crack origins were traced, 87 (a little over 15%) involved corrosion and/or fretting. Hoeppner's conclusions closely agree with those of Campell and Lahey; however, as implied by the earlier statement that the technical community does not have the entire story, this percentage is probably much higher.

3.1.2 Economic impact of corrosion

The health and reputations of customers, operators, and designers of aircraft are protected by diligent maintenance. These practices have led to much reduced accident rates and, in other cases, expert piloting, rugged construction, and a little luck have kept incidents from becoming serious or fatal accidents. Putting the latter factors aside, corrosion protection is still costly in its own right. Wanhill (1995) reports that more than one billion dollars is expended annually for corrosion control in worldwide aircraft fleets. Corrosion in civil transports, states Wanhill, accounts for six to eight per cent of direct airframe maintenance costs, and the relative cost in military aircraft stands at 25 to 30% of total maintenance costs. However, other sources, such as the United States Air Force (USAF), suggest that costs of corrosion are much higher than those reported by Wanhill.

The USAF released the first in a series of comprehensive corrosion cost surveys in September 1990 (Cooke *et al.*, 1990), and the report paints an even more critical picture of corrosion costs. Cost analysis showed that the USAF was spending approximately $718 million per year on direct corrosion maintenance. These direct costs include repair, detection, and deterrence and exclude classified weapon systems and intangibles such as readiness losses from aircraft down-time. Since that report was issued, the USAF has surveyed corrosion costs three more times: in 1997, 2001, and 2004. The most recent study (Kinzie, 2004) shows that costs have ballooned to $1.5 billion. Even when factoring inflation into the equation, costs have risen by 50% since 1990. Keep in mind, too, that fleet sizes over this period of time have decreased substantially, with a 35% reduction occurring between 1990 and 1997 alone.

3.1.3 Safety aspect of corrosion

What of the costs in terms of human life? It is difficult to put a number on that, especially on a personal level, but it is indisputable that people die in corrosion-related aircraft accidents. Hoeppner *et al.* (1995) found that at least 81 passengers and crew of 687 general aviation, commercial, and military aircraft died in accidents and incidents in the United States between 1975 and 1993. Sometimes the events have a much worse bark than bite. For instance, Hoeppner's data showed that the most common corrosion-induced failures on aircraft involve landing gear systems. Such landing gear failures seldom result in fatal injuries or destroyed aircraft, as found by Hoeppner, but they do result in parts of the aircraft hitting the ground that have no business being in contact with said ground.

On occasion, the sequence of events in a corrosion failure are much more catastrophic. In 1988, a flight attendant was killed and many passengers were injured during an explosive decompression involving Aloha Airlines Flight 243. In this accident, 18 feet of the upper fuselage crown separated from the aircraft while it cruised at 24000 feet.

The National Transportation Safety Board (NTSB) claimed that the nineteen-year-old aircraft and its passengers were the victim of multiple-site fatigue damage at critical rivet rows in the fuselage lap joints as well as the victim of numerous human errors on the part of the operators, regulators, and designers (NTSB report 1989). Corrosion was excluded as a specific cause, yet corrosion and moisture intrusion into the lap joints played a significant role in degrading the joints' integrity. This environmental and structural degradation facilitated fatigue cracking.

The accident caused a flurry of activity in the Federal Aviation Administration (FAA), and the issue of 'aging aircraft' flew forward and perched atop the list of concerns in air safety. The result was a massive effort to evaluate the structural integrity of the world's air transport fleet with much of the energy being focused on corrosion (even though corrosion was not listed as an official cause in the Aloha accident).

Lincoln of the USAF defined an aging aircraft as, 'aircraft that have overflown their design service lives, that have corrosion problems, that have widespread fatigue problems, and that have numerous repairs many of which are not damage tolerant' (Lincoln, 1994). Ironically, the term 'aging aircraft', having enjoyed nearly two decades in the limelight, has fallen out of favor in some circles to the point that the funding of research into these important issues has faded, too – but not because the problems have been solved.

In the following sections, we will discuss some of the most threatening forms of corrosion, their effects on structure, and their existence (or lack thereof) within existing fatigue design. To close the chapter, we will briefly discuss the importance of holistic structural integrity and the broad safety and economic benefits afforded by proper corrosion structural effects management.

3.2 Types of corrosion and their impact on aircraft structure

Many types of corrosion affect common materials used in aircraft structure, and depending on which reference you consult, it is possible to list anywhere from seven to fourteen varieties. Those frequently encountered include:

- general attack;
- pitting;
- intergranular/exfoliation;
- galvanic;
- crevice;
- filiform;
- erosion corrosion;
- environmentally-assisted cracking (EAC):
    - corrosion fatigue,
    - stress corrosion cracking,
    - hydrogen embrittlement;
- fretting fatigue.

Many of the common types of aircraft corrosion, although mechanistically different, are similar in terms of their potential effects on structural integrity. From a structural-effects view, 'parent' groups of corrosion morphology appear to exist. These 'parent' groups, namely general attack, pitting, and intergranular/exfoliation, have been classified for the purpose of this study as structurally significant and unique corrosion forms.

Many, but not all, of the corrosion types on the above list are discussed in the following pages.

3.3 Structurally significant and unique corrosion forms

The first group of corrosion types to be discussed include general attack, pitting, and exfoliation.

3.3.1 General attack

General corrosion attack 'causes the metal to be consumed uniformly over the entire surface that is wetted with the corrosive environment' (De Luccia, 1991). In principle, this is an easy concept to understand, but in practice, such a corrosive state is almost impossible to obtain. The uniform distribution of corrosion is a simplifying assumption, used in cases where corrosion is widespread and the level of corrosion is not extremely varied throughout the corroded region.

General corrosion attack also weakens metals by introducing hydrogen (which embrittles the metal), by introducing load redistribution causing secondary bending stresses, or by introducing subtle, local stress risers (owing to surface roughness), which accelerate fatigue crack nucleation and growth.

In practice, however, structural effects analysis for general corrosion attack is usually boiled down to a net-section material loss. This material loss is translated into an increase in net-section stress used in evaluating structural capability against criteria such as design ultimate and limit load, residual strength, and crack growth. This simplified assumption can lead to unconservative results, particularly in crack growth/fatigue analyses (Brooks *et al.*, 2001). The manner in which this can occur will be discussed later in this chapter.

3.3.2 Pitting corrosion

Pitting corrosion often occurs when a structure is undergoing general corrosion attack. The corroding medium attacks the entire surface, more particularly the surface impurities and weak spots, causing numerous isolated pits in the structure's surface. However, pitting corrosion can be distinguished from general corrosion by the near absence of measurable material thinning, and, in many cases, by the severe degradation in a structure's integrity relative to general corrosion, particularly in tension-dominated components.

Several pitting models (Lindley *et al.*, 1982; Kawai and Kasai, 1985; Kondo, 1989; Mills *et al.*, 2002; Mills *et al.*, 2004; Crawford *et al.*, 2004) have been proposed over the decades, particularly in conjunction with fatigue analysis. This is perhaps not surprising as pitting corrosion is often easily discovered after a structure fails, and pitting corrosion typically leads to serious degradation of structural capability, particularly in single-load-path, highly loaded components. Pitting does its damage by providing local stress risers that can greatly accelerate crack formation and short crack growth. This is not limited to fatigue crack formation; stress corrosion cracks can also be generated from pits if the load, material, and chemical environment is supportive of this failure mode. The chemical influence of pitting can also cause embrittlement of the material at the pit surface, a factor that can also enhance crack nucleation and short crack growth.

Unfortunately for the aerospace industry, most metals currently used in today's structures are susceptible to pitting corrosion under typical service conditions. Structural aluminums and high-strength steels are vulnerable to pitting, particularly when protective coatings break down. For instance, high strength steels are often protected with cadmium plating. If the plating were not present, the material would likely undergo a general attack. But, with cadmium plating in place, any local break in the coating can rapidly lead to pitting. This is far more threatening to many structures.

High-strength steels are selected for use in some components because the components carry very high stresses. These ultra-high strength materials are also known for having good damage resistance in the proper conditions, meaning that they can resist the formation of fatigue cracks at their nominal operating stress. What history shows us, though, is that structures made from these materials have a very low damage tolerance in the presence of pitting. In the presence of pits, fatigue cracks may form very quickly. The use of the term 'damage tolerance' here is not to be confused with the damage tolerance design paradigm described later, the latter of which has essentially become synonymous with crack growth analysis and inspection and has little bearing on the vulnerability of a structure to corrosion.

The following examples illustrate the impact of pitting on structural capability. In the mid-1980s, cracks were detected in critical wing carry-through structure of two separate strike aircraft (Mills *et al.*, 2001). In both cases, the cracks formed at corrosion pits; the initial fracture was intergranular (attributed to stress corrosion) before developing into a fatigue crack. In one aircraft the cracking started from three separate pits and joined to form one large crack. Had these problems not been found by inspection, subsequent fracture of the structure would have resulted in the loss of the aircraft. The structure in question exhibits extremely limited damage tolerance.

Landing gear has already been highlighted in this chapter as being a problem area for aircraft. One does not have to search far to find an example of a landing gear failure. In 1998, the main landing gear spindle on a military transport aircraft failed due to corrosion pits (Huffman, 2000) that were approximately 0.102 mm deep. Fatigue only progressed to a depth of 0.152 mm before fast fracture occurred. This is another classic example of low damage tolerance. This case is very similar to that of the wing structure example provided above: both components were made from high-strength steel, both were susceptible to pitting, and both were subject to failure at relatively small and difficult-to-inspect crack sizes. Where the two cases differ is in component criticality.

These sorts of dangerous issues extend into the general aviation community too, not just the military and airlines. In 1990, a pilot and crew were killed when the right wing outboard of the engine nacelle separated from their Aero Commander 680 while performing a geological survey. The aircraft entered an uncontrolled descent and crashed into a field near Hassela, Sweden. Investigations revealed that the wing failed due to corrosion pits that nucleated fatigue cracks in the lower spar cap, part of the primary load-carrying structure in the wing (SCAA, 1991). This failure was one of many for this aircraft type. Swift notes (Swift, 1995) that the chronic diseases of corrosion and/or fatigue have influenced most of the 24 documented in-flight wing failures in the Aero Commander family of flying machines.

As with the crash in Sweden, other failures in the Aero Commander were centered in the wing spar. Since the troubled wing spar was not easily inspectable, reports of corrosion problems did not initially come from routine maintenance but rather from crash investigations.

3.3.3 Intergranular and exfoliation corrosion

Intergranular corrosion, particularly its special subset, exfoliation, is a well-known cause of failures of aircraft structure, especially in high-strength aluminum alloys. These alloys use certain elements and heat treatments to significantly raise their ultimate tensile and yield strengths and achieve the impressive strength-to-weight ratios demanded by the aircraft industry. Unfortunately, increased strength often has inherent trade-offs in the form of decreased ductility, increased notch sensitivity, and increased corrosion susceptibility.

The 7xxx-series alloys contain copper and zinc, and these alloys are particularly susceptible to intergranular attack, because of the high galvanic couples between grain bodies and boundaries produced by the heat-treatment. The unstable precipitates formed by the peak-aged (maximum strength) heat-treatments cause the copper and/or zinc in solution to accumulate at grain boundaries and leave an adjacent precipitate-free zone near the boundary. Since aluminum is anodic to copper in the galvanic series, the grain boundaries preferentially corrode. Over-aged tempers have greatly reduced intergranular susceptibility in certain alloys, but these tempers are still vulnerable to pitting.

Exfoliation corrosion, by ASTM definition, is: 'corrosion that proceeds laterally from the sites of initiation along planes parallel to the surface, generally at grain boundaries, forming corrosion products that force metal away from the body of the material, giving rise to a layered appearance' (ASTM, 1986, p. 283).

The flake-like morphology of exfoliation corrosion makes the thinning of the material cross section very conspicuous and, as such, exfoliation in aircraft structure has typically been treated as a threat to structural integrity in much the same manner as general attack, namely: area loss. It was to be 1990 before research began to emerge in earnest that addressed exfoliation–fatigue interactions. Some focused on the crack propagation rates (Chubb *et al.*, 1991; Chubb *et al.*, 1995; Koch *et al.*, 1995; Baldwin *et al.*, 1997; Mills 1997) of cracks growing through exfoliated regions, while others focused on the influences of exfoliation on the formation of fatigue (Mills, 1995; Sharp *et al.*, 1998; Sharp *et al.*, 2000). These later studies verified in the laboratory what the US Navy learned in the field some 30 years prior (Shaffer *et al.*, 1968) – that exfoliation can influence structural response by using many of the same mechanisms as pitting corrosion.

In the mid-1960s, the Navy was busy combating corrosion and cracking problems in 7075-T651 aluminum extrusions that formed the wing spars of air–sea rescue aircraft. Studies identified pitting and intergranular degradation (exfoliation) as the culprits in reducing fatigue life in the exfoliated material up to 70%. Metallographic examination of the post-test specimens showed that the corroded laminar paths were preferential sites for fatigue crack nucleation. The exfoliation problems originated deep inside rivet holes that suffered pitting attack under crevice corrosion conditions, and the resulting corrosion damage was undetectable by normal visual observation. In all cases, the intergranular corrosion cracks propagated from corrosion pits, and as many as 20 cracks were detected (via ultrasonic inspection) emanating from a given area.

The danger associated with treating exfoliation as simple area loss was illustrated by Sharp *et al.* (2000) and by Mills *et al.* (2004) using refined modeling techniques. The message from the experiments was simple: by using area loss only, fatigue life predictions were over predicted (too long) by a factor of four relative to the test lives of coupons containing exfoliation varying from 2 to 30% thickness loss. Only by including the effects of pitting in the fatigue life models could the experimental lives be accurately predicted. Incidentally, pit-like morphologies were found at all of the crack nucleation sites.

3.4 Other forms of corrosion attack

Many other forms of corrosion affect aircraft structure, as shown in the earlier list, and many have unique mechanisms driving them. However, their structural influences bear much similarity to the three mechanisms discussed in the previous section. For instance, galvanic corrosion arises when dissimilar materials are in contact and cause greatly accelerated corrosion rates for the susceptible material in the pair, but the physical damage that occurs may take the form of general attack, pitting, or exfoliation.

Likewise, crevice corrosion as a type is characterized more by the geometric configuration of the corroding component(s), such as a lap joint, than the damage morphology, which again is typically of the general attack, pitting, and/or exfoliation variety. The advancement of corrosion damage in crevice configurations is also accelerated because of the occluded environment.

Filiform corrosion is quite common in aircraft structure and is characterized by light blistering of the paint and miniature 'mole tunnels' between the coating and the metal. When looking at the damage caused to the structure itself, one sees miniature rivulets running across the surface, seldom more than a few μm deep. Structurally, this damage is typically benign, and it is regarded as being mostly cosmetic, in that it affects coating appearance.

3.5 Insidious synergisms

Some of the corrosion mechanisms on the list are truly synergistic, which makes them even more insidious, as they very efficiently render expected 'safe-lives' and 'crack growth lives' irrelevant, even with a host of safety factors standing guard. Mechanisms frequently encountered include environmentally assisted cracking, such as corrosion fatigue, stress corrosion, and hydrogen embrittlement, and fretting fatigue.

3.5.1 Corrosion fatigue

Corrosion fatigue, by the ASTM standard definition (ASTM, 1986, p. 180), is 'the process in which a metal fractures prematurely under conditions of simultaneous corrosion and repeated cyclic loading at lower stress levels or fewer cycles than would be required in the absence of the corrosive environment.'

The study of corrosion fatigue mechanisms has focused on both mechanical and chemical alteration of crack formation and propagation. The synergistic nature of corrosion fatigue can take many forms – some obvious and some not so obvious. For instance, consider these non-standard definitions:

- Corrosion-nucleated fatigue – the process in which physical corrosion damage (e.g., exfoliation, pitting) and/or chemical damage (e.g., embrittlement) accelerates the formation of fatigue cracks in a component or structure.

- Prior-corrosion fatigue – occurs when a propagating fatigue crack is influenced by a prior corroded region. This can result in acceleration of growth due to the increased stresses associated with mechanical thinning. A propagating fatigue crack in a prior-corroded region may also be synergistically altered by chemical means if an aggressive chemical environment is still present.
- Corrosion induced fatigue via load transfer – occurs when corrosion damage or environmental degradation in a structure causes load to be transferred to nearby structure or to alter the load transfer path within an affected structural detail. The increased stresses or strains associated with the transfer may promote fatigue cracking.

The concepts surrounding this last definition may seem obscure, but the review of case studies, such as the Aloha accident (NTSB, 1989), helps reveal some of the ways in which this failure mode works. In the Aloha accident, lap joints that were intended to be fastened both by cold bond adhesive as well as rivets suffered from the degradation of the adhesive, partly from manufacture and partly from environmental ingress and corrosion. The loss of load transfer from the adhesive concentrated stresses at the fasteners and greatly accelerated the onset of fatigue cracking.

3.5.2 Stress corrosion cracking

Stress corrosion cracking (SCC) leads to some of the most common cracking problems faced by aircraft designers, maintainers, and operators. Three things are necessary for production of stress corrosion cracks; the absence of any of these ingredients means SCC will not occur. These ingredients are:

- a susceptible material,
- a corrosive environment,
- a sustained tensile stress.

Susceptible materials are no strangers in aircraft construction; heightened corrosion susceptibility is traditionally a hallmark of the compromise made when seeking high strength-to-weight ratios. For example, in 1943, Alcoa introduced what was to become one of the most well known aircraft aluminum alloys and tempers of all time, 7075-T6. Initially, the product form was thin sheet, which exhibited less SCC susceptibility, but as products started to take the form of thick plate, extrusions, and forgings, the number of stress corrosion failures increased dramatically (ASM, 1985). Wallace and Hoeppner state that at one point, 90% of aluminum alloy stress corrosion failures could be traced to 2024-T3, 7075-T6, and 7079-T6 (Wallace and Hoeppner, 1985), a fact that would certainly provide evidence that use of these alloys meets the criterion in point one above.

Of the three necessary ingredients for SCC, it would appear that material susceptibility is the easiest to control. The second ingredient, a corrosive environment, is a moving target in terms of definition. For some materials and grain orientations (relative to the direction of sustained tensile stress), high humidity air could meet the criterion, where in other materials and orientations, it might take something more aggressive like wet/dry cycles of sump water to trigger SCC. Perhaps the best defense here is the use of corrosion protection schemes, such as coatings, but even those degrade with time.

The third ingredient, sustained tensile stress, is even harder to control or, for that matter, to characterize. Many sources for this stress exist in a structure. Some may be induced by a component manufacturing process, such as forging, others may induced by assembly of the structure, still others may be induced as part of the normal stresses in the structure caused by the weight of the aircraft.

The beauty of SCC, if such a thing exists, is that the 'simple' elimination of any one of the three ingredients means SCC will not occur. Perhaps that is why a considerable amount of effort has gone into developing methods that reveal the degree of susceptibility of a material. One of the first hints pursued by researchers was the fact that SCC in aircraft aluminum alloys often followed intergranular paths. Thus, intergranular attack susceptibility has long been recognized as a precursor to SCC susceptibility in high-strength aluminum alloys. Because of the link between intergranular attack and SCC, researchers have developed methods to determine the susceptibility of high-strength aluminum alloys to intergranular attack, particularly exfoliation (Romans, 1969; Sprowls *et al.*, 1972). One of the more common methods, the EXCO test, was later adopted into an American Society for Testing and Materials (ASTM) standard that has seen revision up through 1990 (ASTM 1990). The 7xxx-series alloys contain copper and zinc, and these alloys are particularly susceptible to exfoliation because of the high galvanic couples between grain bodies and boundaries produced by the heat-treatment. The unstable precipitates formed by the peak-aged (maximum strength) heat-treatments cause the copper and/or zinc in solution to accumulate at grain boundaries and leave an adjacent precipitate-free zone near the boundary. Since aluminum is anodic to copper in the galvanic series, the grain boundaries preferentially corrode leading to intergranular corrosion.

One of the most effective weapons for combating SCC, in terms of material susceptibility has been through modifications in heat treatments. For instance, the over-aged, T7, temper applied to 7xxx-series alloys is the most common modification (Lifka and Sprowls, 1972), and this method is successful in reducing the amount of segregation of precipitates in the matrix (Smith, 1993). The over-aged treatment tends to reduce strength in these alloys by 15% because of the larger precipitate size, which sometimes makes substituting corrosion resistant, T7x components in place of corrosion-prone structure difficult.

3.5.3 Fretting fatigue

Fretting fatigue is a failure mode often overlooked in aircraft structural integrity, particularly as a cause of accidents. Fretting is both a corrosion and a wear mechanism in which mating surfaces nominally at rest undergo small-amplitude oscillatory motion. Oxide debris generated by this process can severely damage a surface by gouging and pitting.

From a structural standpoint, fretting can have much in common with pitting; however, the effects of fretting on fatigue are often much harder to analyze than other types of corrosion owing to the existence of a contact stress state. In some components, contact stresses can be substantial and complex and provide strain gradients akin to a very severe stress concentration at a notch or hole. Fretting problems prove a constant battle in compressor disks and turbines, particularly at dovetail joints between blades and disks where contact stresses are high during operation. This is a serious issue as damage tolerance in these components is often low and consequence of failure great.

3.5.4 Role of surface integrity

An important point about all of these 'insidious synergisms', as they were labeled, is that often violation of surface integrity is a key ingredient to precipitating failure. Refer, for instance, to the definition provided earlier for 'corrosion-nucleated fatigue.' Although the violation of surface integrity is not a necessary condition, there are still many instances where corrosion fatigue, stress corrosion, and fretting fatigue get a boost from the stress-concentrating effect of pit-like discontinuities. In other words, failure modes and mechanisms often mix, and the process that leads to failure is probably not the result of a single factor. One of the most common mistakes in a failure analysis is to classify a fatigue fracture emanating from a corrosion pit as a 'fatigue failure,' as was found repeatedly by Hoeppner *et al.* (1995). It should be obvious how ignoring the root cause could lead one to pursue the incorrect solution set.

A classic example of this (Mills *et al.*, 2004), is the unexplained cracking of a major structural component in a large transport aircraft. The problem was so perplexing, and the existing analytical solutions so unworkable, that the plan of action was wholesale replacement of the component – no minor undertaking to say the least. 'Durability' analysis of the structure showed it to have an 'infinite life.' However, failure was occurring in the fleet at

40000 h. 'Damage tolerance' analysis of the structure resulted in a 6000-h inspection interval. Not only was this number inaccurate, as evidenced by the fleet failures, but also the inspection burden would have been substantial. However, further study revealed that the cracks were emanating from corrosion pits. The inclusion of corrosion pits in the durability analysis produced a solution very similar to the actual failure lives in service and allowed for the development and implementation of an NDT solution to detect for pits to screen for components that needed rework, or in the worst case, replacement. This is a substantial deviation from the original plan to replace all of these components throughout the fleet, a plan that was a direct result of the inability of the 'fatigue only' analysis to produce a realistic simulation of the failure process.

3.6 Design paradigms and corrosion structural effects

For military and commercial aviation industries, to include all varieties of fixed-wing, rotary-wing, and power plants, a variety of structural design philosophies have evolved, some of which have become synonymous with structural management philosophies.

This section will briefly examine the strengths and weaknesses of each of these philosophies, particularly related to corrosion structural effects, and will introduce an evolved, holistic, design and sustainment philosophy to combat the growing concern of corrosion in aircraft.

The philosophies of interest include:

- safe-life,
- fail safe,
- damage tolerance,
- holistic structural integrity.

We will now consider each individually.

3.6.1 Safe-life

Britain's de Havilland Comet began commercial jet service in 1952. The aircraft and the service it provided had captured the world's attention, including that of US aircraft manufacturers, whom it now appeared, were well behind their competition. The aircraft's reputation was tarnished, though, when less than two years after the jet started its record-breaking service, two Comets disintegrated in flight within three months of each other.

The investigation of these crashes quickly focused on fatigue and, after eliminating a wing failure as the cause, a full scale fatigue test of the fuselage soon revealed that the fatal failure sequence was caused by cracks emanating from countersunk fastener holes immediately adjacent to severe radii corners at windows and escape hatches (Wanhill, 2002).

From this investigation, it appeared likely that further fatigue failures would occur as engineers continued to use higher-strength alloys that did not exhibit similar increases in fatigue resistance. Also, it became clear that life estimates would be necessary for fatigue critical structures in aircraft.

From this requirement, safe-life fatigue design was born, albeit history has repeatedly shown this name to be a bit of a misnomer. Older definitions of safe-life meant that the structure has been evaluated to be able to withstand the repeated loads of variable magnitude expected during its service life without detectable cracks. In more recent times, the definition has evolved to mean 'that number of events such as flights, landings, or flight hours, during which there is a low probability that the strength will degrade below its design ultimate value due to fatigue cracking (FAA 1998).

In safe-life design, structure and components are certified by analysis and by laboratory testing (including full-scale structure). These tests are usually conducted in controlled, benign environments. The accelerated nature of the testing usually means that corrosion does not appear as a failure mode in the laboratory. Thus, the impact of corrosion damage is not assessed. Designers and regulators are aware of this and attempt to protect the structure and the people using them with safety factors, sometimes very substantial safety factors. However, the effects of corrosion on structure are extremely varied, and the response of the structure to such damage is dependent on a host of factors such as operational stress level, usage severity, material strength and fracture properties, availability of alternate load paths, and so on. The end result is that many structures are retired well before they need to be retired; yet, others fail well before the expiration of their rated 'safe-life.'

It is difficult to imagine a wider range of scenarios emanating from one design and maintenance philosophy, and this can be directly attributed, in part, to the fact that corrosion has not been assessed, even though tools now exist that can eliminate this problem. Since, under safe-life, corrosion has not been assessed, corrosion cannot be allowed to exist on the structure; the standing order is to remove corrosion from everywhere that it is seen. In some cases, this may be a very prudent decision, such as in a critical rotor component on a helicopter. Indeed on such components, it is likely that the component will be replaced, not repaired. But the safe-life philosophy is indiscriminant of components.

One of the primary reasons that corrosion is the proverbial thorn in the side of safe-life designs is that safe-life analyses are stress- or strain-based.

Thus, as long as stresses or strains do not exceed a certain value, then it is envisioned that cracks will not form in the projected life of the part. However, as discussed previously, corrosion pits and fretting scars form stress risers sufficient to encourage the formation of cracks.

This is certainly not a new realization. In 1917, Haigh (1917), working with brasses, was busy conducting the first corrosion fatigue studies. Of more eventual significance to the aircraft industry, the 1920s found Moore (1927), and McAdam (1927) observing degraded performance in high-strength aluminum, then called duralumin. The 1927 study conducted by Moore showed that prior corrosion had a larger effect on the fatigue endurance of duralumin specimens. Moore found that the cyclic endurance limit or duralumin tube was reduced as much as 50% when the material was corroded before fatigue testing.

McAdam published his work in the very same journal as Moore and added some insight into the corrosion/fatigue interactions when he wrote the following passage: 'Owing to surface irregularities and inclusions in the specimen, the actual stress is always more or less greater than the nominal stress.'

Thus, it was recognized early on that surface integrity is an important factor in the fatigue performance of an alloy, and McAdam's comment became the subject of many later prior corrosion/fatigue studies including work by the USAF in the 1960s (Harmsworth, 1961; Gruff and Hutcheson, 1969). A brief search of the literature will reveal that studies into these phenomena continue today – some 90 years past the first contributions of Haigh.

Another significant area of concern in the safe-life philosophy lies within what the name itself implies: that a certified safe-life for the component exists, and that this safe-life has accounted for all foreseeable threats to structural integrity. This implication leads to a lack of directed inspections; after all, there is no perceived need to inspect something that has a certified crack-free life. Inspections that do occur are merely inspections of opportunity. These inspections quite possibly lack the rigor of fully understanding component criticality and the damage sought, the optimum methods for NDT, the interval at which to conduct the inspection, and the reliability of the inspection result. None of these critical factors are inherent to the system.

Aviation has a lengthy history of failed 'safe-life' components such as wing attachment fittings (SCAA, 1991), engine discs (NTSB, 1991), turboprop impellers (McLeod, 1979), propellers (NTSB, 1993), and landing gear (Hoeppner *et al.*, 1995). This list of components contains some of those most absolutely critical to safe operation of an aircraft. The failures of these components, with the exception of landing gear, have an increased likelihood of leading to serious or fatal accidents.

To help address the problems related to failures of safe-life designed major airframe components, the USAF started its Aircraft Structural Integrity Program (ASIP) in 1958 (Lincoln, 1985). With this new program, design started to shift in the 1960s towards fail-safe design, which is discussed next. However, even with the obvious inadequacies of safe-life design, many components, such as landing gear and commercial aircraft engines and helicopter rotors, continue to be designed with this criterion.

3.6.2 Fail-safe design

Fail-safe design was essentially an extension of the safe-life concept (it continues to be used today, but it is not a stand-alone design methodology in the USAF and in FAA Part 25 regulations for commercial transports). In these regulatory environments, fail-safe designs still need to meet damage tolerance requirements. To quote Wanhill (2002): '. . . a fail-safe design concept does not by itself constitute a fail-safe design. Inspectability is equally important . . .'.

Although safe-life had been an improvement in design philosophies, fatigue failures still abound. The principle of fail-safety was to provide redundant load paths as back-ups in the event of localized failure. The FAA's (2005) accepted definition is as follows: 'fail safe is the attribute of the structure that permits it to retain its required residual strength for a period of unrepaired use after the failure or partial failure of a principal structural element'.

Goranson (1993) explains that fail-safe has had a decent but imperfect record in commercial jet aircraft. Structural damage, including corrosion, has been sustained many times without catastrophe. Goranson illustrates some shortcomings in fail-safe design, especially in aging transport structures: 'crack initiation in adjacent, redundant members is likely and similar unless the load paths are totally independent or significantly different. Thus, accepting the existence of the circumstances that necessitated redundancy also means accepting that the redundancy is not very effective in some instances to provide desired structural reliability.'

This failure scenario has occurred in Air Force fighter aircraft as well, highlighted by the 1973 loss of an F-4 Phantom II due to fracture of its 'fail-safe' wing structure. This crash made it painfully clear that structure could not be truly 'fail-safe' without inspection (Lincoln, 1985).

This realization was paralleled in the commercial industry as airline operators, already flying aircraft beyond typical lives, were expected to find cracks that were, according to Goranson, 'far from obvious.' To find cracks, however, was no easy task. Knowing how to look, where to look, and how often to look provided some serious challenges to operators. In the early 1970s, the Air Force tackled these problems by radically modifying ASIP to incorporate inspection intervals based on crack growth rates and critical crack sizes (Lincoln, 1985). The resulting structural integrity philosophy was called 'damage tolerance.'

3.7 Damage tolerance

3.7.1 Fundamentals of the concept

The concept of damage tolerance incorporated, for the first time in fatigue design, the possibility that cracks exist in the structure of aircraft. The FAA (2005) defines this design requirement as follows: 'Damage tolerance is the attribute of the structure that permits it to retain its required residual strength for a period of use after the structure has sustained a given level of fatigue, corrosion, accidental, or discrete source damage.'

On the surface, this may sound like fail-safe design in that the structure must be able to sustain damage without failure. However, implementation of damage tolerance essentially requires directed inspection. For inspection to be as useful as possible, engineers and operators need to know how often to look. Cracks in a damage-tolerant structure are viewed as a preexisting condition, and the goal of this fatigue design tool is to predict how long it will take a crack to grow from its assumed initial size to its critical size in the component. Inspection intervals are then set to monitor for such cracks well before they are predicted to become critical.

The success of damage tolerance does not hinge only on knowing when to inspect structure. We also need to know what we are looking for, where to look for it, and how to look. This is no small task, as knowing what we are looking for involves understanding possible failure modes, knowing where to look requires identifying critical and otherwise susceptible structure, knowing how to look involves understanding NDT systems and associated probabilities of detection, and knowing when to look requires the understanding of damage mechanics and propagation.

Despite this mouthful of complex issues, Goranson (1993) reports the incorporation of damage tolerance into the USAF's ASIP reduced hull losses (destroyed aircraft) by an impressive 80%. However, as Lincoln pointed out, this concept still had its limitations in that USAF damage tolerance certifications evaluated monolithic (single) cracks in structure (Lincoln, 1994). In contrast to the monolithic cracks assumed and analyzed under the damage tolerance paradigm, structural degradation often takes very different forms, such as multiple-site damage (MSD) and multiple-element damage (MED), both of which are subsets of widespread fatigue damage (WFD).

The propensity of corrosion and wear to greatly accelerate crack nucleation under fatigue loading can be a major influence on the formation of

WFD and its subsets of MSD and MED in a structure. The literature supports that the stress concentrating effect of corrosion and wear scars not only reduces the time to crack nucleation, it also reduces the overall variability in fatigue life. This makes it more likely that corrosion in aircraft structure, if widespread across many structural details, will lead to MSD, MED, and WFD.

3.7.2 Assessing the influence of corrosion fatigue in damage tolerance today

As stated above, damage tolerance design and maintenance practices in aerospace are not structured to handle corrosion. In the realm of damage tolerance, corrosion considerations are usually limited to crack propagation acceleration from corrosion fatigue. For example, this approach was used when the USAF certified structure in the KC-135 (originally designed as fail-safe in the late 1950s) as damage tolerant. Crack growth rates of the structural alloys exposed to water were used to determine inspection intervals.

The lifing of structure and the determination of inspection intervals seldom incorporate corrosion influences beyond corrosion fatigue for some interesting reasons. Recall that the damage tolerance assumes that cracks already exist in a structure. For instance, many fatigue-critical locations on the KC-135 are assumed to already contain cracks that are 1.25 mm in length. These assumed initial cracks are larger than typical corrosion details (e.g., pits), so corrosion is deemed unimportant in these locations.

The other area where corrosion fits into current damage tolerance applications is in the effect of prior corrosion damage on crack propagation rates. Under these circumstances, the thinning of material translates to higher net section stress and higher crack growth rates.

3.7.3 Potential pitfalls of damage tolerance regarding corrosion

Those who resist becoming more holistic and integrating corrosion analyses into the damage tolerance framework – and there are many – almost always argue that the assumed initial crack sizes in a damage tolerance analysis are considerably larger than any possible physical corrosion damage. This logic overlooks a number of interesting and crucial factors about how corrosion affects structure and about how we use damage tolerance from a practical sense. Such arguments thus far have provided an effective, if unwise, block to technology development and integration. Consider the following arguments for augmenting damage tolerance analysis (DTA) to include state-of-the-art corrosion structural effects models.

First, relatively few components in an aircraft structure are actually managed as fracture critical. The most critical locations (based on crack growth life from an assumed starting crack size) in these components are known as control points. Recall that the heart of damage tolerance is inspection, and each location deemed critical must be inspected. As such, practical economic limits apply to the number of components treated with the full protection afforded by damage tolerance. For the components that are selected for management by DTA, the implementation of a more holistic methodology might provide only minimal enhancements in terms of safety and cost, unless of course these components are designed in a manner that makes them vulnerable to WFD.

Second, the starting crack size for an analysis at a damage tolerance control point is not always of size much larger than typical corrosion details. The crack size most-often associated with damage tolerance is 1.25 mm. This is an order of magnitude larger than many corrosion pits, for instance, and corrosion structural effects models suggest that a pit of this size has minimal influence on the propagation rate of a co-located fatigue crack. However, crack sizes associated with control points at cold-worked fastener holes (and many fighter structures use cold-working at critical locations of tension-dominated structure), are sometimes as small as 0.125 mm, (the 'minimum allowed' in Lincoln, 1985). Neither of these initial crack sizes are physically based, and this fact could provide another chapter's worth of material, but suffice it to say at this point that the effects of pitting, for example, on cracks of this size are real and detrimental (this also says nothing of the effectiveness of the cold-working in the presence of corrosion damage). The end result is that even the conservative numbers produced by a DTA in such a location could rapidly become unconservative in the presence of pit-like corrosion morphologies.

Third, where control points are not concerned, the rest of the structure is managed and analyzed quite differently. Crack sizes associated with so-called 'durability components,' for instance, are much smaller than those associated with damage tolerance components. Durability analyses (more correctly called reliability assessments) are supposed to provide an estimate of the reliability of the system and its 'economic life.' Durability crack sizes, like the crack sizes used for control points at cold-worked fastener holes, are small enough that geometric corrosion effects are very real and significant. Exclusion of corrosion from durability analyses ensures that a more accurate economic life will never be developed for a component or airframe. It is interesting, if not puzzling, that:

- durability analyses at the design stage are supposed to give a measurement of the aircraft's economic life,
- the economic life is often determined in practice by runaway corrosion costs, and
- the original durability analysis did not consider corrosion.

Fourth, exclusion of corrosion from analyses ensures that the industry will never move from the 'find and fix' mentality associated with this degradation mechanism. The lack of understanding of corrosion structural effects has led to this mentality. What makes the situation even more unstable is that NDT for corrosion is becoming more effective year after year to the point now that sensor-based technologies are available, some of which claim to be able to detect even minute pitting *in situ*. Enhanced NDT coupled with unchanging management strategies (i.e., 'find and fix') will rapidly lead to exploding maintenance costs and increased aircraft downtime. This can be disastrous to both military and commercial operators. The only way to effectively implement advanced NDT for corrosion is with concurrent development of analytical capabilities. Anything less will be unaffordable and will inherently undermine the great advances in NDT.

Fifth, the fracture critical locations on the aircraft have recurring inspection intervals set by analysis assuming an initial crack size that is tethered to a defined probability of detection (POD) of that crack at a defined confidence level. Since the life calculated for the part is intimately tied to the POD of the chosen NDT method, much benefit is gained by advancing NDT technologies that have sufficient POD and confidence at successively smaller crack sizes. As mentioned repeatedly in this chapter, the current damage tolerance crack size in redundant ('fail safe') structure in fixed-wing aircraft is often given as 1.25 mm. At these sizes, the geometric effects of corrosion are greatly diminished (although not eliminated). However, if you reduce the damage tolerance crack size to, say, 0.75 mm or smaller, the effects of corrosion become quite pronounced – so pronounced in fact that the safety factor of two applied to the inspection interval may become unconservative. This possibility could lead to a couple of conditions: either we could adopt the new NDT but not update our analyses, at which point we are in danger of catastrophic failure, or we could accept that analytical capabilities have not kept pace with NDT and, as discussed above, not be able to effectively implement the enhanced NDT.

Sixth, damage tolerance analyses were, as mentioned earlier, structured around monolithic cracks. They were not, as Swift (1994) points out, set up to handle MSD, MED, and WFD.

> Since the Aloha accident it has been the contention of the airworthiness authorities that managing structural safety in the presence of WFD or MSD

> is not reliable with current in-service inspection sensitivity. In fact, the FAA no longer allows continued inspection of known problem areas as an alternative to fixing the problem itself. The issue is, then, how can the monitoring period be technically viable when MSD cannot be reliably found before it has already reduced residual strength capability below regulatory levels.
>
> The concern appears to be confined to just a few people who fully understand the implication of MSD. The others appear to think that small MSD can never be a problem because they have been able to tolerate much larger cracks in the past. This is true but it is not the issue. The issue is that the airplane is designed to tolerate certain lead crack sizes. The inspection program is based on these lead cracks sizes, and MSD has significant effect on residual strength at these sizes.

This issue still poses a serious problem. Absent from Swift's observations was a discussion on what often leads to MSD, MED, and WFD. Corrosion and fretting are capable contributors to the initial onset and propagation of these insidious failure modes. Understanding all of these effects, both individually and synergistically, is crucial to a more holistic view of structural integrity.

Seventh, DTA has evolved in the industry to essentially be synonymous with crack growth analysis. As with safe-life, a substantial disconnect exists between the assumptions in the lifing paradigms and the physical process of failure. We pay no mind to how the cracks got there, which leads us to repeat our mistakes and to develop incomplete or even improper solutions to problems that arise.

The need for a more holistic design philosophy is not a recent epiphany. In fact, significant works in the literature began to appear (Hoeppner, 1971) even before Damage Tolerance became law in the USAF and FAA. This next section just touches on the elements of holistic structural integrity, a philosophy that seeks in part to reduce the threat of corrosion to structural integrity.

3.8 Holistic structural integrity

The theme that courses through the previous discussions on traditional aircraft lifing methodologies is that these methodologies consider only the cyclic domain (fatigue). This holds true for both damage tolerance and safe life methodologies. However, corrosion does not honor this practice.

What is needed is a design and sustainment process that properly accounts for the corrosion/fatigue interactions prevalent in aircraft structures and is flexible enough to efficiently augment existing design and sustainment methodologies (particularly inspection-based methodologies such as damage tolerance).

Consider these words by Macferren (1994):

> Imagine an ideal engineering situation. We would know all the types of damages that might cause our product to fail. We would understand all failure modes, the mechanisms contributing to those modes, and the driving forces behind each mechanism. We would understand synergism between mechanisms. We would have equations characterizing each mode that are based on equations characterizing each basic failure mechanism, including synergism. We would have equations characterizing the applied loads: mechanical, thermal, chemical, and so on. We also would understand how all quantities vary; for example, we would have equations characterizing the differences in how the product will be used. From this information we would calculate the probability of failure by any mode and the overall probability of failure. Most engineers think such a scenario is impossible – and unnecessary.

The previous pages in this chapter ought to have dispelled the notion that such engineering power is unnecessary. As for being impossible – technology has advanced substantially even in the few short years since Macferren penned the above quote. Although we may not have the complete control of every possible failure mode, synergism, and probability of failure, we do stand ready as a community to implement new tools that can greatly enhance the state-of-the-art beyond safe-life and damage tolerance as they are known today.

In 1998, Brooks and Simpson (1998) stated that technologies have matured sufficiently to allow improved incorporation of corrosion and age-degradation effects into a systematic assessment framework. The integration of these holistic technologies into current fleet management infrastructures is feasible. Combining the impacts of damage accumulation from operational stress excursions of a structure with the impacts of damage accumulation caused by environmental exposure while at rest or in operation is key to this holistic concept. Brooks and Simpson showed examples of tasks that would be added to (note: not replace!) the existing tasks in the USAF ASIP in order to make it more holistic in assessing and managing corrosion.

A key ingredient to such holistic assessments, the accurate modeling of corrosion structural effects, requires two principal elements:

1. simulating how corrosion damage morphologies change over time, and
2. simulating how various corrosion forms interact, either in series or in parallel, with mechanical loads to affect the fatigue and static strength response of a material and structure.

Frameworks and analytical models have been developed in recent years that provide these elements. Chapter 8 is dedicated to corrosion structural effects models, so little detail is given here. A wealth of model development has been channeled towards general attack, pitting, and exfoliation (see earlier sections in this chapter dedicated to these failure modes for sample references) with concerted efforts coming from the US, Canada, and Australia. These models have been used to successfully assess structure and provide solutions for problems that were historically unsolvable using other assessment techniques. One such example has been cited previously from Mills *et al.* (2004), and this example was one of many explored during the USAF Corrosion Fatigue Structural Demonstration program, a sizeable advanced development effort that spanned 2000 to 2003.

3.8.1 Benefits of holistic structural integrity

A number of benefits of holistic structural integrity efforts have been identified, particularly where corrosion is concerned.

First, corrosion, corrosion-fatigue, or fatigue damage is often hidden and can only be estimated through expensive means: teardowns, advanced NDT techniques, or data from historical fleet problems. Holistic assessments provide a way to estimate that hidden damage and/or supplement the expensive techniques with analytics.

Next, by superimposing the predicted effects of corrosion and fatigue on the current health of a structure (as determined by NDT or estimated using holistic modeling), it now becomes possible to determine a physically based estimate of when a structure will reach its limit state. This allows for better maintenance options. As stated earlier, the current industry standard for corrosion maintenance is the 'find and fix' approach, which is enormously expensive and may do more harm than good in certain structure.

Also, holistic assessments can help provide targets and detection limits for NDT. What do we really need to be able to find to stay safe?

Finally, holistic assessments provide the metrics that allow structure to be better classified in a range from non-critical to critical, which enables NDT resources and monetary funds to be used in appropriate areas of the aircraft.

3.8.2 Augmenting management of structure

The previous discussions on safe-life and damage tolerance should make it apparent that there is often a considerable difference between the physical discontinuities assumed to be present in a structure versus those that actually are present. These two common design paradigms represent the extremes of assumptions: one being 'crack free' and the other being that sizeable cracks exist from the point of manufacture. Other than crack growth models used to support damage tolerance, no physically based modeling of the fatigue process is used for design.

With this in mind, it should be clear that the actual effects of corrosion on structure could be quite different than the effects produced within the

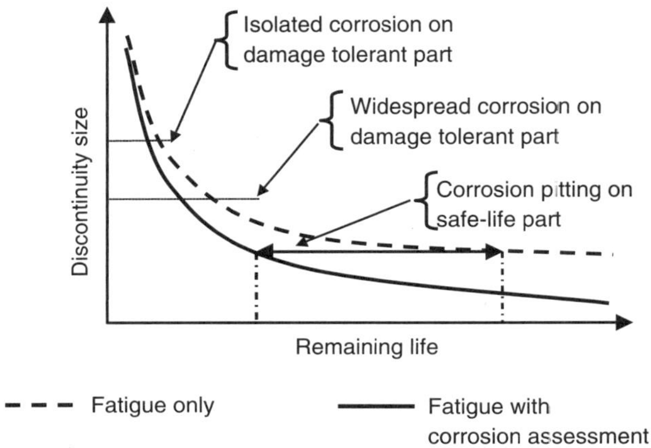

3.1 Notional chart indicates residual life impact of corrosion as a function of initial discontinuity size.

rules of that design methodology. As described earlier in this chapter, one of the most common criticisms of corrosion structural effects by the DTA community is that most assumed initial crack sizes are substantially larger than corrosion details. Recall that in many cases it has been shown that the geometric influence of corrosion damage on the propagation of larger cracks is negligible in the case of pitting or simplified as thickness loss in the case of general attack and exfoliation. This effect quickly amplifies in severity and complexity at smaller crack sizes, as shown in Fig. 3.1. The plot shows notional trends of residual life versus discontinuity size assuming no corrosion (traditional DTA) and with corrosion. At larger crack sizes, the influence of corrosion is minimal, so if this end of the curve represents a damage tolerance control point in a critical area, it is quite possible that incorporating corrosion into the current framework would have little influence on management of that location. On the other hand, the lower end of the curve, representing small initial discontinuities and long residual life is greatly influenced by corrosion. These crack sizes might be those associated with analyses of cold-worked holes or of so-called durability items.

The power of this particular chart is that it can serve double duty by also illustrating the influence of corrosion within the safe-life paradigm. The label of the *y*-axis may be simply changed to read 'stress' rather than 'initial discontinuity size'. The reason this works is that components that are undergoing high stresses have many competing failure modes at work, and cracking occurs so rapidly that there is little practical concern for corrosion. At low stresses, however, corrosion can be devastating with losses in fatigue life sometimes being measured by orders of magnitude.

3.8.3 Elements of holistic structural integrity

Holistic structural integrity is a versatile tool that can positively influence all the activities associated with aircraft engineering, such as design, fabrication, sustainment, and fleet disposition. A variety of tasks have been identified that illustrate the types of analyses and technologies that must be integrated to produce a holistic process. These tasks are:

1. criticality assessment,
2. usage and environment characterization,
3. life and residual strength analyses,
4. nondestructive inspection (NDI),
5. data management and fusion, and
6. process documentation.

The prominence of the individual tasks will naturally vary depending on the phase of life that a system is in. It is possible that a product may be in every phase of life simultaneously. Take the Lockheed F-16 as an example: its long production run and great numbers has certain assets being retired at the same time new ones are being produced. Further to that, mission types (thus load spectra) have continuously evolved to the point that they bear no resemblance to those envisioned during design. This one example shows the need to have a living, closed-loop process. Just because a structure has been designed once and the asset has been in the field for 20 years does not mean that circumstances surrounding its use will not change. The following discussions briefly touch on the first three points.

Criticality assessment is an important first step to holistic analysis. The idea behind it is to establish consequence of failure associated with a component or system and then to ascertain the likelihood of such a failure occurring. Aircraft have many fracture-critical components. However, all fracture-critical parts are not created equal. A host of factors play in determining the susceptibility to failure. Such factors are:

- material,
- geometry,
- product form,
- manufacturing and joining methods,
- environment (stress, chemical, thermal),
- protection systems,
- inspectability,
- maintainability, and
- alternative load paths.

Consider a criticality assessment of a component using corrosion as a failure mode. Material selection can have very profound influences on corrosion susceptibility. Likewise, part geometry and product form can dictate which areas of a component are susceptible and to what types of corrosion the part is susceptible. Riveted structure versus bonded versus welded would have even more corrosion implications. The magnitude and direction of residual stresses, severity of external loads, and ease of access for corrosive environments (and presence of protection schemes) influence the chance that corrosion will occur and the structural response if it does occur. Inspectability and maintainability also enter the equation. Is the damage size that could cause failure smaller than can be reliably detected by existing NDI methods? If damage is detected, can it be effectively mitigated?

Alternative load paths are another important consideration in component criticality. The most critical structures often do not have alternative load paths, so components have to be quite strong. This high strength requirement can lead to selection of materials that may have poor toughness qualities and abhorrent corrosion resistance, thus elevating the risk associated with the subject components.

Usage and environment characterization is a very important aspect of holistic analysis. Clearly, one cannot adequately assess the life of a component without a good understanding of the loads in the system. Similarly, chemical environments can seriously impact crack growth rates, set up EAC conditions (such as stress corrosion), or induce corrosion-nucleated fatigue. Thermal loads and cycles also must be considered, as general material response can be altered by variations in temperature, and temperature-related failure mechanisms, such as creep, may become active.

All of these variations in loads, temperature, and chemical environment can affect component susceptibility (thus criticality), and certainly are key to successful life and residual strength analyses. The point here is that none of these 'tasks' being discussed are independent and all must be integrated into a closed loop process.

To help illustrate the importance of load spectra in determining criticality and determining vulnerability of the structure to corrosion, fatigue life predictions were made for a fictitious part made from the same material and subject to identical levels of damage. The only difference was in the load spectra applied in the models.

The first spectra simulated a lower wing for a C-130 Hercules. Figure 3.2 shows that with no corrosion, the part lasted 50000 spectrum flight hours. The addition of pitting but with no thickness loss cut the projected life by more than half (18200 h). Thinning to 30% material loss further reduced capability to 4150 h. Compare this with the same levels of damage only using a spectrum for the E-8C JSTARs aircraft. This load spectra is for the upper wing, as opposed to the lower wing as with the C-130. Here, the most severe corrosion case lasts 97400 spectrum hours. All other cases with less

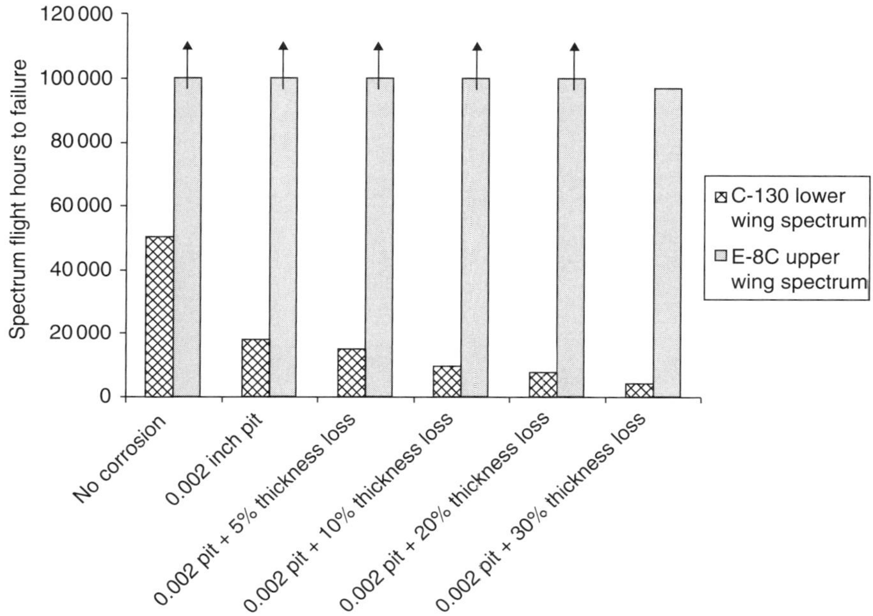

3.2 Sample analyses show the importance of understanding load spectra in fatigue analysis. Here, two identical part geometries with identical amounts of corrosion were subjected to substantially different spectra with vastly different results. Corrosion may be created equal, but corrosion structural effects are not.

corrosion would last well over 100 000 spectrum hours, so the lives in those cases are shown with arrows beyond 100 000 (Fig. 3.2). It is critical to understand that the model of the corrosion damage has not changed. So in this case, the exfoliation may have been created equal, but the structural effects definitely were not equal.

Holistic life and residual strength analyses make use of five broad categories of elements that are necessary for arriving at a solution. These elements are summarized in Fig. 3.3 and include examples of the types of factors that influence each element. Also depicted in the figure is the notion that the type of information needed for an analysis is influenced, in part, by the overarching design philosophy and by certification requirements. At a top level, these elements are part of any structural analysis including failure analysis. However, what separates the holistic approach from previous methodologies is the goal to increase the realism in the analysis, to become physically based, and to account for physical effects and failure modes that have previously been covered by safety factors or not covered at all.

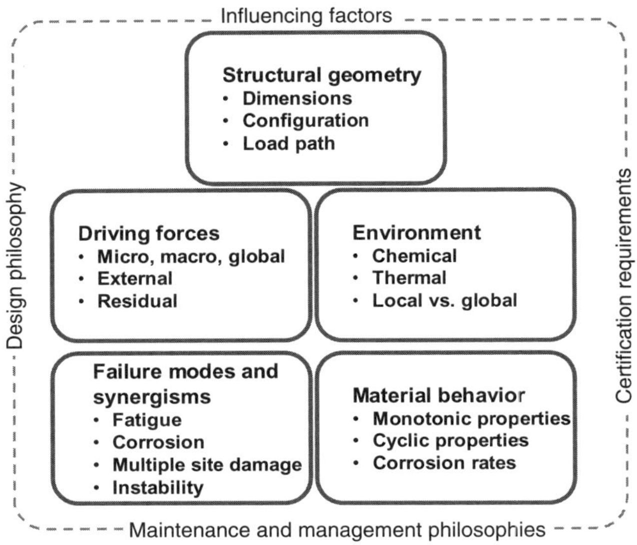

3.3 Primary elements of a holistic structural analysis including the effects of corrosion.

3.9 Conclusions

In this chapter we have provided an in-depth look at corrosion effects on structural integrity, with particular focus on aircraft systems. The discussion touched on economic and safety issues, types of corrosion and their deleterious influences on structure, and the analysis of corrosion within existing and evolved, holistic design and maintenance philosophies.

Existing structural integrity methodologies tend to treat corrosion inadequately. The safety factors put into place in safe-life prove to be unreliable. The assumptions in damage tolerance that large assumed crack sizes 'cover' the influence of corrosion might be valid in some cases, but there are many cases where that assumption is invalid or just unaffordable.

The day is coming soon, for instance, when designers of propeller systems and rotorcraft components will be required to certify these formerly 'safe-life' and vulnerable components as damage tolerant. Such a major shift in philosophy will clearly require effective corrosion structural effects analysis, particularly for pitting. Fortunately, such tools are becoming available, and with an attendant holistic view of structural integrity, we have before us a golden opportunity to greatly enhance safety of these systems along with the safety of major airframe components, many of which are starting to experience WFD (an eventuality not formerly considered by damage tolerance and its 'rogue flaw' assumptions).

A final point is that the holistic approach is adaptable to current structural assessment methodologies to make them more robust, realistic, and capable. Holistic structural integrity is not a revolution, but it is the logical next step in the evolution of aircraft structural integrity.

3.10 References

AMERICAN SOCIETY FOR METALS (ASM) (1985), Metals Handbook, 9th ed., Vol. 13, *Corrosion*, American Society for Metals, Metals Park, OH, 584–609.

AMERICAN SOCIETY FOR TESTING AND MATERIALS (ASTM) (1986), *Compilation of ASTM Standard Definitions,* 6th ed., American Society for Testing and Materials, Philadelphia.

AMERICAN SOCIETY FOR TESTING AND MATERIALS (ASTM) (1990), 'ASTM standard G34-90, standard test method for exfoliation susceptibility in 2xxx and 7xxx series copper containing alloys', American Society for Testing and Materials, Philadelphia, (1990).

BALDWIN J, MILLS T and PAUL C (1997), 'Statistical analysis of fatigue behavior of aluminum alloys in the presence of prior corrosion', in Cook R and Poole P, *Fatigue in new and aging aircraft*, Proceedings of the 19th symposium of the International Committee on Aeronautical Fatigue, Edinburgh, Scotland.

BROOKS C, HONEYCUTT K and PROST-DOMASKY S (2001), 'Monitoring the robustness of corrosion and fatigue prediction models', *2001 USAF aircraft structural integrity program conference*, San Antonio, Texas.

BROOKS C and SIMPSON D (1998), 'Integrating real time age degradation into the structural integrity process', *Proceedings, NATO RTO's workshop 2 on fatigue in the presence of corrosion*, Corfu, Greece, North Atlantic Treaty Organization, Research and Technology Organization, p. 22-1 to 22-13.

CAMPELL G and LAHEY R (1984), 'A survey of serious aircraft accidents involving fatigue fracture', *International Journal of Fatigue*, **6**(1), 25–30.

CHUBB J, MORAD T, HOCKENHULL B and BRISTOW J (1991), 'The effect of exfoliation corrosion on the fatigue behavior of structural aluminum alloys', *Structural Integrity of Aging Airplanes*, 87–97.

CHUBB J, MORAD T, HOCKENHULL B and BRISTOW J (1995), 'The effect of exfoliation corrosion on the fracture and fatigue behavior of 7178-T6 aluminum', *International Journal of Fatigue*, **17**(1), 49–54.

COOKE G, VORE P, GUMIENNY C, COOKE JR. G, LUNSFORD E and KEALY H (1990), 'A study to determine the annual direct cost of corrosion maintenance for weapon systems and equipment in the United States Air Force', USAF contract #F09603-89-C-3016.

CRAWFORD B, LOADER C and SHARP P (2004), 'The effect of pitting corrosion on the position of aircraft structural failures,' *Proceedings, structural integrity and fracture 2004*, Brisbane, Australia.

DE LUCCIA J (1991), 'The corrosion of aging aircraft and its consequences', AIAA-91-0953, AIAA 32nd Structures, Structural Dynamics, and Materials Conference, Baltimore, MD, USA.

FEDERAL AVIATION ADMINISTRATION (FAA) (1998), 'Damage-tolerance and fatigue evaluation of structure', *Advisory Circular 25.571-C.*

FEDERAL AVIATION ADMINISTRATION (FAA) (2005), 'Fatigue, fail-safe, and damage tolerance evaluation of metallic structure for normal, utility, acrobatic, and commuter category airplanes', *Advisory Circular 23-13A*.

GORANSON U (1993), 'Damage tolerance – facts and fiction', *17th Symposium of the International Committee on Aeronautical Fatigue*, Stockholm, Sweden.

GRUFF J and HUTCHESON J (1969), 'Effects of corrosive environments on fatigue life of aluminum alloys under maneuver spectrum loading', *AFFDL Technical Report 70-144*, 521–537.

HAIGH (1917), 'Experiments on the fatigue of brasses', *Journal of the Institute of Metals*, **18**, 55–86.

HARMSWORTH C (1961), 'Effect of corrosion on the fatigue behavior of 2024-T4 aluminum alloy', *ASD Technical Report 61-121*.

HOEPPNER D W (1971), 'Corrosion fatigue considerations in materials selections and engineering design', *Corrosion Fatigue*, NACE-2, 3–11.

HOEPPNER D W, GRIMES L, HOEPPNER A, LEDESMA J, MILLS T and SHAH A (1995), 'Corrosion and fretting as critical aviation safety issues: Case studies, facts, and figures from US aircraft accidents and incidents', in Grandage J and Jost G, *Estimation, Enhancement and Control of Aircraft Fatigue Performance*, Warrington, EMAS.

HUFFMAN J (2000), 'HC-130 left rear main landing gear spindle failure investigation', Internal Report to the US Coast Guard.

KAWAI S and KASAI K (1985), 'Considerations of allowable stress of corrosion fatigue (focused on the influence of pitting)', *Fatigue Fracture of Engineering Materials Structures*, **8**(2), 115–127.

KINZIE R (2004), '2004 USAF direct costs of corrosion,' online at http://www.corrdefense.org/ReferenceLibrary.aspx.

KOCH G, HAGERDORN E and BERENS A (1995), 'Effect of preexisting corrosion on fatigue cracking of aluminum alloys 2024-T3 and 7075-T6', *Final Report to Flight Dynamics Directorate*, USAF Research Laboratory, August 1995.

KONDO Y (1989), 'Prediction of fatigue crack initiation life based on pit growth', *Corrosion Science*, **45**(1), 7–11.

LIFKA B and SPROWLS D (1972), 'Significance of intergranular corrosion on high-strength aluminum alloy products,' *Localized corrosion-cause of metal failure,* ASTM STP 516, American Society for Testing and Materials, 120–144.

LINCOLN J (1985), 'Damage tolerance – USAF experience', in Salvetti A and Cavallini G, *Durability and damage tolerance in aircraft design*, Proceedings of the 13th Symposium of the International Committee on Aeronautical Fatigue, Pisa, Italy, 265–295.

LINCOLN J (1994), 'Challenges for the aircraft structural integrity program', in Harris C, *NASA conference publications 3274 part I, FAA/NASA International Symposium on Advanced Structural Integrity Methods for Airframe Durability and Damage Tolerance*, 409–423.

LINDLEY T, MCINTYRE P and TRANT P (1982), 'Fatigue crack initiation at corrosion pits', *Metals Technology*, **9**, 135–142.

MACFERRAN D (1994), 'Towards a postulate-based methodology for developing specifications and failure criteria', Dissertation, University of Utah.

MCADAM, JR D (1927), Corrosion-fatigue of non-ferrous metals, *Proceedings of the ASTM,* **27**(2), 102–125.

MCLEOD R (1979), 'First stage impeller failure: Rolls Royce Dart 514–7 engine S/N 12023, Quebec Fairchild F-27, C-FQBL, 29 March 1979', *Engineering Report LP 77/79.*

MILLS T (1995), 'The effects of exfoliation corrosion on the fatigue response of 7075-T651 aluminum plate', Thesis, University of Utah.

MILLS T (1997), 'The combined effects of prior-corrosion and aggressive chemical environments on fatigue crack growth behavior in aluminum alloy 7075-T651', Dissertation, University of Utah.

MILLS T, CLARK G, LOADER C, SHARP P and SCHMIDT R (2001), 'Review of F-111 structural materials', *Defence Science and Technology Organisation Technical Report*, DSTO-TR-1118, Australia.

MILLS T, SHARP P, LOADER C (2002), 'The incorporation of pitting corrosion damage into F-111 fatigue life modeling, *Defence Science and Technology Organisation Research Report*, DSTO-RR-0237, Australia.

MILLS T, HONEYCUTT K and BROOKS C (2004), 'Demonstration of an holistic structural integrity process using corrosion/fatigue interactions from laboratory experiments and field experience', *Proceedings, 6th International Aircraft Corrosion Workshop*, Solomon's Island, MD.

MILLS T, HONEYCUTT K, BROOKS C, SHARP P, LOADER C and CRAWFORD B (2004), 'Development and demonstration of an holistic structural integrity process using the initial discontinuity state concept for 7050-T7451 aluminum', *2004 USAF Aircraft Structural Integrity Program Conference*, Memphis, TN.

MOORE R (1927), 'Effect of corrosion upon the fatigue resistance of thin duralumin', *Proceedings of the ASTM*, **27**(2), 128–152.

NATIONAL TRANSPORTATION SAFETY BOARD (1989), 'Aloha Airlines, flight 243, Boeing 737-200, N73711, near Maui, Hawaii, April 28, 1988', *Aircraft Accident Report,* NTSB AAR-89/03, Washington DC.

NATIONAL TRANSPORTATION SAFETY BOARD (1991), 'Compressor disk pitting linked to JT8D failure', *Aviation Equipment Maintenance,* October 1991.

NATIONAL TRANSPORTATION SAFETY BOARD (1993), 'Aircraft accident report, in-flight loss of propeller blade and uncontrolled collision with terrain, Mitsubishi MU-2B-60, N86SD, Zwingle, Iowa, April 19, 1993', *Aircraft Accident Report*, NTSB/AAR-93/08.

ROMANS H (1969), 'An accelerated laboratory test to determine the exfoliation corrosion resistance of aluminum alloys', *Materials Research and Standards*, **9**(11), 31.

SHAFFER I, SEBASTIAN J, ROSENFELD M and KETCHAM S (1968), 'Corrosion and Fatigue Studies of Extruded 7075-T6 Spar Caps', *Journal of Materials*, **3**(2), 400–424.

SHARP P, COLE G, CLARK G and RUSSO S (1998), 'The influence of corrosion on aircraft structural integrity', *Proceedings, 21st Congress of the International Council Aeronautical Sciences*, Melbourne, Australia.

SHARP P, MILLS T, RUSSO S, CLARK G and QIANCHU L (2000), 'Effects of exfoliation corrosion on the fatigue life of two high-strength aluminum alloys', *Proceedings, 4th Joint DoD/FAA/NASA Conference on Aging Aircraft*, St. Louis, Missouri.

SMITH W (1993), *Structures and Properties of Engineering Alloys*, McGraw-Hill, Inc., 176–229.

SPROWLS D, WALSH J and SHUMAKER M (1972), 'Simplified exfoliation testing of aluminum alloys', *Localized corrosion – cause of metal failure,* ASTM STP 516, 38–65.

SWEDISH CIVIL AVIATION ADMINISTRATION (SCAA) (1991), 'Aircraft Accident Report, Report C 1991:25, Accident 1990-07-25, Hassela, X Ian, Issue SE-FTP 56/90.'

SWIFT S (1995), 'The Aero Commander chronicle', in Grandage J and Jost G, *Estimation, Enhancement and Control of Aircraft Fatigue Performance,* Warrington, EMAS, 507–530.

SWIFT T (1994), 'Widespread fatigue damage monitoring – issues and concerns', in Harris C, *NASA conference publications 3274 part II, FAA/NASA international symposium on advanced structural integrity methods for airframe durability and damage tolerance*, 829–870.

WALLACE W and HOEPPNER D (1985), 'Aircraft corrosion: Causes and case histories', *AGARD Corrosion Handbook.*

WANHILL R (1995), 'Aircraft corrosion and fatigue damage assessment', *Proceedings of the 1995 USAF Aircraft Structural Integrity Program Conference*, San Antonio, TX.

WANHILL R (2002), 'Milestone case histories in aircraft structural integrity', *National Aerospace Laboratory of the Netherlands*, Technical Publication, NLR-TP-2002–521.

4

Effect of corrosion on the mechanical behaviour of aircraft aluminum alloys

S. G. PANTELAKIS, University of Patras, Greece; and A. T. KERMANIDIS, University of Thessaly, Greece

Abstract: A brief overview of aircraft aluminum alloys, along with a discussion of their susceptibility to corrosion and the various types of corrosion damage is provided. The significance of the effect of corrosion on mechanical behaviour under static and fatigue loading conditions is demonstrated. For this purpose, experimental results concerning the tensile and fatigue behaviour of pre-corroded aluminum specimens are presented. The results, which are supported by metallographic observations, are discussed in terms of the synergetic effect of corrosion damage and corrosion-induced hydrogen embrittlement of the material. The fatigue crack growth and fracture behaviour of pre-corroded aluminum alloys is also examined. Experimental results demonstrate the essential influence of prior corrosion exposure on the material's damage tolerance performance. Corrosion, being a time-dependent and diffusion-controlled process degrades the material properties in a local scale. To describe the fracture behaviour of pre-corroded aluminum alloys, the concept of local fracture toughness is introduced. A mechanical model for assessing the local fracture toughness is presented and incorporated into a fatigue crack growth code for fatigue life assessment of pre-corroded material under irregular loading.

Key words: aluminum alloy, corrosion modeling, hydrogen embrittlement, mechanical behaviour, damage tolerance, aircraft.

4.1 Introduction

The Aloha incident in 1988 warned the aviation industry of the dangers caused by the structural degradation of aging aircraft components and indicated that corrosion of aircraft structures is a problem far more widespread than anticipated.[1] Corrosion of metallic airframes is correlated to the degradation mechanisms that affect the structural reliability, durability, integrity and hence the safety of aircraft. Corrosion is also an economic problem since repair and maintenance procedures undertaken to diminish the effects of corrosion are typically over-conservative due to the lack of reliable methodologies to predict the future effects of corrosion. As a result there is an increasing need for the development of analytical tools for the evaluation of the effects of corrosion on structural behaviour in order to enable the continuous safe and cost-efficient operation of aircraft.[2] The development of such methodologies requires a comprehensive understanding and characterization of corrosion damage mechanisms and relies on the existence of sufficient experimental data. However, limited experimental data exist in published literature on the mechanical performance of corrosion-damaged aircraft aluminum alloys.

In this chapter, a series of experimental results showing the influence of existing corrosion on the mechanical behaviour of aircraft aluminum alloys is presented. In the first part, the static as well as fatigue and damage-tolerance behaviour of the corroded material is examined and, in the second part, analytical models are proposed, which take into account the effect of corrosion for the prediction of the fatigue and damage-tolerance behaviour of corrosion-damaged components. Firstly, a brief literature review is performed on the corrosion susceptibility of aircraft aluminum alloys.[3]

4.2 Corrosion behaviour of aircraft aluminum alloys

4.2.1 Corrosion behaviour of heat-treatable aluminum alloys

2xxx series alloys

Copper is the alloying element in all 2xxx alloys, which also contain magnesium and/or manganese. The 2xxx series alloys are precipitation hardened and after final fabrication, the alloys are solution heat treated and quenched. They are then frequently used in the as-quenched and naturally aged (at room temperature) condition (T3 or T4 tempers), or artificially aged to increase strength via formation of Al–Cu–Mg strengthening precipitates (T6 or T8 tempers). In common with all aluminum alloys, constituent particles play a key role in pit initiation, owing to the galvanic interaction at the particle/matrix interface. Because of the copper content, 2xxx alloys tend to be more susceptible to pitting and general attack than other alloys. During dissolution of Al–Cu–Mg particles, regions of metallic copper are formed on the alloy, thereby promoting accelerated galvanic attack and increasing the pitting severity relative to non-copper-containing alloys.[4] Copper can serve to initiate new pits or can assist in the propagation of already-established pits. Depending on temper and processing specifics, the intergranular (IG) corrosion and stress-corrosion cracking (SCC) resistance of 2xxx alloys can vary significantly. The basic mechanism is galvanic interaction between the grain boundary region and the interior grain matrix.[5,6] A similar scenario exists for SCC and exfoliation corrosion. Pitting susceptibility is related to the copper content. Lower-copper alloys (2036, 2008) exhibit significantly better pitting resistance than higher-copper alloys, such as 2024. The variation in IG corrosion susceptibility as a function of copper content is less pronounced. Effects of quench rate and aging time, as discussed previously, are more dominant factors in determining IG corrosion susceptibility.

6xxx series alloys

The 6xxx series alloys contain primarily magnesium and silicon, some with additions of copper and/or manganese (<1.2%). They are strengthened by precipitation hardening, with the primary precipitate being Mg_2Si. These alloys generally have good corrosion resistance especially to pitting corrosion. Those that contain copper have somewhat poorer corrosion resistance. In common with most commercial aluminum alloys, Al–Fe–X constituents play a dominant role in pit initiation. Intergranular corrosion resistance can vary. Low copper alloys, such as 6061, exhibit good resistance to all forms of corrosion. Alloys with an excess of silicon (that is, more than is needed to form Mg_2Si), tend to be more susceptible to IG corrosion. Copper is added to some 6xxx alloys to augment strength. Examples include 6013, 6056, and 6111. Although pitting susceptibility is only slightly worse than it is for copper-free 6xxx alloys, IG corrosion susceptibility can be significantly worse. The severity is dependent on copper content and temper. Unlike 2xxx alloys, where the IG corrosion susceptibility can be decreased by artificial aging to peak strength, 6xxx alloys often require over-aging beyond peak strength in order to significantly decrease IG corrosion susceptibility.

7xxx series alloys

The 7xxx alloy series all contain zinc (<9%). In addition, nearly all of these alloys contain magnesium (<4%), and many of them contain copper (<3%). They can be generally divided into two categories. The first category consists of the copper-containing, high-strength alloys (e.g., 7075, 7050, 7055). These alloys are available in various forms (sheet, plate, extrusion) and are used extensively in the aerospace industry. The second category is low-copper alloys (e.g., 7005, 7029). These alloys offer somewhat better resistance to general corrosion and pitting (owing to low copper) and are used in structural applications and automotive applications such as bumpers. As in most aluminum alloys, in the 7xxx alloys, constituent particles play a dominant role in pit initiation. The severity of pitting is dependent on the copper content. Lower-copper alloys are somewhat less susceptible to pitting corrosion. While the basic pitting performance and mechanism are similar to those for other aluminum alloys, the IG corrosion, exfoliation, and stress-corrosion cracking (SCC) mechanisms of this class of alloys is far more complex.

Aluminum-lithium alloys, which belong to the 8xxx series, present similar corrosion resistance to many 7xxx series alloys.[3]

4.2.2 Hydrogen embrittlement of aircraft aluminum alloys

Investigations on the corrosion behaviour of aluminum alloys have revealed that hydrogen produced during the corrosion process may diffuse into the material and lead to concentration and trapping of hydrogen at vulnerable sites. This depends on the alloy system.[7–10] Investigations on hydrogen embrittlement of Al–alloys from the 7xxx series. A metastable aluminum hydride has been considered responsible for the brittle intergranular fracture of Al–Zn–Mg alloys subjected to stress corrosion cracking in water vapour.[11] However, the evaluation of the above hydride has been difficult owing to its high instability. The preferential decohesion of grain boundaries containing segregated magnesium is a different explanation for the intergranular fracture of these alloys. Moreover, thermal desorption has been successfully used to study hydrogen diffusion and trapping in Al–Cu, Al–Mg_2Si[12] and Al–Li alloys[13] and recent investigations have provided evidence of a corrosion-induced hydrogen embrittlement mechanism for the 2024 alloy.[9,14]

4.3 Effect of corrosion on the mechanical behaviour of aircraft aluminum alloys

During its service life, an aircraft structure gradually accumulates corrosion damage, thus revealing the need for quantifying the effects of 'prior corrosion' on mechanical behaviour. For aged aircraft that have exceeded their design life this becomes more important since the combined effect of existing corrosion with mechanical loads (e.g. fatigue), can lead to structural integrity degradation.[15] Hence, residual strength assessment of aged aluminum aircraft structures requires a thorough consideration of the material's mechanical performance degradation due to corrosion, which is presently taken into account by limiting the effects of corrosion to a decrease in the thickness of the structural member,[16] as well as an increase of the probability for the onset of fatigue cracks.[17,18]

In this part of the work, experimental results are provided on the effect of prior corrosion on the tensile, fatigue and damage tolerance performance of aircraft aluminum alloys. The experimental data are compared with results for uncorroded material and are discussed with regard to the existing corrosion damage mechanisms under the viewpoint of a synergetic effect of corrosion and corrosion-induced hydrogen embrittlement.

4.3.1 Effect of corrosion on the tensile behaviour of aircraft aluminum alloys

Several investigations[19–21] have provided evidence that corrosion exposure of aluminum alloys from the 2xxx, 6xxx and 8xxx series leads to a degradation of the alloy's mechanical properties and a significant material embrittlement. A summary of the experimental investigations on the tensile behaviour of pre-corroded aircraft aluminum alloys performed in the works[19–21] is presented in the forthcoming paragraph, the results are discussed in terms of the synergetic effect of corrosion and corrosion-induced hydrogen embrittlement.

Experimental procedure and results

Tensile experiments performed[19–20] on the 2024, 2091, 6013 and 8090 aluminum alloys revealed a systematic degradation of tensile properties of these alloys after exposure to varying corrosive environments.[22–26] The tensile experiments were performed according to the specification ASTM E8m-94a.[27] The temper conditions were T351 for 2024, T6 for 6013, T81 for 8090 and T3 for 2091 aluminum alloy. In Figs 4.1 and 4.2, characteristic experimental results are displayed, showing a significant material embrittlement, which was reflected in a dramatic reduction of tensile ductility and strain energy density for the alloys under consideration. In the figures, the residual tensile properties (L-direction) are given as a percentage of the respective properties of the reference material. Exposure of the same alloys in an exfoliation corrosion environment for different lengths of time indicated gradually increasing tensile property degradation with exposure time (Figs 4.3 and 4.4). Complementary to the tests,[19–20] a set of experiments performed for alloy 2024,[21] reproduced the results indicating tensile property degradation with increasing corrosion exposure (Fig. 4.5). In the figure, tensile strength, yield strength, elongation to failure and strain energy density are symbolized with R_m, S_y, A_{50} and W, respectively. The results show that tensile ductility is drastically degraded even after a short exposure time of 20 min and is reduced by 85% after an exposure of 96 h. Further investigation of the influence of corrosion on tensile behaviour and the role of hydrogen in the observed material embrittlement was performed on the aluminum alloys 2024-T351 and 6013-T6.[21]

To investigate the effect of hydrogen on the observed tensile behaviour, three test series were performed. Firstly, a number of tests were conducted on samples after mechanical removal of the corroded areas in order to

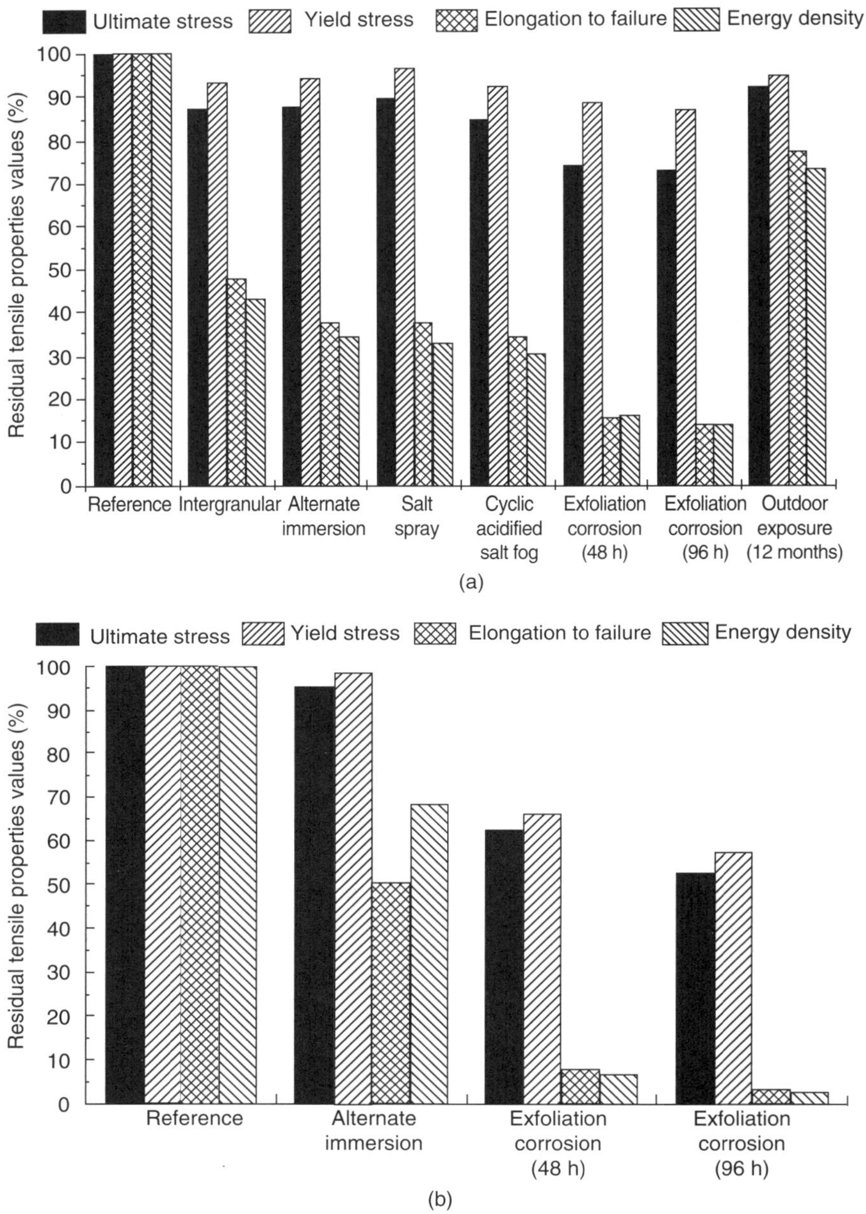

4.1 Tensile property degradation in different corrosive environments in (a) 2024 aluminum alloy and (b) 6013 aluminum alloy.

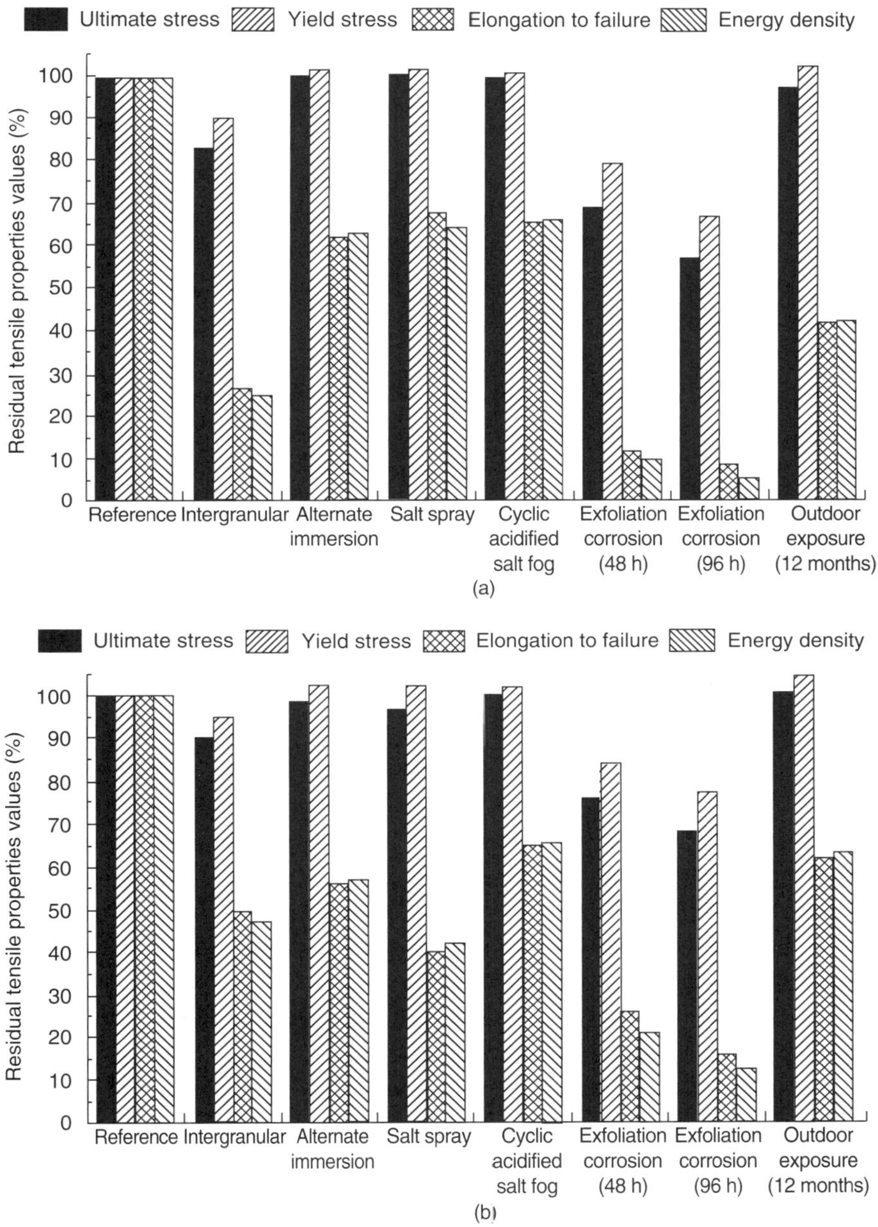

4.2 Tensile property degradation in different corrosive environments for (a) 8090 aluminum alloy and (b) 2091 aluminum alloy.

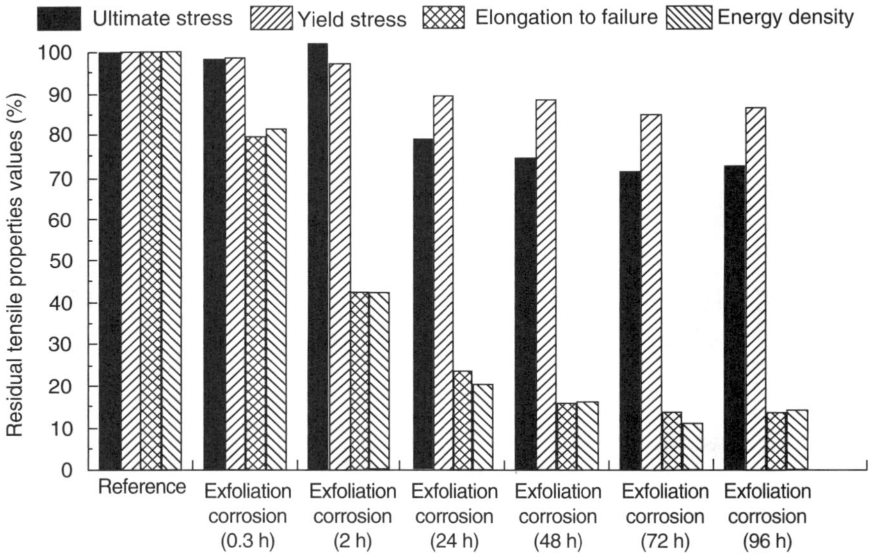

(a)

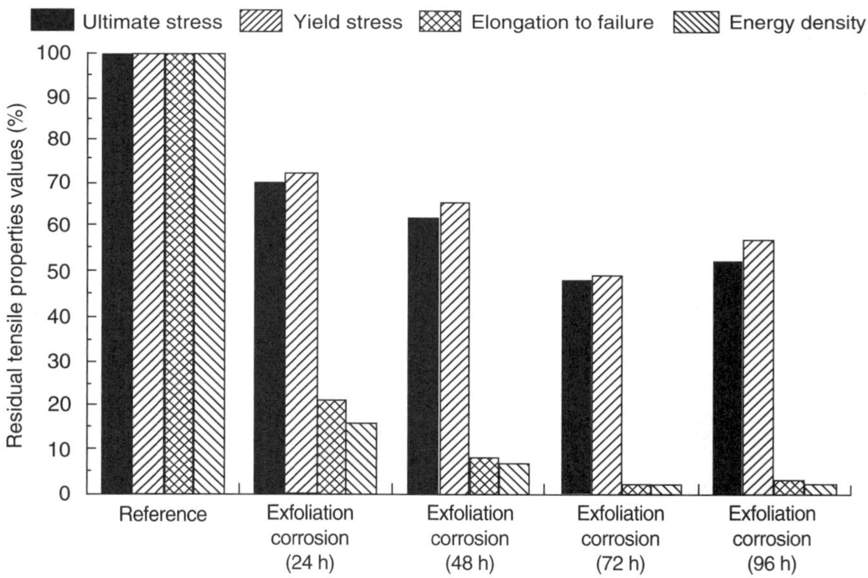

(b)

4.3 Tensile property degradation in EXCO environment for (a) 2024 aluminum alloy and (b) 6013 aluminum alloy.

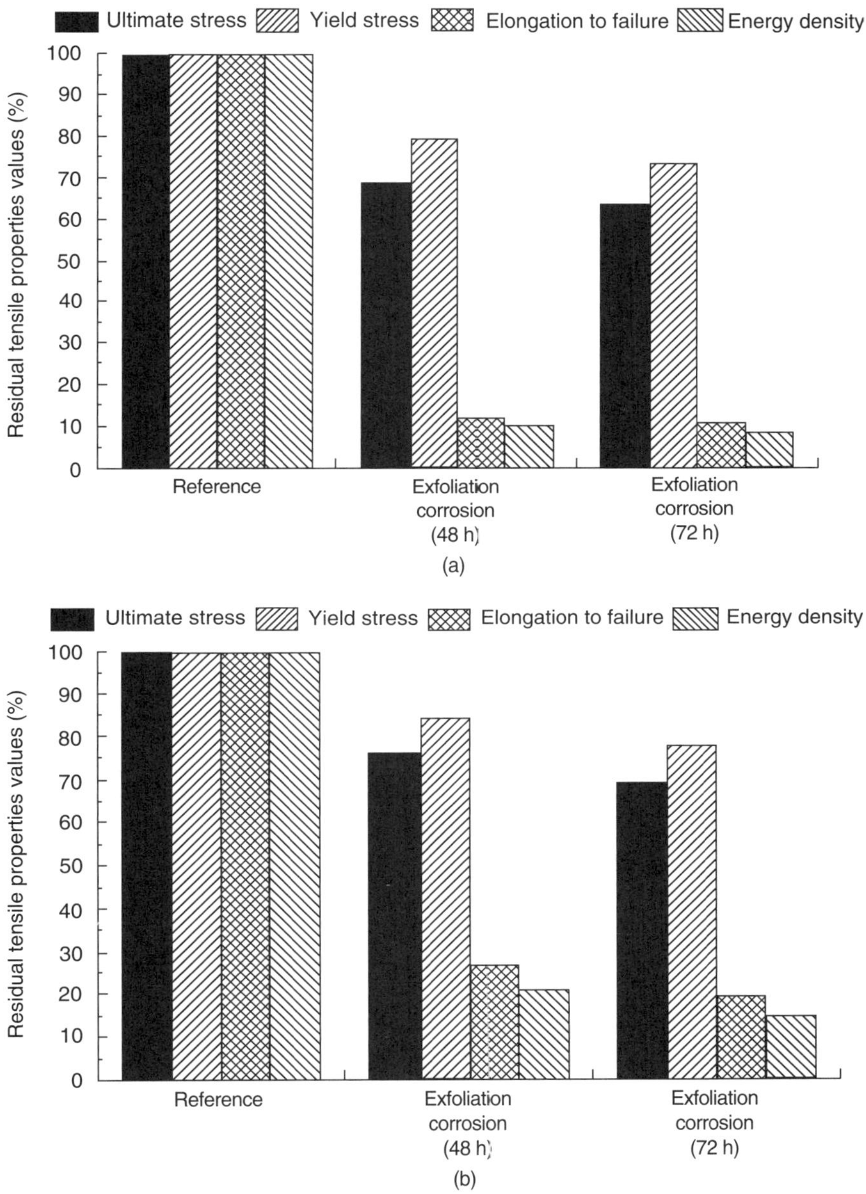

4.4 Tensile property degradation in EXCO environment for (a) 8090 aluminum alloy and (b) 2091 aluminum alloy.

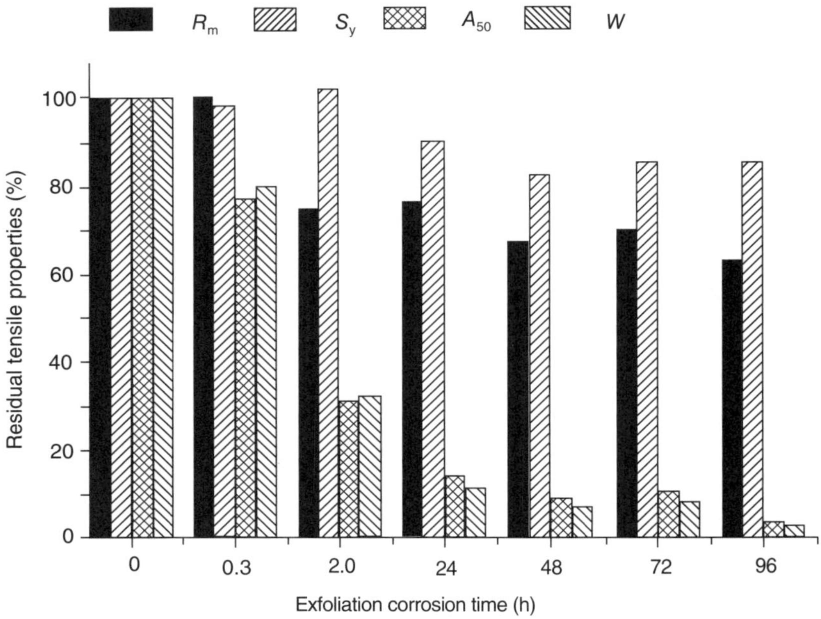

4.5 Gradual tensile property degradation with increasing exposure time in EXCO environment in 2024 aluminum alloy.

Table 4.1 Chemical composition (wt%) of aluminum alloys 2024-T351 and 6013-T6

Aluminum alloy	Si	Fe	Cu	Mn	Mg	Cr	Zn	Ti	Ni	Zr	Al
2024	0.10	0.18	4.35	0.67	1.36	0.02	0.07	0.03	–	0.01	Balance
6013	0.25	–	0.90	0.35	0.95	–	–	–	–	–	Balance

evaluate the effect of corrosion on the uncorroded core material. Secondly, pre-corroded samples were pre-heated at various temperatures and hydrogen measurements were performed in order to identify possible hydrogen trapping sites. Finally, tests were performed after heat treatment at the temperatures, which correspond to the hydrogen trapping sites,[28] for both 2024 and 6013 alloys. This series would indicate whether the trapped hydrogen is responsible for the observed material embrittlement. Since heat treatment influences the microstructure of the material, tests on heat-treated uncorroded samples were also conducted to allow comparison. The chemical compositions of the alloys 2024-T351 and 6013-T6 are shown in Table 4.1. In Tables 4.2 and 4.3 the test series performed are shown.

Table 4.2 Tensile tests performed on the Al 2024-T351 aluminum alloy

Test series description	Corrosion exposure before tensile test	Specimen direction	Number of tests performed
Tensile tests after exfoliation corrosion with mechanical removal of corroded areas	Exfoliation corrosion (exposure times: 24, 48 and 72 h)	L	10
Tensile tests for exfoliation corrosion with heat treatment after corrosion	Exfoliation corrosion 24 h (heat treatment temperatures: 150, 350, 490, 540 °C)	L	12
Tensile tests for exfoliation corrosion with mechanical removal of corroded areas and heat treatment	Exfoliation corrosion 24 h (heat treatment temperature 540 °C)	L	3

Table 4.3 Tensile tests performed on the Al 6013-T6 aluminum alloy

Test series description	Corrosion exposure before tensile test	Specimen direction	Number of tests performed
Tensile tests for exfoliation corrosion with mechanical removal of corroded areas	Exfoliation corrosion 48 h	L	2
Tensile tests for exfoliation corrosion with heat treatment after corrosion	Exfoliation corrosion 24 h (heat treatment temperatures: 200, 350, 510 °C)	L	12

In Fig. 4.6, the results of tensile properties for alloys 2024 and 6013 after mechanical removal of the corroded specimen layers are shown. The mechanical removal of corroded surfaces was performed by taking into account the depth of corrosion attack determined by metallographic inspection.[21] In Fig. 4.6, M1 stands for mechanical removal of all the corroded areas of the specimen and M2 for mechanical removal of corroded layers only at the front and back surface of the specimen, respectively. As shown, mechanical removal of the corroded areas leads to an appreciable recovery of yield and ultimate tensile stress. Yet, tensile ductility was not recovered, indicating that the observed embrittlement is a volumetric phenomenon.

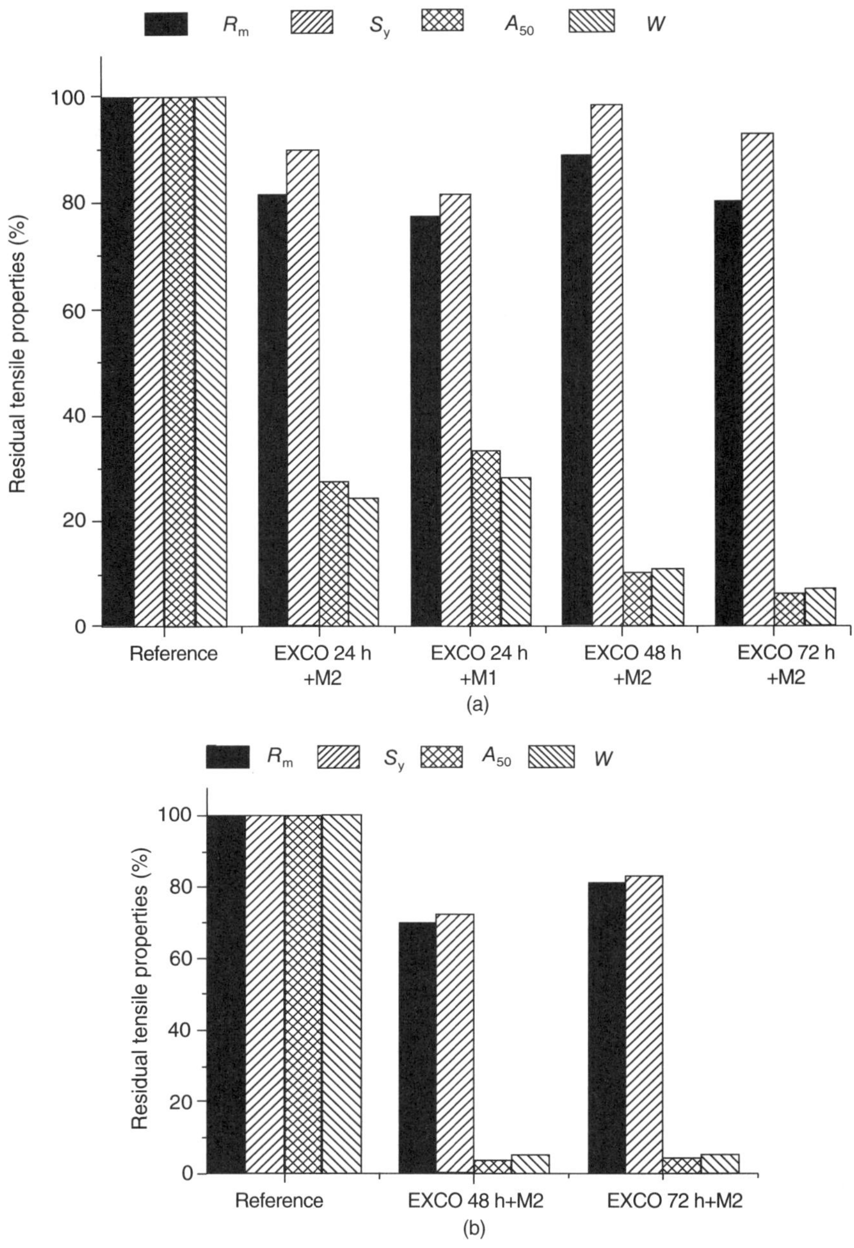

4.6 Tensile behaviour following exfoliation corrosion and mechanical removal of the corrosion-induced surface damage for (a) alloy 2024 and (b) alloy 6013.

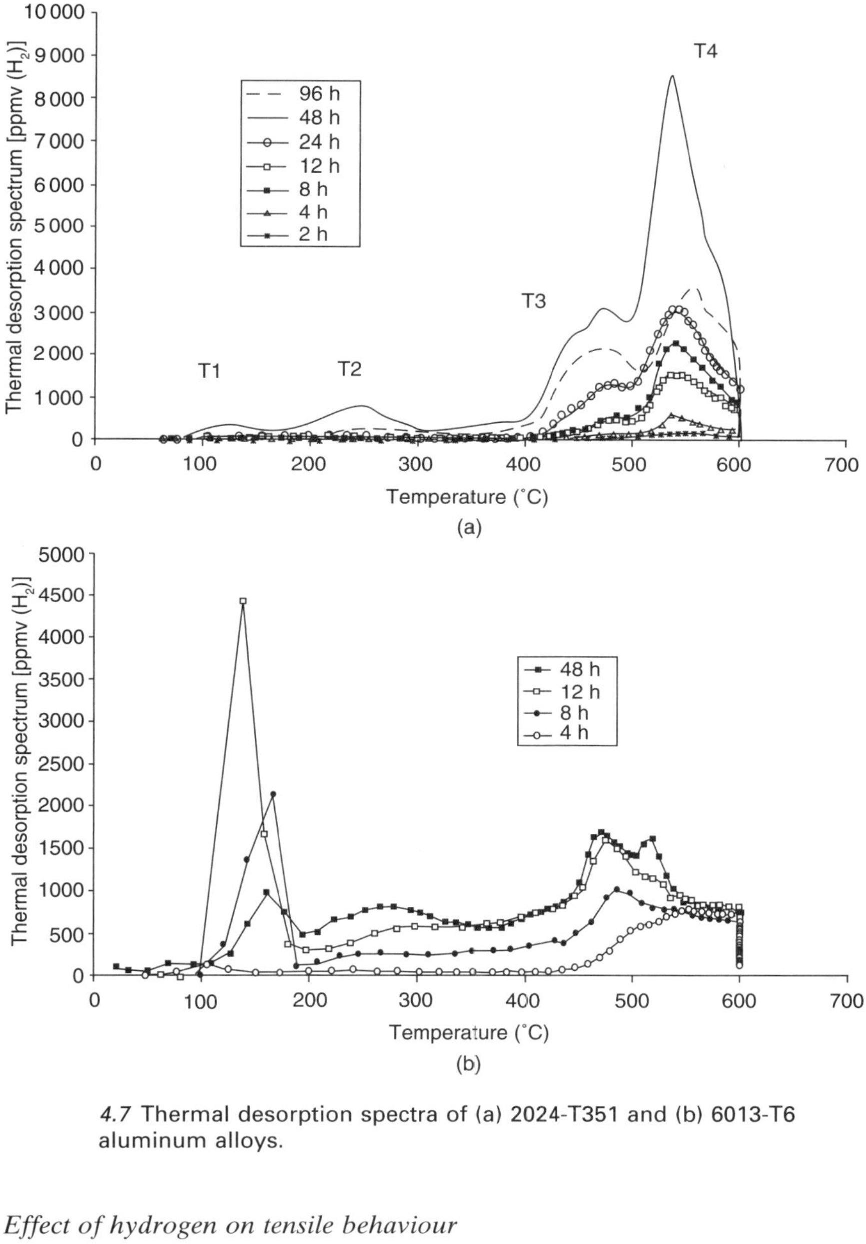

4.7 Thermal desorption spectra of (a) 2024-T351 and (b) 6013-T6 aluminum alloys.

Effect of hydrogen on tensile behaviour

The thermal desorption spectra shown in Fig. 4.7 provide evidence for the existence of multiple hydrogen trapping sites for the investigated alloys, corresponding to the curve peaks in Fig. 4.7. They correspond to the temperatures of 150, 350, 490 and 540 °C for the 2024 alloy and 200, 350 and

510 °C for the 6013 alloy, with T4 having the highest activating temperature and, hence, corresponding to the strongest trap.

A series of pre-corroded specimens were heat treated at these temperatures in order to release the corresponding trapped hydrogen. Tensile ductility was then observed to recover towards the level of the uncorroded material (Fig. 4.8). The results shown in Fig. 4.8 can be used to quantify the effect of each trapping site on the tensile ductility. Heating of corroded 2024-T351 at 540 °C after machining of the corroded surface layers and corrosion edge notches has resulted in a full recovery of the tensile ductility (Fig. 4.8a). This finding clearly suggests that the reason for the reduction of tensile ductility is volumetric corrosion-induced hydrogen embrittlement. Fig. 4.9 shows the hydrogen content of the corroded specimens as a function of exfoliation corrosion time.[8] The hydrogen content is presented for temperatures corresponding to the irreversible trapping sites T2, T3 and T4 as a percentage of the saturated value. Also plotted in Fig. 4.9 is the strain energy density reduction using the results of Fig. 4.5. It is clear from the figure that the reduction in tensile ductility is related to the hydrogen content and that hydrogen saturation leads to an ultimate value of tensile ductility. One can also observe that even a small amount of hydrogen can lead to dramatic reduction in tensile ductility. The results of Fig. 4.9 cannot be used to evaluate the effect of each trapping site on the tensile ductility. However, it has been shown[8] that the larger amount of hydrogen is trapped in the trapping site T4, which has a hydrogen release temperature of 500 °C for 2024 alloy, a temperature that coincides with the dissolution of the Al_2Cu precipitate. Thus, it seems reasonable to associate trapping site T4 with this phase, which is the main strengthening phase of the 2024 alloy. From the technological point of view, this means that full recovery of tensile ductility for corroded 2024 alloy is not possible, since it would require the dissolution of its strengthening phase.

4.3.2 Effect of corrosion on the fatigue behaviour of aircraft aluminum alloys

Several investigators have experimentally studied the effect of corrosion on the fatigue performance of aluminum alloys.[29–34] Amongst them, Bray *et al.* have shown that prior corrosion pitting reduced the fatigue strength of aluminum alloys 2024-T3 and 2524-T3 by approximately 40%.[31] The effects of pre-existing localized surface pitting corrosion on the fatigue lives of alloy 7075 have been investigated.[33–34]

In the following section, fatigue results of pre-corroded 2024-T351 aluminum alloy in selected laboratory corrosive environments are presented. Moreover, the role of anodizing protective layer on the fatigue performance of the samples is also discussed.

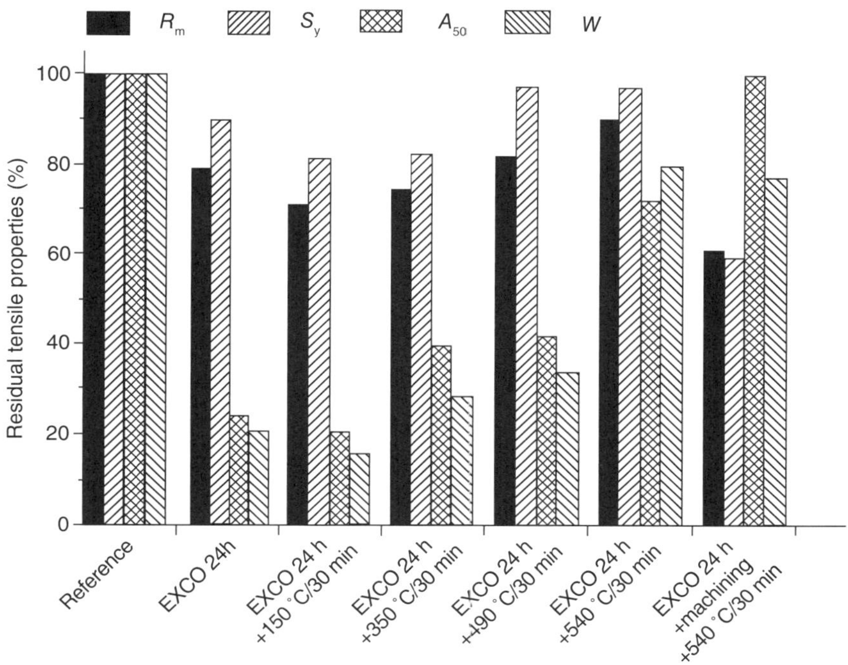

(a)

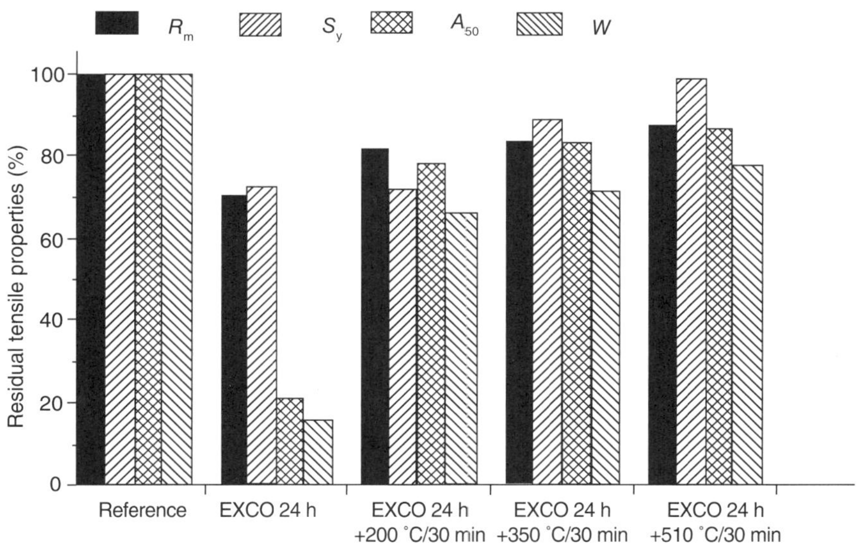

(b)

4.8 Tensile behaviour following exfoliation corrosion and thermal treatment for the removal of trapped hydrogen for (a) 2024 and (b) 6013 alloy.

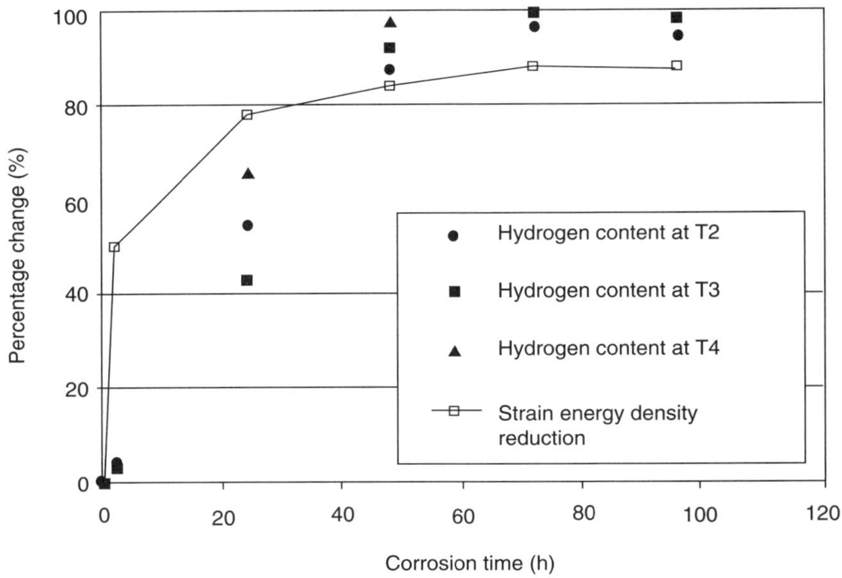

4.9 Hydrogen content at different trapping sites and strain energy density reduction as a function of corrosion time for 2024 alloy.

Experimental investigation and results

The experiments conducted for the 2024-T351 aluminum alloy with the chemical composition shown in Table 4.1, included fatigue tests to obtain the S–N curves. The alloy was received in bare, sheet form of 1.6 mm nominal thickness. Machining of the specimens was made according to the specification ASTM E 466-82.[35] All specimens were cut in longitudinal (L) orientation relative to the rolling direction. To investigate the protective effect of anodization and sealing against corrosion a number of specimens were first coated with a hard anodization layer and then sealed according to the MIL-A-8625E specification[36] before exposure to the corrosive environment. The anodization layer was 50 μm thick. Before mechanical testing, the specimens had been subjected to exfoliation corrosion for 36 h according to the ASTM G34-90 specification and then cleaned according to ASTM G34-90. The fatigue tests performed to obtain the S–N curves with the respective corrosion exposure before each test series are given in Table 4.4. The stress ratio used for the tests was $R = 0.1$ with a frequency of 25 Hz. The tests were carried out according to the ASTM E 466-82 specification.

Metallographic corrosion characterization of specimens exposed to exfoliation corrosion solution for 48 h has shown the coexistence of pitting and

Table 4.4 Fatigue tests at $R = 0.1$ performed on 2024-T351 aluminum alloy specimens to obtain S–N curves

Material: Al 2024-T351	Corrosion exposure before test	Maximum stress at $R = 0.1$ (MPa)	Frequency (Hz)	Number of tests performed
Bare, as-received	None	175, 200, 225, 250, 300, 325	25	16
	Exfoliation corrosion for 36 h	100, 125, 150, 175, 200	25	10
Anodized and sealed	None	135, 150, 175, 200, 230, 250	25	10
	Exfoliation corrosion for 36 h	100, 125, 135, 150, 175, 210	25	11

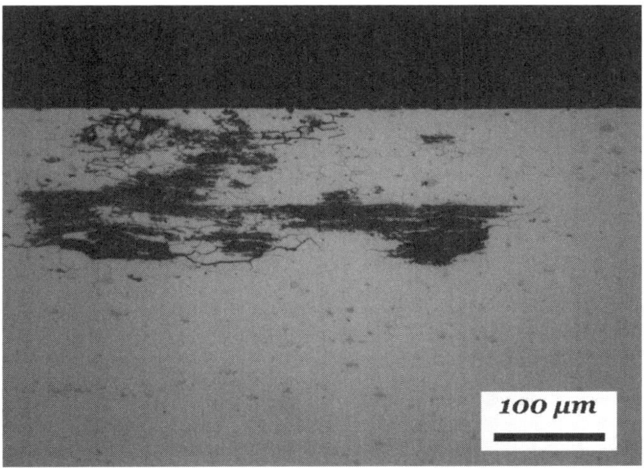

4.10 SEM photograph of the cross section of 2024 specimen subjected to 48 h in an exfoliation corrosion environment.

intergranular corrosion (Fig. 4.10). The presence of corrosion pitting and intergranular corrosion essentially facilitates the onset of fatigue cracks and, hence, reduces the fatigue life of the corroded specimens appreciably. In Fig. 4.11, the effect of a 36-h exposure to exfoliation corrosion solution of bare 2024-T351 aluminum specimens on the fatigue life of the specimens is demonstrated. As expected, the corrosion attack results in a significant drop in the material's fatigue life, which increases with the decrease of fatigue stress. The fatigue endurance limit drops from 175 MPa for the

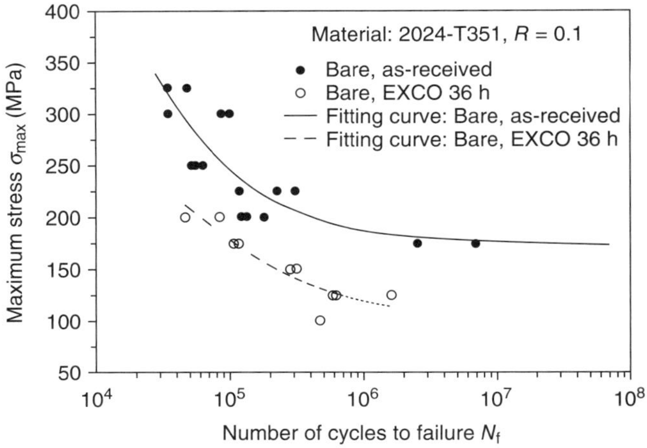

4.11 S–N curves for corroded and uncorroded 2024-T351 aluminum alloy (as-received specimens).

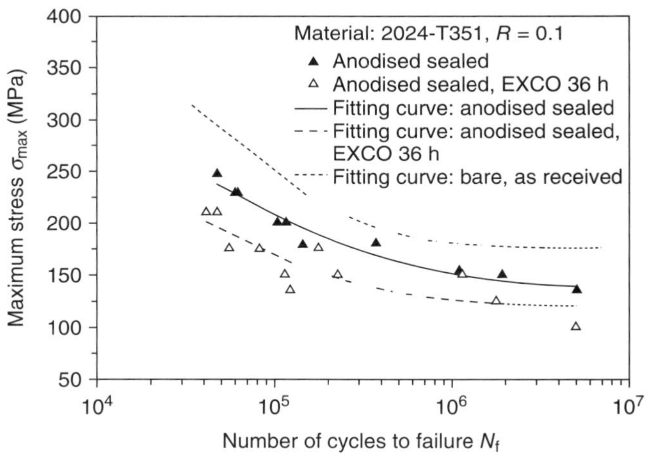

4.12 S–N curves for corroded and uncorroded 2024-T351 aluminum alloy (anodized and sealed specimens).

uncorroded material to 95 MPa for the pre-corroded specimens. Fitting curves for both uncorroded and corroded material were derived using a regression analysis. The fitting curves are also displayed in Fig. 4.11.

The influence of anodization and sealing on the fatigue life of the exfoliated material is shown in Fig. 4.12 in comparison to the S–N curve for the bare material. As expected, the fatigue lives for the anodized specimens are less severely degraded through the corrosion attack than the unprotected

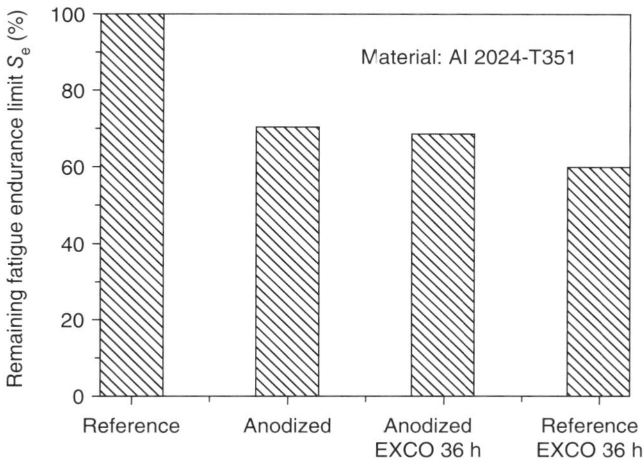

4.13 Fatigue endurance limit after exfoliation corrosion solution for 36 h of bare, as-received and anodized and sealed Al 2024-T351 specimens.

bare 2024 specimens. The fatigue endurance limits for the cases investigated here are compared in Fig. 4.13. The fatigue endurance limit in Fig. 4.13 refers to $N_f = 5 \times 10^6$ cycles, 'Reference' and 'Anodized' refer to bare as-received and anodized sealed material, respectively. With 'EXCO 36 h', the specimens, which were pre-corroded in exfoliation corrosion solution for 36 h, are characterized. As shown in the figure, depending on the process before the fatigue test, the fatigue endurance limit may drop up to 40% compared to the reference material. There is a remarkable reduction of 55 MPa in the fatigue endurance limit of the corroded anodized and sealed specimens when compared to the endurance limit of the as-received material. The obtained reduction results mainly from the anodization process. As discussed above, the classical interpretation of the above results in Fig. 4.11–4.12 relates the drop of fatigue life to the essential reduction of the fatigue crack initiation phase owing to the occurrence of corrosion notches. With regard to the identification of trapped hydrogen in the pre-corroded material and the resulting embrittlement discussed in the previous paragraph, an effect of hydrogen embrittlement on the fatigue life reduction of the corroded specimens cannot be excluded. It is noticeable that the anodization process itself decreases the fatigue life appreciably. This result may be explained by considering that the anodization process can be the cause for the occurrence of material surface notches, which are expected to lead to a reduction of the fatigue crack initiation phase. Recall, that due to the electrolytical nature of the anodization process the specimens are subjected to an active, hydrogen-rich environment at a temperature of 37 °C for the duration of the process (40–60 min). Further investigations will show whether hydrogen has been trapped in the material during anodization.

4.3.3 Effect of corrosion on the damage tolerance behaviour of aircraft aluminum alloys

As described in the previous sections, structural degradation due to existing corrosion has been considered as a load capacity decrease of the structural component. Furthermore, it has been recognized that the occurrence of corrosion damage in the form of pitting or multiple-site damage facilitates the onset of fatigue cracks[29–34] and hence reduces fatigue life. The role of corrosion on multiple-site damage scenarios and aircraft structural integrity has been correlated to the onset of multiple site damage (MSD) from corrosion pits.[31] On the other hand, it was found that there is no significant effect of prior exfoliation corrosion on the fatigue crack growth rate of 2024-T351 specimens.[37] Despite the efforts made and although essential for damage tolerance design, the effect of corrosion on damage tolerance characteristics has not yet been sufficiently understood. Moreover, limited investigations have focused on the influence of pre-existing corrosion on the damage tolerance performance of aluminum alloys characterized by fracture toughness and fatigue crack growth resistance.

In this part of the work, the fatigue and damage tolerance behaviour of pre-corroded 2024-T351 aluminum alloy specimens has been investigated and discussed in relation to the synergetic effect of corrosion and corrosion-induced hydrogen embrittlement. The experiments performed included fatigue crack growth tests and fracture toughness tests on pre-corroded 2024 aluminum alloy.

Experimental procedure and results: Fatigue crack growth tests

The experiments were conducted on bare 2024-T351 aluminum alloy of 1.6 mm thickness and included fatigue crack growth tests for variable stress ratio R on corroded and uncorroded material. An overview of the tests is given in Table 4.5. The selected fatigue stress ratios were $R = \sigma_{min}/\sigma_{max}$ 0.01, 0.1, 0.5 and 0.7 and the test frequency was 20 Hz. In order to evaluate the R effect, the stress range $\Delta\sigma$ was kept constant at a σ_{max} between 101 and 180 MPa. Machining of the specimens was made according to ASTM E 647-93 specification.[38] All specimens were cut in longitudinal (L) orientation relative to the rolling direction. The experimental investigation included uncorroded and pre-corroded specimens in an exfoliation corrosion environment for 36 h according to ASTM standard G34-90. After exposure, the specimens were cleaned according to ASTM G34-90. During all fatigue

Table 4.5 Performed fatigue crack growth tests

Corrosion exposure	Stress ratio $R = \sigma_{min}/\sigma_{max}$	Frequency (Hz)	Maximum stress σ_{max} (MPa)	No. tests
None	0.01	20	109	2
EXCO 36	0.01	20	109	2
None	0.1	20	101	2
EXCO 36	0.1	20	101	2
None	0.5	20	180.4	2
EXCO 36	0.5	20	180.4	1
None	0.7	20	176.1	2
EXCO 36	0.7	20	176.1	2

Table 4.6 ΔK and K_{max} values at critical crack length

Corrosion exposure	Stress ratio $R = \sigma_{min}/\sigma_{max}$	Stress intensity factor range ΔK (at α_{cr}) (MPa m^{½})	Maximum stress intensity factor K_{max} (at α_{cr}) (MPa m^{½})
None	0.01	29.74	30.04
EXCO 36	0.01	22.87	23.09
None	0.1	32.54	36.16
EXCO 36	0.1	25.28	28.09
None	0.5	38.74	77.48
EXCO 36	0.5	16.94	33.88
None	0.7	11.70	40.32
EXCO 36	0.7	11.25	37.54

crack growth tests, crack growth was recorded via a DC potential drop measurement method. The fatigue crack growth tests are described in Table 4.5 and the obtained results are summarized in Tables 4.6–4.7 and Fig. 4.14–4.18. In Table 4.6 the values of the stress intensity factor range ΔK and the maximum stress intensity factor value K_{max} referring to the critical crack length are given. Calculation of ΔK and K_{max} was made with implementation of the equation:[38]

$$K = \beta\sigma\sqrt{\pi\alpha} \qquad [4.1]$$

In the interpretation of results, the effect of corrosion on the correction factor β in Eqn. (4.1) was not taken into account. In Table 4.7 the number of cycles corresponding to specimen failure is displayed together with the critical crack length α_{cr} defined at a time step that corresponds to 3 s before specimen failure. The critical crack length was defined in this manner

Table 4.7 Fatigue crack growth data of corroded and uncorroded specimens

Corrosion exposure	Stress ratio $R = \sigma_{min}/\sigma_{max}$	Maximum stress (MPa)	Fatigue life N_f [cycles; (mean values)]	Critical crack length α_{cr} [mm; (mean values)]
None	0.01	109	46 640–47 460 (47050)	20.02–16.46 (18.24)
EXCO 36	0.01	109	42 700 (42 700)	12.59 (12.59)
None	0.1	101	81 620–69 100 (75 360)	24.8–23.2 (24.0)
EXCO 36	0.1	101	58 640–51 960 (55 300)	15.59–21.31 (18.45)
None	0.5	180.4	43 960–49 820 (46 890)	19.4–35.73 (27.56)
EXCO 36	0.5	180.4	30 560–29 560 (30 060)	9.72–10.92 (10.32)
None	0.7	176.1	213 540 (213 540)	14.17 (14.17)
EXCO 36	0.7	176.1	139 880–121 260 (130 570)	11.75–13.67 (12.71)

because the data recording frequency made the measurement of the actual value at fracture unfeasible. The critical crack length corresponds to the last data point of the curves in Fig. 4.14a–4.18a.

In Fig. 4.14–4.18 the fatigue crack growth curves together with the crack growth rate measurements versus stress intensity factor range ΔK are displayed for the corroded and uncorroded specimens at various stress ratios.

From the results, the deteriorating effect of corrosion on fatigue life of pre-corroded specimens is revealed for all stress ratios examined. Fatigue life reduction, corresponding to fatigue crack propagation cycles, is promoted by the existence of localized corrosion sites acting as stress concentration locations. The reduction in fatigue life ranges from 12% ($R = 0.01$) to 38.8% ($R = 0.7$) (Table 4.7). The effect of corrosion on fatigue crack propagation is more pronounced for small values of R. The observed behaviour can be explained by considering the mechanisms of partial embrittlement of the specimens as a result of corrosion exposure. For small values of ΔK damage accumulation mechanisms seem not to be influenced to the same degree by the partially embrittled material. At higher tensile stresses (higher ΔK values) embrittlement may have a more pronounced effect on the ability of the material to deform plastically ahead of the crack tip.

In Table 4.7 the effect of corrosion damage on the critical crack length to failure is illustrated. The critical crack length of the corroded specimens is reduced by a magnitude of 45% compared to the respective value of the uncorroded specimens, while the stress intensity factor values

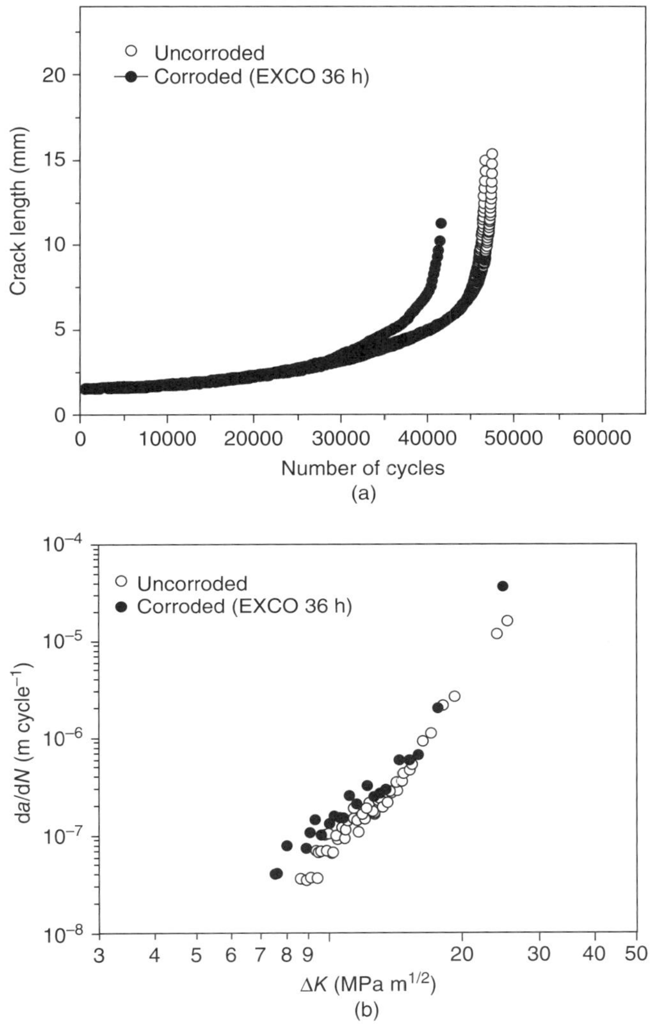

4.14 (a) Crack growth curves and (b) fatigue crack growth rates of corroded and uncorroded specimens at $R = 0.01$.

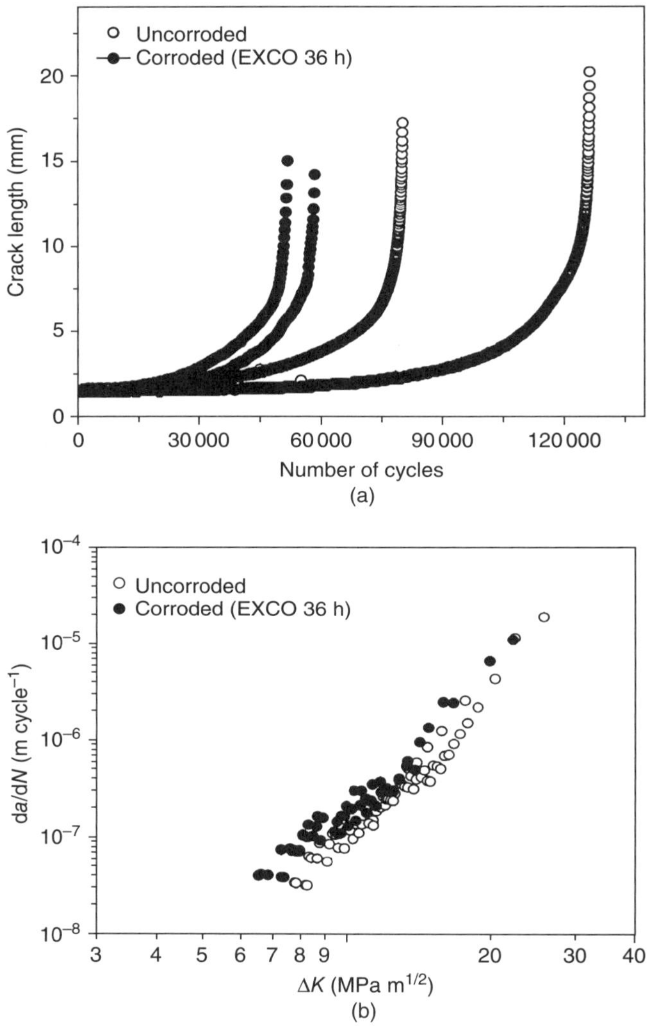

4.15 (a) Crack growth curves and (b) fatigue crack growth rates of corroded and uncorroded specimens at $R = 0.1$.

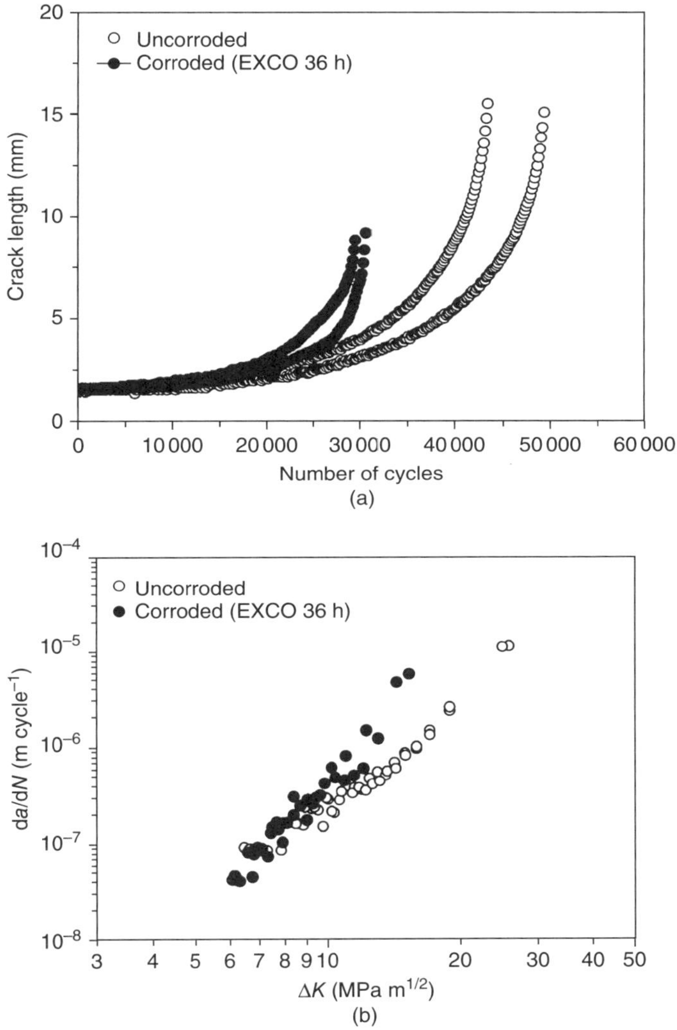

4.16 (a) Crack growth curves and (b) fatigue crack growth rates of corroded and uncorroded specimens at $R = 0.5$.

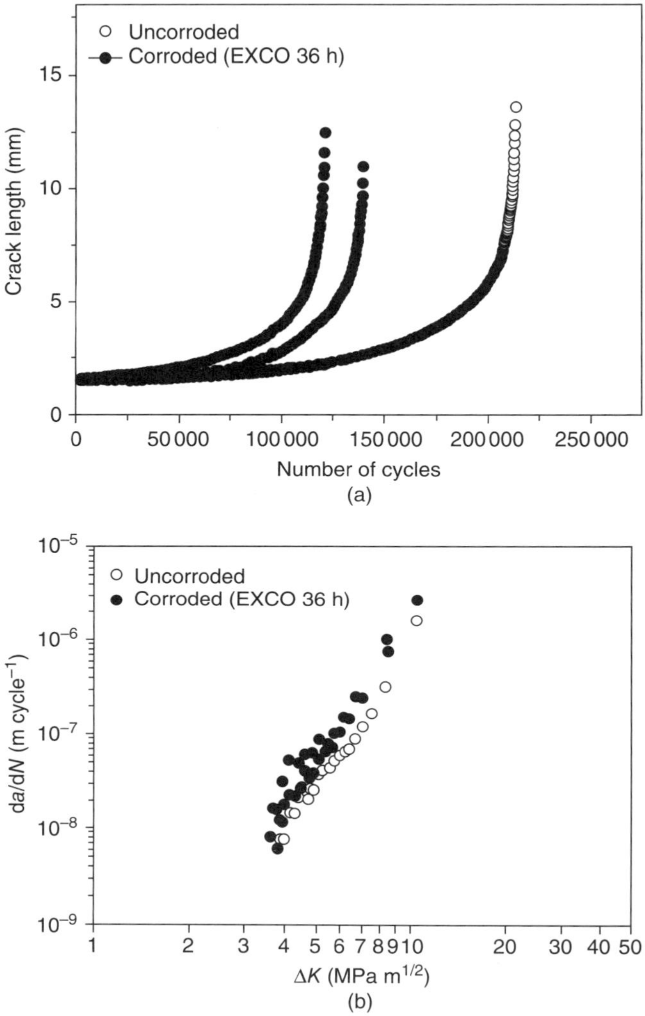

4.17 (a) Crack growth curves and (b) fatigue crack growth rates of corroded and uncorroded specimens at $R = 0.7$.

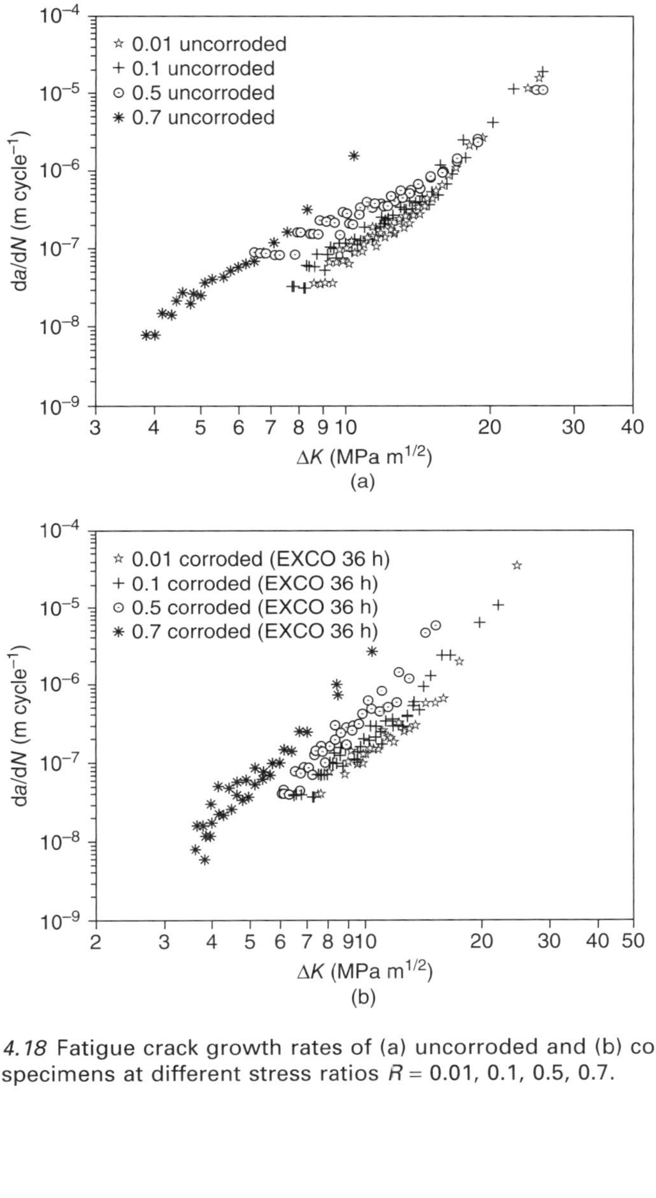

4.18 Fatigue crack growth rates of (a) uncorroded and (b) corroded specimens at different stress ratios $R = 0.01, 0.1, 0.5, 0.7$.

corresponding to α_{cr} are reduced up to 34% relative to the uncorroded material (Table 4.6). The stress intensity factor reduction is attributed to the reduction of the critical crack length owing to fracture toughness degradation of the corroded alloy.

From the experimental findings obtained in Figs 4.14–4.18, it can be observed that crack propagation resistance at high ΔK values suffers a significant degradation compared to the medium ΔK region where corrosion attack does not have the same impact. The latter observation is supported by the calculated coefficients of the Paris equation. The coefficients for the corroded and uncorroded specimens, derived from the da/dN–ΔK plots for stress intensity ranging from 10–18 MPa $m^{1/2}$ are compared in Table 4.8. The coefficients do not differ appreciably. This is consistent with previous results.[37] At stage III crack growth (high ΔK), the calculated crack growth rates are high and cannot be described by the Paris law. At this stage, the Paris equation cannot be used to assess the influence of corrosion on fatigue crack growth.

The obtained decrease in fatigue life of corroded specimens is attributed to the reduction of the critical crack length at failure. The latter is dependent on the material's fracture toughness. These parameters are influenced by the partially embrittled material due to corrosion. The fatigue crack growth behaviour described above is not easy to explain by means of the classical interpretation of corrosion. Fractographic analyses[14] on a 1.6 mm thick sample subjected to exfoliation corrosion exposure for 24 h confirm the embrittlement of the material during the corrosion exposure. The

Table 4.8 Paris coefficients of corroded and uncorroded material

Corrosion exposure	Stress ratio $R = \sigma_{min}/\sigma_{max}$	n	Mean value	C	Mean value
None	0.01	3.09	3.13	7.38×10^{-11}	4.69×10^{-11}
		3.16		2.0×10^{-11}	
	0.1	3.35	3.21	5.56×10^{-11}	9.38×10^{-11}
		3.07		1.32×10^{-10}	
	0.5	3.0	3.12	1.94×10^{-10}	1.60×10^{-10}
		3.23		1.26×10^{-10}	
	0.7	3.6	3.6	9.25×10^{-11}	9.25×10^{-11}
		–		–	
EXCO 36 h	0.01	–	2.85	–	4.88×10^{-11}
		2.85		4.88×10^{-11}	
	0.1	3.16	3.11	1.62×10^{-10}	1.5×10^{-10}
		3.05		1.38×10^{-10}	
	0.5	3.04	3.21	2.38×10^{-10}	4.37×10^{-10}
		3.38		1.99×10^{-10}	
	0.7	3.16	3.47	5.03×10^{-10}	2.9×10^{-10}

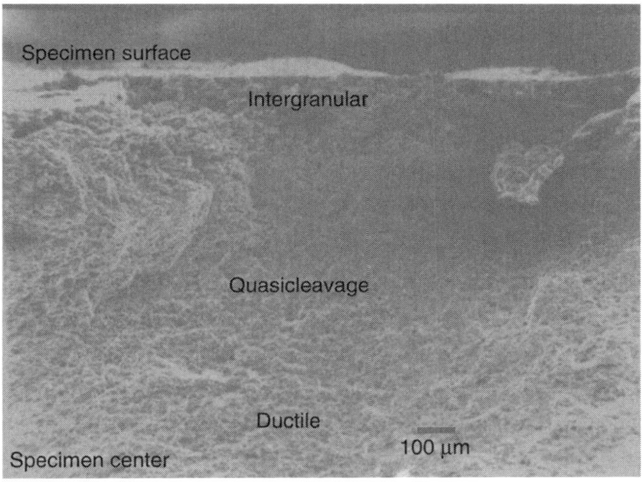

4.19 Embrittled fractured zones in 2024 specimen.

Table 4.9 Fracture toughness tests on 2024 alloy

Material: Al 2024-T351	Corrosion exposure	Direction/number of tests
Bare, as-received	None	L/2
Bare, corroded	Exfoliation corrosion for 36 h	L/2
Anodized and sealed	None	L/2
Anodized and sealed, corroded	Exfoliation corrosion for 36 h	L/2
Anodized and sealed, corroded	Exfoliation corrosion for 48 h	L/2

appearance of the fracture surface varies from the surface of the sample to the interior. At the surface, the fracture is intergranular as shown in Fig. 4.19. Immediately below, there is a zone of quasi-cleavage fracture and, further below, the fracture turns to entirely ductile as shown in Fig. 4.19. The results underline the need to consider the fatigue crack growth behaviour of corroded 2024 as the result of the synergetic effect of corrosion and corrosion-induced hydrogen embrittlement.

Fracture toughness tests

The fracture toughness tests were performed on bare and anodized aluminum 2024 center cracked panels according to ASTM E 561-94 standard.[39] In Table 4.9, the fracture toughness tests carried out according to ASTM E

561-94 specification are presented. The results of the fracture toughness tests together with a characteristic *R*-curve behaviour of corroded and uncorroded bare panels are displayed in Fig. 4.20. Exposure of the bare panels to exfoliation corrosion for 36 h caused a decrease in fracture toughness of 27%. Although this reduction in fracture toughness is appreciable, it is far less than the reduction of tensile ductility of the tensile tests presented in 4.3.1. As expected, the fracture toughness of the anodized specimens was much less affected by the corrosion attack than the fracture

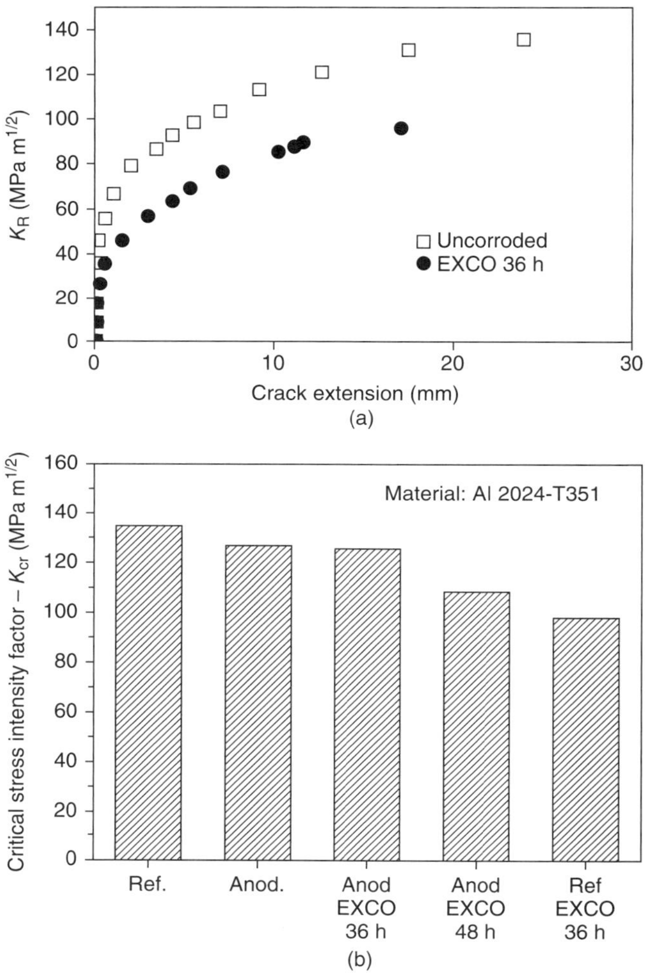

4.20 (a) *R* curve behaviour of uncorroded and corroded 2024 center-cracked panel and (b) results of the performed fracture toughness tests.

toughness of the bare samples. As shown in Fig. 4.20b, the fracture toughness value of the anodized specimens remained practically unchanged after corrosion exposure for 36 h and suffered a drop of 14.4% after corrosion exposure for 48 h. The measured decrease in the fracture toughness value following the anodization process was 7%. Further studies will show whether this reduction is explicitly the result of material surface notches caused by the anodization process or partially the result of hydrogen, which has possibly been trapped in the material during anodization.

4.4 Modeling the corrosion effect on damage tolerance characteristics

The considerable requirement to keep aging airframes in service as long as possible is facilitated through damage tolerance design concepts having periodic maintenance to assess the state of corrosion and fatigue damage in the airframe. The usual practice for removing corrosion damage allows up to 10% of the component thickness to be ground away; otherwise the component must be replaced or repaired. Since fatigue cracks will initiate preferentially from any surface pits, including corrosion pits, the presence of corrosion reduces the fatigue life of the component. Hence, it is desirable to have analytical tools that will allow an assessment of the effect of this corrosion damage on the remaining fatigue life of the component. These tools can be used to identify corrosion damage that needs to be removed and to establish appropriate inspection intervals.

In the final part of this chapter, analytical methodologies are presented, which may be used to account for the influence of corrosion damage when assessing the damage tolerance characteristics of aged aircraft components. The proposed methodologies can be implemented to predict the influence of existing corrosion damage on residual strength and fatigue crack growth behaviour in the aged material.

4.4.1 Fracture-mechanical model for the prediction of residual strength

The significance of corrosion-induced hydrogen embrittlement on the fracture toughness of a corroded aluminum alloy has not yet been adequately recognized, and is still not considered when evaluating the structural integrity of aged aircraft components. At present, there are neither experimental data nor an established experimental methodology for assessing the fracture toughness values of a corroded and hydrogen-embrittled area of the alloy. Nevertheless, the significant information gained from a series of intensive recent investigations on the issue of the corrosion induced hydrogen embrittlement of 2024 alloy[7–10,14,19–21] allows for the suggestion of a simplified corrosion and hydrogen embrittlement process, which may be explained as the physical background for evaluating the corrosion-induced hydrogen-embrittlement effects on the material's fracture toughness. During the exposure of the alloy to the exfoliation solution, corrosion attack occurs first at the material surfaces and develops gradually from pitting into intergranular attack. Pits and cracks represent channels through which the corrosive medium can reach the uncorroded material, so that further corrosion develops. Because the diffusing hydrogen is the product of the corrosion electrochemical reactions, it is obvious that with increasing corrosion exposure time, additionally to the hydrogen that is produced by the continuing corrosion process at the material surface, further hydrogen production sources evolve at the developing front of the corrosion-attacked surface layer, which increases in depth towards the material interior. Hence, hydrogen diffuses beyond the corrosion-attacked surface layer and, thus, causes material embrittlement that is not damaged by corrosion. The embrittled zone lying between the corrosion-attacked material surface layer and the typical ductile fracture surface of the 2024 alloy obtained in Fig 4.19, is supporting the view described above for the interactive corrosion and hydrogen embrittlement damage process. As the mentioned damage processes are diffusion controlled, their effect on the material's fracture toughness is expected to be time-dependent and local. Hence for evaluating the structural integrity of aged and corroded components, there is the need to introduce a new, location and time-dependent quantity, that will be referred to as local fracture toughness and should be distinguished from the macroscopic fracture toughness of the material which can be measured by means of the previously described tests.

For evaluation of the local fracture toughness of a material area damaged by the corrosion processes discussed above, a fracture-mechanical model will be introduced. When the fracture toughness panel is immersed in an exfoliation corrosion solution, corrosion and hydrogen-embrittlement damage accumulates gradually in the material for the duration of the immersion. The damage initiates at the material surfaces, which are attacked by the corrosive environment and then propagates to the alloy interior, driven by diffusion-controlled mechanisms. It is important to mention that, according to previous experimental results,[10] diffusion of hydrogen in a thin-rolled sheet occurs preferentially through the thin sides of the sheet and the through-thickness cuts and is very low through the upper and lower large rolled surfaces. In Fig. 4.21, a detail of the crack tip of a corroded fracture toughness specimen and the damaged area surrounding it, is shown schematically. Based on previous findings,[10] the main paths for hydrogen diffusion are the open crack surfaces. To model the crack opening process by considering the effects of the existing corrosion and hydrogen-

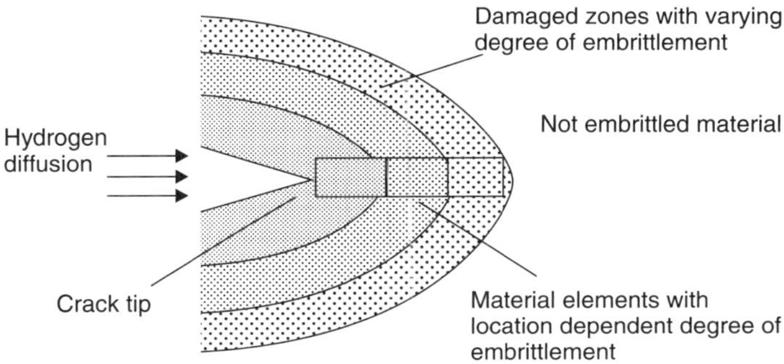

4.21 Location-dependent degree of embrittlement at the crack tip.

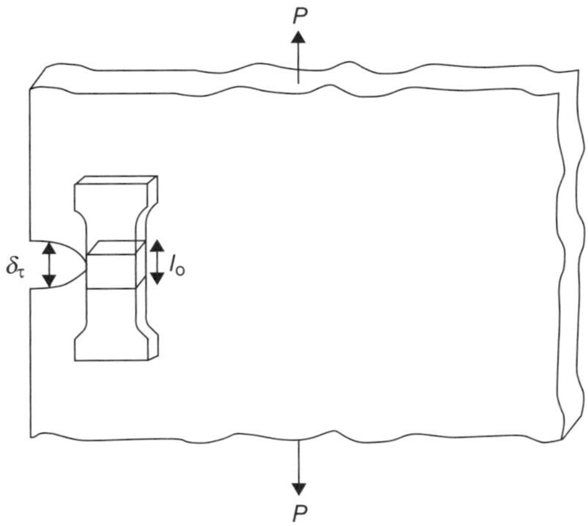

4.22 Material element at crack tip under mode I monotonic loading.

embrittlement process, one may assume a sequence of prospective material elements with a varying degree of damage and, hence, a varying degree of embrittlement. Since the underlying damage processes are diffusion controlled, the severity of the corrosion and hydrogen-induced damage in each of these elements depends on location and time.

The model takes into consideration the material elements ahead of the crack tip shown in Fig. 4.22. They are damaged to different degree of embrittlement. It is well known that, for a stationary crack subjected to monotonic loading, when the scale of plasticity is small, the Dugdale model provides the following results:[40]

$$\omega_s = \frac{\pi}{8}\left(\frac{K}{\sigma_F}\right)^2 \quad [4.2]$$

$$\delta_t = \frac{K^2}{E\sigma_F} \quad [4.3]$$

where ω_s is the monotonic plastic zone size, δ_t is the crack opening displacement, $K = \beta\sigma\sqrt{\pi\alpha}$ is the stress intensity factor, σ_F is the material's yield strength, β is the correction factor and E is the Young's modulus. Within the plastic zone size (0, ω_s) the material elements are subjected to different values of plastic stretch $\delta = \left(\frac{x}{\omega_s}\right)$, where x is the distance from the crack tip. When the panel is subjected to a uniaxial tensile load perpendicular to the crack, it can be assumed that the material element at the immediate vicinity of the crack tip is deformed by $\varepsilon_{pl} = \frac{\delta_t}{l_0}$ with l_0 being the characteristic length of the material element ahead of the crack tip. Then from eqn. [4.3] it follows

$$\varepsilon_{pl} = \frac{K^2}{E\sigma_F l_0} \quad [4.4]$$

It is reasonable to assume that fracture in the immediate vicinity of the crack tip occurs when the maximum crack opening displacement δ_t, at the crack tip reaches a critical value δ_{cr}.[41] By replacing K in Equation [4.4] by its critical value K_{cr}, the corresponding critical plastic deformation ε_f can be obtained as:

$$\varepsilon_f = \frac{K_{cr}^2}{E\sigma_F l_0} \quad [4.5]$$

where K_{cr} is the material's fracture toughness. By making use of Eq. [4.5] for corroded and uncorroded material the local fracture toughness of the corroded material is derived as:

$$K_{cr}^{cor} = K_{cr}\sqrt{\frac{\varepsilon_f^{cor}\sigma_F^{cor}}{\varepsilon_f\sigma_F}} \quad [4.6]$$

It has been suggested[17–18] that the dependency of yield strength and elongation to fracture reduction with the exfoliation corrosion exposure time, may be expressed by the set of empirical equations:

$$\sigma_F^{cor}(t) = \sigma_F\left(1 - Pt^m\right) \quad [4.7]$$

$$\varepsilon_f^{cor}(t) = \varepsilon_f\left[1 - \bar{\varepsilon}_f\left(1 - e^{-\beta t}\right)\right] \quad [4.8]$$

where $P, m, \bar{\varepsilon}_f$ and β are case specific constants with m and $\beta < 1$. By making use of these expressions, equation [4.6] takes the form:

$$K_{cr}^{cor} = K_{cr}\sqrt{[1-\bar{\varepsilon}_f(1-e^{-\beta t})](1-Pt^m)} \qquad [4.9]$$

or

$$K_{cr}^{cor} = K_{cr}\sqrt{\lambda} \qquad [4.10]$$

where the corrosion factor λ stands for:

$$\lambda = [1-\bar{\varepsilon}_f(1-e^{-\beta t})](1-Pt^m) \qquad [4.11]$$

The corrosion factors of the material range between $\lambda = 1$ for the uncorroded material and a minimum corrosion factor λ_{min} referring to the material which has reached maximum embrittlement for the case under consideration.

4.4.2 Fatigue crack growth prediction

For the fatigue crack growth prediction of components including corroded areas, the LTSM-F code is presented in the following section. It consists of two modules, the first of which supports the simulation of irregular service spectra by equivalent stress cycles, while the second serves as a basis for the calculation of fatigue crack growth, accounting also for overload interaction effects. The fatigue crack growth retardation model was first introduced by Pantelakis *et al.*[42] and further upgraded by Kermanidis *et al.*[43] In the model, plastic deformation accounts for damage that accumulates after each cycle of loading.[42] An overload is considered as an irregularity in the loading, which interrupts the constant response of the material and introduces inhomogeneity to its behaviour. The same approach may be applied to service spectra. Service spectra include a sequence of irregularities in the load. They lead to a varying degree of plasticity of the damaged material directly ahead of an existing crack. In this work, the retardation model is modified to also describe constant stress amplitude fatigue crack growth; prediction of the overload effect on fatigue crack growth is also considered. For the simulation of the irregular service spectrum, the rainflow counting method[44] has been modified. The LTSM-F code[45] is summarized below.

Stress spectrum simulation

Implementation of the model for the assessment of fatigue crack growth under service spectra requires the transformation of the loading spectrum from Fig. 4.23a to a sequence of distinguished reversed stress cycles as shown in Fig. 4.23b. Plastic deformation is considered as an accumulating

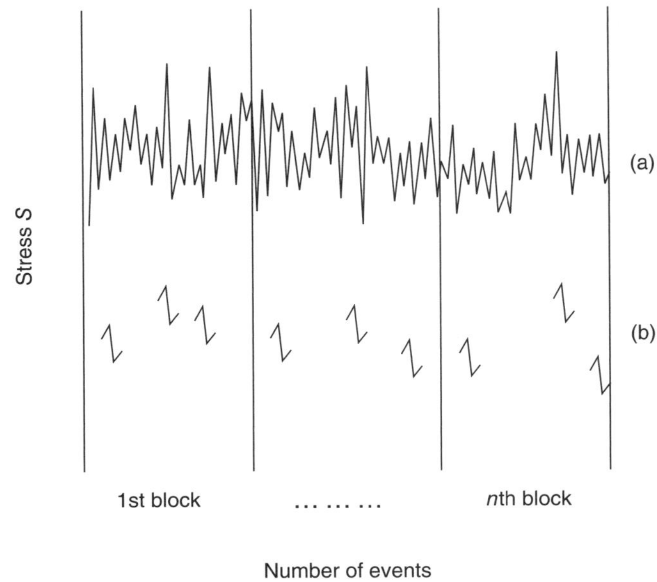

4.23 (a) Typical aircraft stress spectrum discretized into *n* loading blocks. (b) Equivalent spectrum consisting of full, distinguished cycles.

damage after each load event, while any additional damage effect due to unloading or compressive events is considered to be negligible. The implementation of the rainflow counting method can satisfy the above requirements. However, classical rainflow algorithms do not account for the relative position of each event within the service fatigue spectrum. This may be interpreted as the *a priori* assumption of linear damage accumulation during fatigue. In the LTSM-F code, the rainflow algorithm[44] has been modified. The service loading spectrum is discretized into *n* loading blocks. Each block consists of a small number of events as shown in Fig. 4.23. Then, the rainflow algorithm is implemented for each of these blocks separately. The introduced modification limits the assumption of linear fatigue damage accumulation within each block. Mathematically, it refers to the approximation of a power function by a sufficient number of prospective, short straight lines with increasing slopes. In addition, this modification allows the consideration of the relative position of the overloads within the entire service spectrum.

Fatigue crack growth calculation

In the fatigue crack growth analysis code,[43] the crack growth rate is described by the equation:

$$\frac{\mathrm{d}\alpha}{\mathrm{d}N}=\left(\frac{\sqrt{A}}{\sqrt{2}K_{cr}}\right)^{2/m}\left(\frac{\pi}{32\sigma_F^2}\right)^{1-\frac{\beta-1}{m}}\Delta K^{2\left(1-\frac{\beta-2}{m}\right)} \quad [4.12]$$

where the quantities

$$C=\left(\frac{\sqrt{A}}{\sqrt{2}K_{cr}}\right)^{2/m}\left(\frac{\pi}{32\sigma_F^2}\right)^{1-\frac{\beta-1}{m}} \quad [4.13]$$

and

$$n=2\left(1-\frac{\beta-2}{m}\right) \quad [4.14]$$

are constant and depend on the material and the stress ratio. Equation [4.12] is of the same form as the Paris law and may be applied for fitting constant stress amplitude fatigue crack growth curves. Fitting parameters are the quantities A and β. For all metals, the parameter m range is 0.5–0.7[46] and may be taken to be constant. For the assessment of the crack growth behaviour after an overload, a retardation factor λ is defined as

$$\lambda=\frac{(\mathrm{d}\alpha/\mathrm{d}N)_{\mathrm{ret}}}{(\mathrm{d}\alpha/\mathrm{d}N)_{\mathrm{nom}}} \quad [4.15]$$

where $(\mathrm{d}\alpha/\mathrm{d}N)_{\mathrm{ret}}$ is the crack propagation rate following the overload while $(\mathrm{d}\alpha/\mathrm{d}N)_{\mathrm{nom}}$ corresponds to the rate without the overload. Using Equation [4.12] for calculating both, $(\mathrm{d}\alpha/\mathrm{d}N)_{\mathrm{ret}}$ and $(\mathrm{d}\alpha/\mathrm{d}N)_{\mathrm{nom}}$, the retardation factor can be derived as

$$\lambda=\left(\frac{\sigma_{F0}}{\sigma_F}\right)^2 \quad [4.16]$$

where S_{y0} and S_y are, respectively, the cyclic yield stress without and with overload plasticity. S_y decreases with the propagation of the crack through the overload plastic zone up to the value $\lambda = 1$ (Fig. 4.24). A linear variation from S_y^* at the crack front to S_{y0} at the region where the plastic zone created by an overload meets the virgin material is assumed as shown in Fig. 4.24. The linear approximation of S_y inside the plastic zone is given by:

$$\sigma_F=\sigma_F^*-\frac{\alpha+(\omega_S/2)-\alpha_0}{\omega_0}\left(\sigma_F^*-\sigma_{F0}\right) \quad [4.17]$$

The calculation of the actual crack length α after application of i fatigue blocks of variable stress amplitudes is based on the relation:

$$\alpha=\alpha_i+\sum_{i=1}^{k}\lambda_i\left(\frac{\mathrm{d}\alpha}{\mathrm{d}N}\right)_{\mathrm{nom}} \quad [4.18]$$

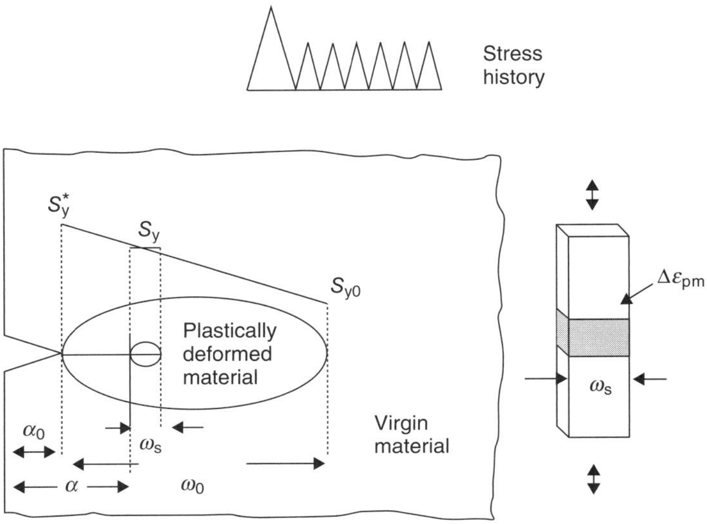

4.24 Basic considerations of the LTSM-F model.

where α_i is the crack length before application of the ith block, k the number of cycles involved in the block i and $(d\alpha/dN)_{nom}$ can be estimated from the constant amplitude crack growth rule.

Based on the described model an assessment of the influence of corrosion damage on the crack growth behaviour of the material can be performed. For this, the local fracture toughness K_{cr}^{cor} value determined by equation [4.10] and the yield stress value of the corroded material σ_F^{cor} is implemented. Accordingly, the crack growth rate in the corrosion damaged material may be assessed by using equation [4.12] in the form:

$$\left(\frac{d\alpha}{dN}\right)^{cor} = \left(\frac{\sqrt{A}}{\sqrt{2}K_{cr}^{cor}}\right)^{2/m}\left(\frac{\pi}{32\sigma_F^{cor^2}}\right)^{1-\frac{\beta-1}{m}} \Delta K^{2\left(1-\frac{\beta-2}{m}\right)} \qquad [4.19]$$

For the assessment of crack growth rate following the application of an overload, the retardation factor λ is implemented in the form:

$$\lambda = \left(\frac{\sigma_{F0}^{cor}}{\sigma_F}\right)^2 \qquad [4.20]$$

Finally, calculation of crack growth is performed via the formula:

$$\alpha = \alpha_i + \sum_{i=1}^{k} \lambda_i \left(\frac{d\alpha}{dN}\right)_{nom}^{cor} \qquad [4.21]$$

4.5 Conclusions

A series of experimental results has been presented, which shows the influence of existing corrosion on the mechanical behaviour of aircraft aluminum alloys. The results include static, fatigue, fatigue crack growth and fracture toughness tests on pre-corroded aluminum alloys. In order to allow comparison between the experimental findings, the tests were conducted also for uncorroded material. The basic conclusions drawn are the following:

1. Corrosion-induced degradation of tensile properties occurs gradually with increasing exposure time. Tensile ductility decreases exponentially to extremely low final values.
2. Mechanical removal of the corroded areas restored the yield and ultimate tensile stress but not the tensile ductility. The latter was restored only after thermal treatment of the alloys at temperatures corresponding to hydrogen trapping, suggesting that corrosion of the examined alloys is associated with hydrogen embrittlement.
3. The experimental results have shown an appreciable decrease in fatigue resistance and damage tolerance of the corroded material. The fracture toughness of the corroded material decreases significantly as well. The obtained results were discussed in relation to the synergetic effect of corrosion and corrosion-induced hydrogen embrittlement.
4. It is essential to evaluate a local fracture toughness value, associated with the corroded areas of a structure, which accounts for the synergetic effect of corrosion and the corrosion-induced hydrogen embrittlement. A fracture mechanical model is presented for the assessment of the local fracture toughness of a corrosion damaged material. Finally for the evaluation of the effect of corrosion on the fatigue crack growth behaviour of the aged material a fatigue crack growth model is presented that uses the local fracture toughness value of the corroded material.
5. The results have demonstrated the need to account for the influence of pre-existing corrosion on the material's properties for the reliable fatigue and damage tolerance analyses of components involving corroded areas.

4.6 References

1 R. P. WEI and D. G. HARLOW, 'Corrosion and Corrosion Fatigue of Airframe Materials', US Department of Transportation, Federal Aviation Administration, DOT/FAA/AR-95/76, February 1996, Final Report, National Technical Information Service, Springfield, VA 22161.
2 'Aging of US Air Force Aging Aircraft', *Nat. Mat. Adv. Board Pub.* NMAB-488-2, National Academy Press, Washington, DC, 1997, p. 54.

3 *ASM Handbook, Corrosion Fundamentals,Testing and Protection*, Volume 13 A, 2003.

4 R. G. BUCHHEIT, R. P. GRANT, P. F. HLAVA, B. MCKENZIE, and G. L. ZENDER, 'Local dissolution phenomena associated with S phase particles in alloy 2024', *J. Electrochem. Soc.*, 1997, **144**(8), 2621–2628.

5 *Aluminum: Properties and Physical Metallurgy*, J. E. Hatch, Ed., ASM International, 1984.

6 J. R. DAVIS (Ed), *Alloying: Understanding the Basics*, ASM International, 2001, pp. 352–416.

7 S. G. PANTELAKIS, G. N. HAIDEMENOPOULOS, 'The effect of corrosion and hydrogen embrittlement on the mechanical behaviour of aluminum aircraft alloys', in: J. K. Wessel (Ed.), *Handbook of Advanced Materials*, McGraw-Hill, London, 2002.

8 E. CHARITIDOU, G. PAPAPOLYMEROU, G. N. HAIDEMENOPOULOS, N. HASSIOTIS, V. BONTOZOGLOU, 'Characterization of trapped hydrogen in exfoliation corroded aluminium alloy 2024', *Scripta Mater.* 1999, **41**, 1327–1332.

9 G. N. HAIDEMENOPOULOS, N. HASSIOTIS, G. PAPAPOLYMEROU, V. BONTOZOGLOU, 'Hydrogen absorption into aluminum alloy 2024-T3 during exfoliation and alternate immersion testing', *Corrosion* 1998, **54**, 73–78.

10 E. KAMOUTSI, 'Hydrogen trapping during corrosion of aluminium alloys', PhD Thesis, University of Thessaly, Greece, 2004.

11 C. D. S. TUCK, 'Evidence for the formation of magnesium hydride on the grain boundaries of Al–Zn–Mg alloys during their exposure to water vapour', in: *Proceedings of the Third International Conference of the Effects of Hydrogen on the Behavior of Materials*, Jackson, Wyoming 1980, pp. 503–511.

12 H. SAITOH, Y. IIJIMA, K. HIRANO, 'Behaviour of hydrogen in pure aluminium Al–4 mass% Cu and Al–1 mass% Mg_2Si alloys studied by tritium electron microautoradiography', *J. Mater. Sci.* 1994, **29**, 5739–5744.

13 P. N. ANYALEBECHI, 'Hydrogen diffusion in Al–Li alloys', *Metall. Trans. B* 1990, **21**, 649–655.

14 P. V. PETROYIANNIS, A. T. KERMANIDIS, E. KAMOUTSI, S. G. PANTELAKIS, V. BONTOZOGLOU, G. N. HAIDEMENOPOULOS, 'Evidence on the corrosion-induced hydrogen embrittlement of the 2024 aluminum alloy', *Fatigue Fract. Eng. Mater. Struct.*, 2005, **28**, 565–574.

15 R. J. H. WANHILL, *Aircraft Corrosion and Fatigue Damage Assessment*, NLR Technical Publication TP 94401 L, National Aerospace Laboratory, Amsterdam, 1994.

16 M. E. INMAN, R. G. KELLY, S. A. WILLARD, R. S. PIASCIK, in: *Proceedings of the FAA–NASA Symposium on the Continued Airworthiness of Aircraft Structures*, Virginia, USA, p. 129.

17 J. E. ZAMBER, B. M. HILLBERRY, 'Probabilistic approach to predicting fatigue lives of corroded 2024-T3', *AIAA J.* 1999, **37**(10), 1311.

18 G. H. BRAY, R. J. BUCCI, E. L. COLVIN, M. KULAK, 'Effects of prior corrosion on the S/N fatigue performance of aluminium sheet alloys 2024-T3 and 2524-T3', in: W. A. Van der Sluys, R. S. Piascik, R. Zawierucha (Eds.), *Effects of the Environment on the Initiation of Crack Growth*, ASTM STP 1298, American Society for Testing and Materials, 1997, p. 89.

19 S. G. PANTELAKIS, N. I. VASSILAS, P. G. DAGLARAS, 'Effects of corrosive environment on the mechanical behavior of the advanced Al–Li alloys 2091 and 8090 and the conventional aerospace alloy 2024', *Metallurgy* 1993, **47**, 135–141.

20 S. G. PANTELAKIS, P. G. DAGLARAS, C. A. APOSTOLOPOULOS, 'Tensile and energy density properties of 2024, 6013, 8090 and 2091 aircraft aluminum alloy after corrosion exposure', *J. Theor. Appl. Fract. Mech.* 2000, **33**, 117–134.

21 P. V. PETROYIANNIS, A. T. KERMANIDIS, P. PAPANIKOS, S. G. PANTELAKIS, 'Corrosion-induced hydrogen embrittlement of 2024 and 6013 aluminum alloys', *Theoretical and Applied Fracture Mechanics* 2004, **41**, 173–183.

22 ASTM G34-90, 'Standard test method for exfoliation corrosion susceptibility in 2XXX and 7XXX series aluminum alloys (EXCO Test)' in: *Annual Book of ASTM Standards, Section 3, Metal Test Methods and Analytical Procedures*, ASTM, West Conshohocken, Philadelphia, USA, 1995, pp. 114–119.

23 ASTM G44–94, 'Standard practice for evaluating stress corrosion cracking resistance of models and alloys by alternate immersion in 3.5% sodium chloride solution', in: *Annual Book of ASTM Standards, Section 3, Metal Test Methods and Analytical Procedures*, ASTM, West Conshohocken, Philadelphia, USA, 1995, pp. 157–161.

24 ASTM B117-94, 'Standard practice for operating salt (fog) testing apparatus', in: *Annual Book of ASTM Standards, Section 3, Metal Test Methods and Analytical Procedures*, ASTM, West Conshohocken, Philadelphia, USA, 1995, pp. 1–8.

25 ASTM G85-94, 'Standard practice for modified salt spray (for) testing', in: *Annual Book of ASTM Standards, Section 3, Metal Test Methods and Analytical Procedures*, ASTM, West Conshohocken, Philadelphia, USA, 1995, pp. 351–360.

26 ASTM G50-76, 'Standard practice for conducting atmospheric corrosion tests on metals', in: *Annual Book of ASTM Standards, Section 3, Metal Test Methods and Analytical Procedures*, ASTM, West Conshohocken, Philadelphia, USA, 1995, pp. 185–189.

27 ASTM E8M-94A, 'Standard test methods for tension testing of metallic materials (metric)', *Annual Book of ASTM Standards, Section 3, Metal Test Methods and Analytical Procedures*, ASTM, West Conshohocken, Philadelphia, USA, pp. 81–96.

28 S. G. PANTELAKIS, G. N. HAIDEMENOPOULOS, 'Corrosion and hydrogen embrittlement of aircraft aluminum alloys', in: *Proceedings of the 4th International Conference on New Challenges in Mesomechanics*, Aalborg University, Denmark, 2002.

29 G. S. CHEN, C. M. LIAO, K. C. C. WAN, M. GAO, R. P. WEI. 'Pitting corrosion and fatigue crack nucleation', in: Van Der Sluys W. A., Piascik R. S., Zawieruca R., editors. *Effects of environment on the initiation of crack growth, ASTM STP 1298*. West Conshohocken, PA, USA: American Society for Testing and Materials; 1997. p. 18–33.

30 K. K. SANKARAN, R. PEREZ, K. V. JATA. 'Effects of pitting corrosion on the fatigue behavior of aluminum alloy 7075-T6: modeling and experimental studies'. *Mater Sci Eng A* 2001, **A297**, 223–229.

31 G. H. BRAY, R. J. BUCCI, E. L. COLVIN, M. KULAK. 'Effect of prior corrosion on the S/N fatigue performance of aluminum sheet alloys 2024-T3 and 2524-T3', in: Van Der Sluys W. A., Piascik R. S., Zawieruca R., editors. *Effects of environment on the initiation of crack growth, ASTM STP 1298*. West Conshohocken, PA, USA: American Society for Testing and Materials; 1997. pp. 89–101.

32 S. I. ROKHLIN, J. Y. KIM, H. NAGY, B. ZOOFAN. 'Effect of pitting corrosion on fatigue crack initiation and fatigue life'. *Eng Fract Mech* 1999, **62**(4–5), 425–444.

33 P. S. PAO, C. R. FENG, S. J. GILL. 'Corrosion fatigue crack initiation in aluminum alloys 7075 and 7050'. *Corrosion* 2000, **56**(10), 1022–1031.

34 P. S. PAO, S. J. GILL, C. R. FENG, 'On fatigue crack initiation from corrosion pits in 7075-T7351 aluminum alloy'. *Scripta Mater* 2000, **43**, 391–396.
35 ASTM E466-82, 'Standard practice for conducting constant amplitude axial fatigue tests of metallic materials', *Annual Book of ASTM Standards, Metals – Test Methods and Analytical Procedures*, Section 3, Vol. 03.01, Metals-Mechanical Testing; Elevated Low-Temperature Tests; Metallography, 1994.
36 MIL-A-8625 E, Type III-Class 1, 'Anodic coatings for aluminum and aluminum alloys', Military Specification, Department of Defense, Systems Engineering and Standardization Department, Naval Air Engineering Center, Lakehurst, NJ 08733-5100, April 25, 1, 1988.
37 J. P. CHUBB, T. A. MORAD, B. S. HOCKENHULL, J. W. BRISTOW, 'The effect of exfoliation corrosion on the fatigue behaviour of structural aluminium alloys', *Structural Integrity of Ageing Airplanes*, Springer-Verlag, Berlin, 1991, p. 87.
38 ASTM E647-93, 'Standard test method for measurement of fatigue crack growth tests', *Annual Book of ASTM Standards, Metals – Test Methods and Analytical Procedures*, Section 3, Vol. 03.01, Metals-Mechanical Testing; Elevated Low-Temperature Tests; Metallography, 1994.
39 ASTM E561-94, 'Standard Practice for R-Curve Determination', *Annual Book of ASTM Standards, Metals – Test Methods and Analytical Procedures*, Section 3, Vol. 03.01, Metals-Mechanical Testing; Elevated Low-Temperature Tests; Metallography, 1994.
40 S.-X. WU, Y.-W. MAI, B. COTTERELL, 'A model of fatigue crack growth based on Dugdale model and damage accumulation'. *Int. J. Fracture*, 1992, **57**, 253–267.
41 J. R. RICE. 'Mechanics of crack tip deformation and extension by fatigue'. In *Fatigue Crack Propagation*, ASTM STP 415, 1967, pp. 247–311 (American Society for Testing and Materials, Philadelphia, Pennsylvania).
42 S. PANTELAKIS, T. KERMANIDIS, and D. PAVLOU. 'Fatigue crack growth retardation assessment of 2024-T3 and 6061-T6 aluminium specimens'. *Theor. Appl. Fract. Mech*. 1995, **22**, 35–42
43 A. T. KERMANIDIS, S. G. PANTELAKIS, 'Fatigue crack growth analysis of 2024 T3 aluminum specimens under aircraft service spectra', *Fatigue Fract. Eng. Mater. Struct.*, 2001, **24**, 699–710.
44 M. MATSUISHI., and T. ENDO, 'Fatigue of metals subjected to varying stress'. In: *Proceedings of the Kyushu Branch of Japan Society of Mechanical Engineering, Fukuoka, Japan* (in Japanese) 1968, pp. 37–40.
45 S. G. PANTELAKIS, A. T. KERMANIDIS, and P. G. DAGLARAS, 'Crack growth analysis code for assessing fatigue life of 2219,T851 aluminum specimens under aircraft structure service spectra'. *J. Theor. Appl. Fract. Mech*. 1997, **28**, 1–12.
46 H. O. FUCHS, and R. I. STEPHENS, (1980) *Metal Fatigue in Engineering.* John Wiley & Sons, Inc., New York, NY, USA.
47 A. T. KERMANIDIS, 'Effect of corrosion on the structural integrity of lightweight aircraft structures', PhD Thesis, University of Patras, Department of Mechanical and Aeronautical Engineering, Patras, Greece, (2003)

Part II

Corrosion monitoring, evaluation and prediction

5
Nondestructive testing of corrosion in the aerospace industry

D. S. FORSYTH, TRI/Austin, USA

Abstract: Methods of nondestructive testing (NDT), any test that does not impair the function of the test subject, are described. Owing to the diversity of corrosion damage that occurs on the variety of engineering materials in the aerospace industry, there are a number of NDT methods that have been applied. A clear understanding of the corrosion damage and the effect of the damage on structural integrity is necessary to properly specify an NDT method for corrosion.

Key words: nondestructive testing, radiography, ultrasound, eddy current, enhanced visual testing, corrosion, aerospace industry.

5.1 Introduction to nondestructive testing

Nondestructive testing (NDT) is defined by the American Society for Nondestructive Testing (ASNT) as: 'The determination of the physical condition of an object without affecting that object's ability to fulfill its intended function. Nondestructive testing techniques typically use a probing energy form to determine material properties or to indicate the presence of material discontinuities (surface, internal or concealed)'. For the purpose of this article, the terms nondestructive testing, nondestructive inspection (NDI), and nondestructive evaluation (NDE) will be considered to be equivalent.

In the modern NDT paradigm, the uses of NDT can be broken into several categories where it plays an important role:

- Material property measurements
- Process design for materials manufacturing
- Online process control
- Quality control as various stages of manufacturing are completed

In addition, NDT plays an important role in the continued safe operation of physical assets. For instance, NDT is being used in conventional inspections and in health monitoring, where NDT sensors are embedded or attached to the system being inspected or monitored for defects or damage.

In all cases, the customer must define the requirements of the test, such as the minimum level of acceptability for the property being measured and the characteristics of the material discontinuities to be identified. Given this information, the NDT engineer or experienced technician can choose the appropriate method and develop an appropriate technique for the inspection requirements.

An NDT method is classified according to its underlying physical principle. For example, the common methods are:

- Visual and optical testing (VT),
- Radiographic testing (RT),
- Electromagnetic testing (ET),
- Ultrasonic testing (UT),
- Liquid penetrant testing (PT),
- Magnetic particle testing (MT),
- Acoustic emission testing (AE), and
- Infrared and thermal testing (IR).

An NDT technique defines all the parameters for the application of a specific method to a specific problem. These parameters include the instruments, probes, acceptance criteria, calibration specifications, and much more. ASNT offers a series of handbooks that are a key reference for the practical implementation of NDT. The following sections will briefly describe each of the common methods listed above. Portions of the text are taken (with permission) from Matzkanin and Yolken (2007).

5.1.1 Requirements for corrosion nondestructive testing

NDT is one element in the life cycle management of aircraft. It is integrally related to the methods used to assess structural integrity and risk assessment: these set the requirements for NDT. The aerospace community designs and maintains aircraft by either damage tolerance or safe life approaches. Both of these, in the current operational environment, are based on crack growth and elimination of other damage mechanisms such as corrosion through maintenance and repair. In this environment, the requirements for NDT for corrosion are not well defined. This author has heard from civilian and military regulators that 'we don't fly planes with corrosion', despite all evidence to the contrary.

The difficulty of the current situation, 'find and fix', is that it precludes use of sensitive NDT in instances where operators feel through experience that corrosion is not an issue. If corrosion damage is found at an early stage where there is no immediate safety impact, operators still feel compelled to take maintenance action immediately instead of deferring to a later, more convenient opportunity.

Several programs currently in progress in the United States and other countries are attempting to improve upon this state of affairs by developing analytical capabilities for corrosion damage assessment in aircraft structures. These analytical models will require, as input, quantified corrosion damage data from NDT. The outputs from the corrosion damage assessments will allow the planning of maintenance actions depending on the current state of corrosion, its influence on structural integrity, and its projected growth. The outputs of these programs will then provide more detailed requirements and cost justification for the use of advanced NDT for corrosion.

5.1.2 Visual nondestructive testing

By far, the most common NDT method is visual and optical testing. In many instances, a trained inspector armed with simple tools, such as a flashlight and magnifying glass, can perform a very effective inspection. In quality control, as well as in maintenance operations, visual testing is the first line of defense. When deciding whether to use visual testing, it is important to understand its potential as well as its limitations. If the visual method is not sufficient for the problem at hand, more complex methods must be considered. Using the visual inspection method for enclosed systems can be challenging and possibly ineffective. To enable a technician or engineer to inspect these difficult-to-see areas, a device known as a borescope is often used. Borescopes are essentially miniaturized cameras that can be placed on the end of a fiber optic cable. The camera can then be inserted into regions that are obstructed from direct visual inspection, and the resulting images are viewed in real-time on a video screen by the inspector.

5.1.3 Enhanced visual/optical nondestructive testing

There are a variety of enhanced visual/optical NDT methods available. In terms of corrosion NDT, these methods are generally used to detect and measure deformations on surfaces. These deformations may be caused by pitting on the exposed surface, or by subsurface corrosion damage in built-up structure. There are a number of implementations of instruments based on Moire, electronic speckle pattern interference (ESPI) and digital speckle correlation (Jin and Chiang, 1998), and holography. Other optical surface topography systems have been used for characterization of corrosion damage (Komorowski *et al.*, 1996, Forsyth *et al.*, 1997). Direct optical metrology methods such as laser interferometry and triangulation-based methods have been used in laboratory-type situations for measuring pillowing caused by corrosion in thin aluminum structures (Eastaugh *et al.*, 1998).

5.1.4 Ultrasonic nondestructive testing

Ultrasonic testing (UT) employs an extremely diverse set of methods based upon the generation and detection of mechanical vibrations or waves within test objects. The test objects are not restricted to metals, or even to solids. The term ultrasonic refers to sound waves of frequency above the limit of human hearing. Most ultrasonic techniques employ frequencies in the range of 1 to 10 MHz. The velocity of ultrasonic waves traveling through a material is a simple function of the material's modulus and density, and thus ultrasonic methods are uniquely suited to materials characterization studies. In addition, ultrasonic waves are strongly reflected at boundaries where material properties change and thus are often used for thickness measurements and crack detection. Recent advances in ultrasonic techniques have largely been in the field of phased array ultrasonics, now available in portable instruments. The timed or phased firing of arrays of ultrasonic elements in a single transducer allows for precise tailoring of the resulting ultrasonic waves introduced into the test object.

5.1.5 Eddy current nondestructive testing

Electromagnetic testing (ET), especially eddy current testing, is commonly used to inspect objects throughout their life cycle. Eddy current techniques employ alternating currents applied to a conducting coil held close to the test object. In response, the test object generates eddy currents to oppose the alternating current in the coil. The eddy currents are then sensed by the same coil, separate coils, or magnetic field sensors. Changes in the induced eddy currents may be caused by changes to a material's electromagnetic properties and/or changes in geometry, including the abrupt changes in current flow caused by cracks. Thus, ET methods are highly effective for the detection of cracks present on or below the surface of metallic objects. ET equipment has become extremely portable and is relatively cheap. It is the second most common method specified for NDT of aircraft. Recent advances in eddy current technology include multi-channel portable instruments, allowing faster inspections of large areas, and new magnetic sensors, such as the giant magnetoresistive sensors (GMR) developed for computer hard drives, instead of coils.

5.1.6 Thermographic nondestructive testing

Infrared and thermal testing methods are characterized by the use of thermal measurements of a test object as it undergoes a response to a stimulus. Thermal imaging cameras are the most common sensing method. Passive imaging of machinery or electronics may be used to detect hot spots indicative of problems. Imaging of test objects after the application of energy can be used to monitor the flow of heat in the object, which is a function of material properties as well as boundaries. Flash thermography techniques have been very successful in imaging disbonds and delaminations in composite parts, for example. The high cost of quality thermal cameras was previously a drawback of the infrared (IR) method, but recently these have become significantly less expensive. Another significant recent advancement is the use of mechanical energy to stimulate localized heating at subsurface discontinuities, such as cracks in metals, opening up a new field of application for the IR method.

5.1.7 Radiographic nondestructive testing

Historically, radiography is the next most common NDT method. Significant activity in the field occurred almost immediately after Roentgen's discovery of x-rays in 1895. Early literature notes the ability of radiographs to detect discontinuities in castings, forgings, and welds in metals. Discontinuities such as pores or inclusions in metals are readily detected in many cases. Cracks may also be detected using radiographic techniques, but attention must be paid to orientation and residual stress issues. Radiography continues to be widely used despite the expense and safety implications of the equipment. Recent advances in digital radiography have helped reduce the cost of employing this method by eliminating the use of film.

5.1.8 Additional nondestructive testing methods

There are a number of other NDT methods that have been used for corrosion NDT. These include the magneto-optic imager (MOI), a commercial device that images magnetic fields induced by a sheet current (Thome *et al.*, 1996). Microwave NDT methods have been used to find corrosion under paint layers (Hughes *et al.*, 2003). Terahertz imaging is being used to find corrosion damage under thermal insulation tiles on the space shuttle (Anastasi *et al.*, 2007).

Health monitoring for corrosion is a growing field, with the potential to reduce the impact of disassembly and reassembly of aircraft to enable traditional NDT. Some of the sensor types are direct evolutions of NDT methods, and are simply attached to the structure to be left in place. This topic is covered in chapter 7.

5.2 Data fusion for nondestructive testing

The complexities of corrosion damage, and the fact that inspections for fatigue damage are often carried out on the same structure, means that the maintainer may end up with multiple inspections on the same structure, each with a fraction of the total information needed to make maintenance decisions. Data fusion is required to make sense of the multiple sources of information, whether it is done manually by the inspector or engineer, or whether it is computer assisted.

A simple definition of data fusion is the combination of multiple inputs into one output. Thus, data fusion includes basic systems such as voting (if a majority of inputs are true, the output is true) as well as highly complex systems such as military target tracking or remote sensing using multiple band radars operating in different locations. There are three general categories used to describe the level at which data fusion takes place. Pixel level data fusion describes applications where little or no preprocessing is applied to the data, and the fusion operation acts on the lowest level of the data. Feature level fusion refers to cases where feature extraction has been performed on the data before fusion. Finally, decision level fusion refers to the fusion of data that is carried out after feature extraction and identification on the data inputs. The results reported in this work use pixel-level fusion algorithms. Some applications of data fusion to NDT have been published, most of which are of moderate complexity (for example Gros, 1997 and Gros, 2001) and involve pixel or feature level fusion. This author has previously applied simple data fusion methods on NDT results to identify and measure corrosion in aircraft structures (Forsyth and Komorowski, 2000); the work presented in that paper used NDT from commercially available inspection equipment and applied more advanced fusion techniques to yield quantitative estimates of the thicknesses of individual layers of a two-layer lap joint.

The generic steps required to perform data fusion on NDT data are:

- inspection preprocessing,
- registration of individual inspections on a common coordinate system, and
- data fusion.

It is important to note that the steps of preprocessing and registration are in themselves value-added steps. Even before any data fusion operation has been performed, the NDT data from disparate sources have been brought together on one software platform, and registered on a common coordinate system. This is a significant improvement over most current practices for handling NDT data, and greatly facilitates the use of databases for maintenance planning. It also allows improved inspector interpretation by making comparison between NDT data much simpler. Often the preprocessing step can be used to transform a single NDT data source from the NDT domain to a quantitative measure; for example, ‘Edge of Light’ images can be transformed from brightness levels to images of maximum pillowing deformation (Forsyth and Komorowski, 2000).

The final data fusion algorithm to be used will be specific to the application. Development of these algorithms will only be cost-effective for repetitive inspection situations, such as the common lap splice joint. However, the preliminary steps of data handling, preprocessing, and registration are likely to become more commonly used as fleet maintenance practices are modernized, reducing the costs of implementing data fusion in practical situations.

5.3 Reliability of nondestructive testing for corrosion

When NDT is used as part of the management of risk in the life cycle maintenance of an aircraft, it is imperative to know what is the probability of finding (or equivalently of missing) discontinuities of interest in an inspection. This is usually called the probability of detection (POD). The development of the POD metric was originally directed towards fatigue cracks, but it is important to note that the POD approach is not limited to cracks, and has in fact been applied to other discontinuities such as corrosion loss, impact damage, or delaminations (Forsyth *et al.*, 1998). The current POD approaches present POD as a function of a single metric of damage, for example crack length, as shown in Fig. 5.1. In cases where the corrosion damage of interest can be characterized by a single metric, the conventional

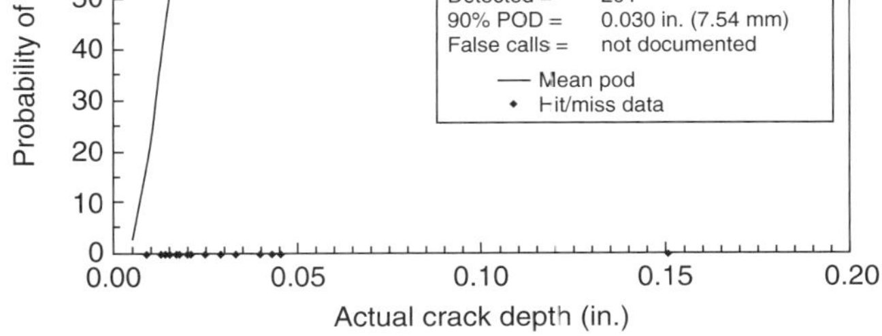

5.1 An example of a POD curve, from the ultrasonic inspection of welds in aluminum for cracks (from Rummel and Matzkanin 1996, used with permission).

POD approaches will be suitable. This may be the case for intergranular cracking and pitting on an exposed surface. In other cases, multi-dimensional damage is not well characterized by a single metric, and a number of approaches have been developed.

Both the USAF (USDO D, 1999) and the United States Federal Aviation Agency (FAA) (Spencer *et al.* 1993a, 1993b and Spencer and Schurman, 1995) have published guidelines that describe in detail the experiments required to estimate the POD of an inspection system. These documents are in the public domain, and can be obtained for free from the respective government agencies as well as the Department of Defense's Advanced Materials, Manufacturing, and Testing Information Analysis Center (AMMTIAC). The USAF MIL-HDBK-1823 is being updated as this book goes to press.

There are a number of useful general statements that can be made about estimating POD. The process of POD estimation requires a number of inspections to be performed:

- using the complete, pre-defined inspection system that is being assessed:
  - including representative equipment, procedures, inspectors, and target parts;
- using parts with discontinuities that represent the discontinuities of interest:
  - or a means to assess the difference between the two, for example, using machined notches or flat bottomed holes can provide a useful measure of capability, but should not be assumed to be representative of cracks or other natural discontinuities; and
- using an inspection procedure and environment typical of the deployed environment:
  - human factors studies have shown that the relationship of factors such as environment (lighting, temperature, etc.), training, experience, motivation and others is not simple and often not intuitive.

It is key to understand the physical parameters that may affect the response of the NDT system to a discontinuity to be able to execute a representative POD estimate. If parts from service, with discontinuities arising from service, are available; this is the optimal situation. However, in most cases, this is not possible. Therefore, every reasonable effort should be made to replicate the service discontinuities as close as possible, or to use engineering judgment as to whether safety factors are needed to account for the difference between the POD experiment and in-service conditions.

Unique approaches to POD for corrosion damage have been developed for the case of corrosion in the internal surfaces of aircraft skin splice joints. Ashbaugh *et al.* (2001) showed that detection of corrosion in this case was a function of both the thickness loss and area affected (Fig. 5.2).

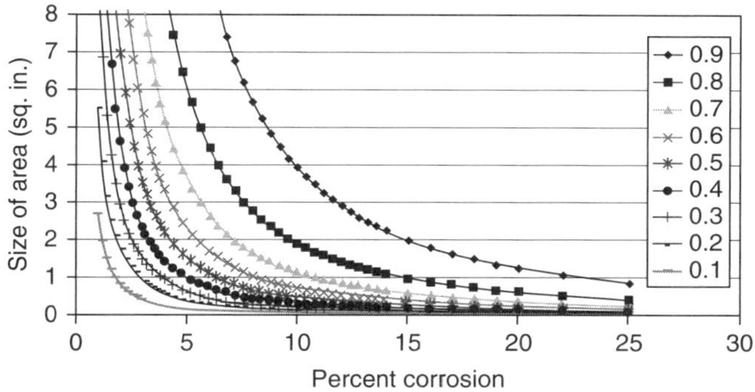

5.2 Data from experiment, showing contours of POD values as a function of both area and thickness loss of damage (from Ashbaugh et al. 2001 used with permission).

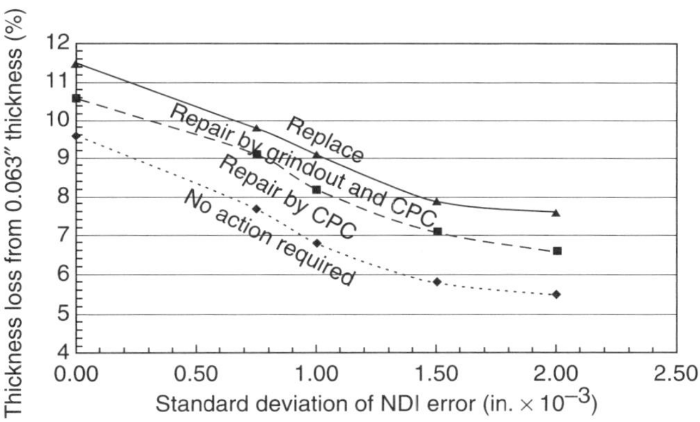

5.3 A graph of the effect of NDT error on the level of damage and subsequent maintenance actions from a K/C-135 case study.

Liao *et al.* (2003) used the error in thickness loss measurement by NDT to calculate the effect of this error on the risk and therefore maintenance actions required. K/C-135 lap splice joints were used as the basis of this case study. As shown in Fig. 5.3, increasing error results in increasing uncertainty about the NDT assessment and, therefore, more severe maintenance actions are required as NDT error increases for the same estimated thickness loss.

5.4 Typical applications of nondestructive testing to corrosion in aerospace systems

The design of modern aircraft is a relatively mature engineering discipline, largely progressing by evolutionary not revolutionary steps. At a high level,

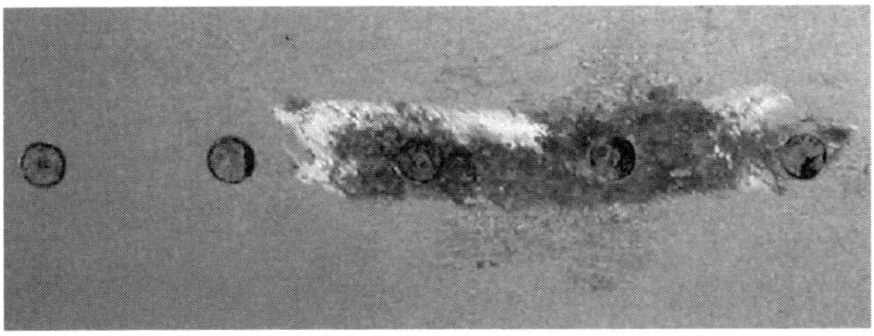

5.4 A photograph of a row of fasteners from the upper wing skin of a Boeing 707, showing exfoliation damage and grinding marks from repairs (from Forsyth *et al.* 2002). The material is 7178-T6 aluminum.

5.5 Metallographic section around a fastener hole in a Boeing 707 wing skin plank, showing multiple layers of intergranular attack from pits in the countersink. The material is 7178-T6 aluminum.

there are strong similarities in structures and engines across rotary aircraft and fixed-wing aircraft from fighters to transports. In this chapter, we describe common design details that may be susceptible to corrosion, and how NDT is applied to these design details.

In order to define our nomenclature for the following, at its simplest, we can think of the design of aircraft structures as consisting of two major components: fuselage and wings. The fuselage typically is constructed of horizontal stringers, circular frames, and skins. The wing is similar, with spars running perpendicular to the fuselage, ribs running parallel to the fuselage, and skins. Horizontal and vertical stabilizer structures are usually similar to the wing construction and nomenclature.

5.4.1 Multi-layer thin structures

First we consider multi-layered thin structures. By thin, we arbitrarily choose 0.250″ as a guide. The key structural elements in this category are the skin

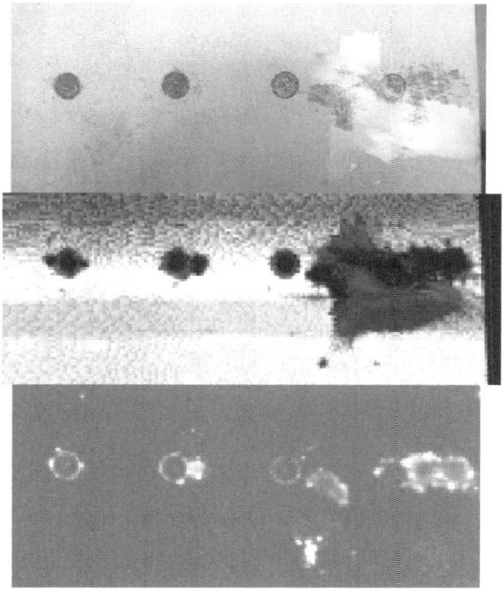

5.6 From top to bottom, a photograph, a UT image, and an IR image of a section of wing plank from a Boeing 707, showing exfoliation damage and grinding marks from repairs. The UT image was assembled from a mechanical scan of resolution 1 mm in both *x* and *y* directions. The IR image is from a 320 × 240 pixel camera. The material is 7178-T6 aluminum.

to skin joints or splice joints, and skin to substructure joints. These joints are formed in a variety of different designs, but in general are riveted. Some early designs, notably the K/C-135, utilize spot welds in some of the splice joints. A sealant or adhesive layer is usually included, but not relied upon for structural capability.

These structures are predominantly manufactured from 2024-T3 aluminum sheet. Significant efforts have been expended by numerous agencies to characterize the corrosion and fatigue behaviors of splice joints made from Al 2024-T3.

Most NDT techniques developed for the detection of corrosion in lap splice joints measure the average thickness loss due to general attack and pitting corrosion. Often these techniques are only capable of measuring this on the faying surface of the first layer. Estimates of the performance of some of these techniques are summarized in Table 5.1. The data in the table is a summary of results from various reports (see Hoppe *et al.*, 2002, Lepine *et al.* 2002).

To date, single frequency ET methods have demonstrated a sensitivity to 10% material loss on the first or second layer of multi-layer specimens, with precision of 2% of total layer thickness, with spatial resolutions limited to

Table 5.1 Estimates of existing NDT capabilities in determining the average thickness loss in thin skin on faying surface of first layer

Technique	Sensitivity	Precision	Spatial resolution
Single frequency eddy current	10% material loss	2%	0.100″**
Ultrasonics pulse–echo	5% material loss*	1%	0.020″
Thermography	10% material loss*	2%	0.100″**
Advanced eddy current	5% material loss	1%	0.050″**

* first layer only.
** spatial resolution is a function of depth of the discontinuity.

probe diameters, which are typically about 0.4″. Multiple frequency techniques can provide better results, by reducing effects such as liftoff. Newer techniques such as pulsed eddy current methods also hold promise.

Pulsed eddy current (P-ET) techniques have shown sensitivity to artificially created flaws with as little as 2% material loss, and the ability to distinguish on which layer the flaw occurs. In actual application to corrosion, a more conservative estimate of the potential of P-ET techniques in the near future would be 5% thickness loss per skin, with an accuracy of 1%.

The use of Hall effect sensors and magneto-resistive sensors in ET probes can improve spatial resolution, but in any case the electromagnetic field is dispersive, and resolution decreases with depth into the material. Experimental data providing approximate limits can be found in reference (Smith and Hugo, 2001).

Ultrasonic pulse–echo measurements are very accurate for damage in the first layer, but reflection and scattering at the interfaces between layers limit the ability of UT to inspect deeper.

It is likely that the enhanced visual techniques mentioned earlier could reliably detect <5% material loss depending on sheet thickness, with a spatial resolution determined by fastener spacing. It is interesting to note that these techniques, unlike others, do not report variations in sheet thickness owing to manufacturing as material lost to corrosion. Their estimates are based on the extent of pillowing due to the accumulation of corrosion products. However, for very thin skins (<0.030″) and dimpled rivets (as opposed to countersunk) it is very difficult to distinguish between corrosion pillowing and pillowing characteristic to the joint construction.

Lap joint pillowing

NDT techniques that can measure the pillowing deformation include enhanced visual/optical techniques, optical metrology, and mechanical measurements. Mechanisms other than corrosion, such as clamping stresses due to the manufacturing process or mechanical overloading during operation, can also cause deformation of riveted joints. This reduces the ability to infer material loss data from pillowing measurements.

Enhanced visual/optical techniques are sensitive to extremely small changes in surface topography. It is likely that many of these techniques could achieve sensitivity to 0.002″ deformation with a precision of 0.001″ in the application of measuring maximum pillowing deformation in an area between rivets in a riveted joint.

Mechanical measurements of deformation are less accurate than optical measurements, but it is likely that they can be performed to a precision of 0.005″ with current techniques.

It should pointed out that in a typical lap joint with rivet spacing of 1″ and skin thickness 0.040″, the average thickness loss of 5% results in 0.006″ pillowing deformation.

Corrosion pitting

Pitting occurring on hidden surfaces is extremely difficult to measure. Some potential has been shown for overall surface roughness owing to pitting by eddy current (Aldrin *et al.*, 2005), Compton back-scattering, high-resolution ultrasonics, and backscattered ultrasonic spectroscopy (Smith and Bruce, 2001). It is considered more likely that pitting characteristics will have to be inferred for each situation based on material data and average material loss measurements.

However, pitting on exposed surfaces could be detected and measured using visual inspections assisted by low-power magnification, or enhanced optical inspections. A sensitivity of 200–500 μm pit size is achievable on cleaned surfaces under favorable conditions with relatively simple equipment. Replication also can be used for precise measurement of the dimensions (including the depth) under a metallurgical optical microscope or a scanning electron microscope. Pit tunneling can occur in common aluminums such as 2024-T3, and thus the overall dimensions can be underestimated by these techniques (Lepine *et al.*, 2002).

The spatial distribution of pitting on an exposed surface can be easily obtained using fluorescent liquid penetrant as well as enhanced optical methods. Also, replication and microscopic examinations as well as boroscopes are capable of measuring with very high accuracy (in the order of 0.001″).

5.4.2 Thick section structure

Typical examples of thick section structure are wing skins, ribs, and spars; and complex forgings used for fittings at locations like wing to body attachment points. These structures are commonly manufactured from 7000 series aluminum alloys, 7075-T6 being common on older aircraft. More corrosion resistant alloys are generally used in modern designs.

Because of the materials and product forms typical of these structures, intergranular corrosion attack is a common problem. This attack begins at pitting at exposed grains, and quickly becomes intergranular in nature. If sustained stresses are present, due to residual stresses or even simple 'weight on wheels' loads, the phenomenon often called 'stress corrosion cracking' can result.

In a number of cases, inspection may be only an interim solution, providing time for material substitution programs. Any pitting on highly loaded forgings of Al 7075-T6 is capable of nucleating intergranular cracking. If these components are exposed for visual inspection, this may be sufficient. Interesting case studies include the C-130 'pork chop' fitting and the C-141 landing gear hub (Brooks *et al.*, 2001).

Corrosion pitting

Pitting attack can occur on thick section components, and can be structurally significant. The same NDT issues apply here as described in section 5.4.1.

Intergranular and exfoliation attack

A number of the 7000 series aluminum alloys that were commonly used in aircraft manufacture are susceptible to intergranular attack. These include 7078, 7079, and 7075 in the T6 temper. The T6 temper was popular due to its high strength. Typical applications on a transport aircraft include upper wing skins, spars, and ribs.

Intergranular attack commonly nucleates at corrosion pits on exposed endgrains. Holes with steel fasteners are a common site, as fretting eventually wears off the coatings designed to isolate the steel from the aluminum. In severe cases, the material bulges around the fastener due to the corrosion product between multiple intergranular cracks. This is called exfoliation.

This type of damage is relatively easy to detect, especially once it has progressed beyond the countersink of the fastener. Even before it is visible as exfoliation, UT and IR methods can readily detect it, but will only measure the top layer and any layers beneath which extend beyond the top layer (Forsyth *et al.*, 2002).

More advanced UT methods use surface waves or reflections to interrogate the volume obscured by fasteners, and can detect smaller areas of exfoliation. These methods are still subject to the phenomenon of top or bottom layers obscuring exfoliation occurring between them.

Depth location of intergranular corrosion

As described above, using UT techniques, it is relatively simple to determine the depth of the first layer of exfoliation that extends beyond fasteners. Accuracy in thick sections should be 0.005″ or better in depth. More advanced UT techniques can interrogate the bottom of the fastener hole and locate the bottom layer. Accuracy of 0.010″ is probably achievable under well-controlled situations.

Extent of intergranular corrosion

Exfoliation that extends beyond the fastener head can be detected by simple UT and IR methods, as well as enhanced visual methods. Sizing for all these methods is limited by probe sizes and in the case of IR methods, diffusion. UT is the most accurate sizing method, and a measurement accuracy of 0.010″ should be achievable. The sensitivity of IR methods is more affected by the depth at which the exfoliation occurs, and this is not well known.

5.4.3 Engine components

Engine components experience a wider variety of operating environments than structures, and they are unlikely to last long enough in typical usage to develop corrosion damage. Erosion corrosion of thermal protection systems may occur, but this is not similar to any of the previous issues discussed in this chapter.

5.4.4 Non-structural systems

Non-structural systems on aircraft such as wiring, control linkages, and hydraulics can also be susceptible to corrosion damage. There is little evidence to suggest that corrosion of actual wiring is an important factor in wire aging. Rather it is the aging of the polymer insulating layers that is significant. There are documented cases of corrosion damage on control linkages influencing aircraft accidents, notably the infamous Aloha Airlines Flight 243 accident in 1988. These parts are normally not layered, thus visual inspection in most cases is sufficient to maintain safety if performed at sufficient intervals.

5.4.5 Coatings

Coatings are not typically subjected to any NDT other than simple visual inspection. This is despite that the lack of use of NDT for corrosion under coatings requires costly removal and repainting at heavy maintenance checks. Microwave NDT and IR methods have been developed for this problem (Hughes *et al.*, 2003), but remain at the precommercial stage without any significant adopter.

5.5 Summary and conclusions

There is a great need for cost-effective and reliable NDT techniques that can be used to ensure the safe operation of aging aircraft. A large number of NDT techniques have been applied to the detection and characterization of corrosion in simulated or real aircraft parts. Table 5.2 summarizes the application areas, capabilities, limitations and metrics for each technique. However, there has been relatively little effort by independent parties to measure the sensitivity and reliability of these techniques under realistic conditions. Therefore, the NDT metrics quoted here are best estimates under the most favorable conditions. Each NDT method has certain capabilities and limitations and often more than one technique is needed to cover the various component and corrosion types encountered in aging aircraft.

For NDT of airframe structures such as lap joints and wing skins, a two-step approach may be the most feasible. A number of organizations (NATIBO, 1998, Alcott *et al.*, 1995, Forsyth *et al.*, 1998) have independently suggested the use of an enhanced visual method as a fast, first-pass inspection to identify suspect areas. This large-area inspection would be followed by a second inspection by eddy current and/or ultrasonic to verify the visual inspections and quantify the detected discontinuities. The increasing availability of robotic scanners and array sensors may improve the speed of acquisition of the more sensitive NDT techniques enough so they can be used on much wider areas and eliminate the proposed 'triage' inspection stage.

The choice of NDT methods for internal components is dependent on the material, shape, location and accessibility of the component as well as the nature of the problem. Often, it may involve a visual inspection followed by a more detailed NDT using either eddy current or ultrasonic techniques.

In terms of corrosion metrics, surface bulging or pillowing can be measured using enhanced optical methods with a precision as low as 0.001″. Calibration with respect to known references is usually needed for these and many other NDT methods. Surface pitting is also detectable after cleaning of the part using enhanced visual inspections and can be measured in terms of surface dimensions with a high accuracy using boroscopes or by producing a replica and examination under a microscope. Pit depth measurements are more difficult but pit distribution can be mapped using

Table 5.2 Summary of NDT techniques for aging aircraft

Technique	Advantages	Limitations	Metrics, resolution
Eddy current	*One side access, can be automated*	*Slow, correlation difficult*	*Multi-layer corrosion, material loss, cracks*
Low-frequency	Equipment available	Specific depth	10% loss, 1 mm crack
Multi-frequency	Various depths	Time consuming	<10% loss
Pulsed/transient	Depth location	At experimental stage	5% loss, cracks
Remote field	Deep flaws	No through-thickness	>1 mm cracks
Magneto-optic	Fast imaging of small areas	Flat surfaces	>>10% loss
Ultrasonic	*Well developed, reliable, automation*	*Coupling fluid needed. Slow, first layer only*	*Thickness, cracks, exfoliation, disbond*
Pulse–echo	Easy, inexpensive	Only first layer	2–5% loss, 2 mm crack
Thru-transmission	Less attenuation	Access to both sides	Disbond, delamination
Spectroscopy	Small flaws detectable	Very slow	Roughness, pits, 8 mic.
Guided waves	Large area coverage	Poor reproducibility	Disbond, corrosion
Surface waves	Very sensitive, simple	Small area, slow	
Non-contact	No coupling needed	Experimental stage	Small cracks, 0.3 mm
Acoustic emission	For hidden parts	Poor reliability	Active flaws
Optical	*Large area coverage, easy*	*Surface features only, interpretation difficult*	*Surface dents, corrosion, pillowing*
Visual	Fast, inexpensive	Not reliable	Paint flaking, dents
Enhanced visual	Provides image	Indirect measurement	3% loss, surface pits
Boroscopes	Precise dimension	Time consuming	Surface pits, 0.25 mm
Liquid penetrant	Total coverage	Cleaning needed	Open cracks, 2 mm
Thermography	*Large areas, fast*	*Interpretation difficult, poor resolution, must paint surface*	*>5% thickness loss on one layer, disbond*
Radiography	*Full volume inspection*	*Safety precautions*	*Density profiles, dimension, internal corrosion*
x-rays	Visible records	Access to both sides needed	Cracks, exfoliation
Compton	One side access only	Inefficient, experimental	Roughness, 0.25 mm
Tomography	Cross-sectional view	Small parts, expensive	Internal dimensions, 0.025 mm
Neutron	Sensitive to hydrogen	Not portable, laboratory tool	Stress, hydrogen embrittlement

enhanced optical methods or by fluorescent liquid penetrant. The latter is also useful during cleaning of corrosion to ensure removal of corrosion pits and products.

Thickness loss due to corrosion is best measured in single-layer structure using ultrasonic pulse-echo method as it provides a direct and accurate measure of the first layer thickness (as low as 2%). However, surface roughness can affect the results and the method is not applicable beyond the first layer. This method is also suitable to locate relatively large exfoliation cracks around fastener holes of thick wing skins. The frequency analysis of ultrasonic echoes (ultrasonic spectroscopy) may provide indications of overall roughness owing to pitting, no information on the resolution is available. Some correlation between the eddy current signal and surface roughness has been reported but further work is needed to confirm this claim. Thickness loss can also be measured using single frequency eddy current method with a detectability of 10% material loss to corrosion. However, the induced field is affected not only by corrosion but also by geometrical changes such as thickness/gap variations as well as the presence of fasteners and substructures. Advanced approaches such as transient or pulsed eddy current and multi-frequency inspections offer significant prospect for improvement.

5.6 References

ALCOTT J, FULTON R, GOAD R, MITCHELL G, MORRIS M, PALLATTO M, RENNELL R, YOUNG M, BRAUSCH J, JABLUNOVSKY G, (1995), 'Development of procedures for using non-destructive inspection equipment to detect hidden corrosion on USAF aircraft', prepared for OC-ALC/TIES under contract F41608-93-D-0649.

ALDRIN J C, SABBAGH H A, SABBAGH E H, MURPHY R K, CONCORDIA M, JUDD D R, LINDGREN E, KNOPP J, (2005), 'Methodology using inverse methods for pit characterization in multilayer structures', in: Thompson D O and Chimenti D E, *Review of progress in Quantitative Nondestructive Evaluation: Volume 25*, AIP Conference Proceedings, 820.

ANASTASI R F, MADARAS E I, SEEBO J P, SMITH S W, LOMNESS J K, HINTZE P E, KAMMERER C C, WINFREE W P, RUSSELL R W, (2007), 'Terahertz NDE application for corrosion detection and evaluation under Shuttle tiles', *Proc. SPIE Int. Soc. Opt. Eng.* **6531**, 1–6.

ASHBAUGH D M, BODE M D, BOYCE K L, SPENCER F W, (2001), 'Corrosion structured experiment', in Thompson D O and Chimenti D E, *Review of progress in Quantitative Nondestructive Evaluation: Volume 21*, AIP Conference Proceedings, 615.

BOSSI R H, IDDINGS F A, WHEELER G C, MOORE P O, (2002), *Radiographic Testing, Nondestructive Testing Handbook (3rd. Ed.)*, Volume 4, American Society for Nondestructive Testing, Columbus, OH.

BROOKS C, HONEYCUTT K, PROST-DOMASKY S, PEELER D, (2001), 'Monitoring the Robustness of Corrosion and Fatigue Prediction Models', in *2001 USAF*

Aircraft Structural Integrity Program Conference, Williamsburg, VA, 11–13 December 2001.

EASTAUGH G F, MERATI A A, SIMPSON D L, STRAZNICKY P V, KRIZAN D V, (1998), 'The effect of corrosion on durability and damage tolerance characteristics of longitudinal fuselage skin splices', in *USAF Aircraft Structural Integrity Program Conference*, San Antonio, 1–3 December 1998.

FORSYTH D S, KOMOROWSKI J P, (2000), 'The role of data fusion in NDE for aging aircraft', in *Nondestructive Evaluation of Aging Aircraft, Airports, and Aerospace Hardware IV, SPIE Proceedings*, **3994**, 47–58.

FORSYTH D S, GOULD R W, KOMOROWSKI J P, (1998), 'Correlation of enhanced visual inspection image features with corrosion loss measurements', in Maldague X P V, *TONE Volume 3: III International Workshop – Advances in Signal Processing for Non Destructive Evaluation of Materials*, ASNT, Columbus, 365–372.

FORSYTH D S, KOMOROWSKI J P, MARINCAK A, GOULD R W, (1997), 'The Edge Of Light enhanced optical NDI technique', *CASJ*, **43**(4), 231–235.

FORSYTH D S, LIU Z, HOFFMANN J, PEELER D, (2002), 'Data fusion for quantitative nondestructive inspection of corrosion damage in aircraft wing structures', in: *2002 United States Air Force Aircraft Structural Integrity Program (ASIP) conference*, Savannah, 10–12 December 2002.

GROS X E, (1997), *NDT Data Fusion*, London: Arnold.

GROS X E, (2001), *Applications of NDT Data Fusion*, Boston: Kluwer Academic Publishers.

HOPPE W, PIERCE J, SCOTT O, (2002), *Automated Corrosion Detection Program Final Report for 07 April 1997–06 October 2001*, United States Air Force report AFRL-MP-WP-TR-2001-4162.

HUGHES D, ZOUGHI R, AUSTIN R K, WOOD N, ENGELBART R, (2003), 'Near-Field Microwave Detection of Corrosion Precursor Pitting under Thin Dielectric Coatings in Metallic Substrate', in Thompson D O and Chimenti D E, *Review of Progress in Quantitative Nondestructive Evaluation: Volume 22, AIP Conference Proceedings*, **657**, 462–469.

JIN F, CHIANG F P, (1998), 'ESPI and digital speckle correlation applied to inspection of crevice corrosion on aging aircraft', *Research in Nondestructive Evaluation*, **10**(2), 63–73.

KOMOROWSKI J P, BELLINGER N C, GOULD R W, MARINCAK A, REYNOLDS R, (1996), 'Quantification of corrosion in aircraft structures with double pass retroreflection', *CASJ*, **42**(2), 76–82.

KOMOROWSKI J P, FORSYTH D S, SIMPSON D L, GOULD R W, (1998), 'Corrosion detection in aircraft lap joints: proposed approach to development of POD data', in: *RTO Workshop on Airframe Inspection Reliability under Field/depot Conditions*, 13–14 May 1998, Brussels, RTO-MP-10, AC/323(AVT)TP/2, 8-1–8-8.

LEPINE B A, MERATI A A, KOURLINE A, FORSYTH D S, KHOMUSI S, (2002), 'Correlating Corrosion Characterization Metrics to Nondestructive Inspections of a 2024-T3 Fuselage Lap Splice', in *United States Air Force Aircraft Structural Integrity Program Conference 2002*, Savannah, 10–12 December 2002.

LIAO M, FORSYTH D S, KOMOROWSKI J P, SAFIZADEH S, LIU Z, BELLINGER N C, (2003), 'Risk analysis of corrosion maintenance actions in aircraft structures', in *Proceedings of the 22nd Symposium of the International Committee on Aeronautical Fatigue*, ICAF2003, Lucerne, Switzerland, May 2003.

MATZKANIN G A, YOLKEN H T, (2007), 'Nondestructive evaluation techniques for detecting hidden corrosion', *AMMTIAC Quarterly*, **2**(2), 3–6.

RUMMEL W D, MATZKANIN G A, (1997), *Nondestructive evaluation capabilities data book*, Advanced Materials, Manufacturing, and Testing Information Analysis Center, United States Department of Defense, AMMT-029.

THE NORTH AMERICAN TECHNOLOGY and INDUSTRIAL BASE ORGANIZATION (NATIBO), (1998), *Corrosion Detection Technologies Sector Study*, Washington.

SMITH R A, BRUCE D A, (2001), 'Pulsed ultrasonic spectroscopy for roughness measurement of hidden corrosion surfaces', *Insight* **43**(3), 168–172.

SMITH R A, HUGO G R, (2001), 'Deep corrosion and crack detection in aging aircraft using transient eddy-current NDE', in *5th Joint NASA/FAA/DoD Aging Aircraft Conference*, Orlando, Florida, 10–13 September 2001.

SPENCER F W, BORGONOVI G, ROACH D, SCHURMAN D, SMITH R, (1993a), *Reliability Assessment at Airline Inspection Facilities Volume I: A Generic Protocol for Inspection Reliability Experiments*, DOT/FAA/CT-92/12, I.

SPENCER F W, BORGONOVI G, ROACH D, SCHURMAN D, SMITH R, (1993b), *Reliability Assessment at Airline Inspection Facilities Volume II: Protocol for an Eddy Current Inspection Reliability Experiment*, DOT/FAA/CT-92/12, II.

SPENCER F W, SCHURMAN D, (1995), *Reliability Assessment at Airline Inspection Facilities Volume III: Results of an Eddy Current Inspection Reliability Experiment*, DOT/FAA/CT-92/12, III.

THOME D K, FITZPATRICK G L, SKAUGSET R L, (1996), 'Aircraft corrosion and crack inspection using advanced MOI technology', in: *Nondestructive Evaluation of Aging Aircraft, Airports, and Aerospace Hardware, SPIE Proceedings* Vol. 2945, Scottsdale, AZ, 03–05 December 1996, 365–373.

USDO D, (1999), *Nondestructive Evaluation System Reliability Assessment, Department of Defense Handbook, MIL-HDBK-1823.*

6
Corrosion prediction in the aerospace industry

J. ULLETT, S&K Technologies, USA

Abstract: This chapter presents a discussion of corrosion growth rates in high-strength aluminum alloys as measured from laboratory and field experiments as well as real aircraft data. Various factors that influence corrosion growth are reviewed, including product form and finish, structural configurations, and local environment. The development of growth rate prediction models is described with a focus on estimating corrosion thinning.

Key words: intergranular and exfoliation corrosion, crevice corrosion, high-strength aluminum alloys, corrosion thinning, probabilistic rate distributions.

6.1 Introduction

The chapter opens by briefly discussing alternative methods to modeling corrosion growth and the objectives of the modeling work featured here. Section 6.2 highlights some of the factors that influence corrosion rate in aluminum alloys. Material influences include not only alloy and temper but product form, manufacturing processes, surface treatments and finishes as well as protective coatings. Environmental factors that affect corrosion rate in aluminum aircraft structure include both the environmental severity of the basing location and the local or micro-environment within the aircraft structure.

In section 6.3, the sources of corrosion damage and repair data utilized in the development of rates for legacy aircraft aluminum alloys for a project sponsored by the United States Air Force (Ullett, 2007) are reviewed. Several forms of corrosion were of interest including crevice, general surface pitting, and intergranular (IG)/exfoliation. Data sources included: laboratory, outdoor exposure, and aircraft corrosion damage and repair data. Data development covered a span of three (laboratory) to seven (outdoor exposure) years. Real aircraft data were available for periods of five to twelve years.

Model development is discussed in Section 6.4. A statistical approach was chosen to represent the variation in corrosion growth rates owing to variation in parameters, such as grain geometry, that cannot realistically be known for aircraft structure. Because the model development was intended for corrosion managers and maintainers of USAF aircraft, corrosion growth values are given in units of inches rather than millimeters. The intended use of the model and data are to provide estimates of the effect of corrosion growth that can be used in concert with failure analysis models in order to evaluate the impact of corrosion on structural integrity and provide guidance for scheduling maintenance.

In section 6.5 future trends for corrosion modeling are briefly discussed, including the use of in situ sensors to more accurately capture micro-climate information relevant to internal aircraft structure. The chapter concludes with suggestions for further reading and a list of references.

6.1.1 Alternative approaches to modeling

Corrosion modeling is of interest to a variety of industries and applications including petroleum pipelines and offshore structure, nuclear waste containment, and structural health management of bridges as well as aircraft structure – the focus of this chapter. Corrosion growth models span complex physico-chemical phenomenological approaches to empirical based approaches. MacDonald and Engelhardt (2003) discuss challenges for both empirical and deterministic models. They point out that empirical models have limitations in accurately extrapolating to far beyond the bounding conditions of the source data. Deterministic models require accurate definition of the pertinent physico-chemical environment over time and over the geometry of the structure. Both empirical and deterministic models have proved to be useful for various industrial problems. In most cases, corrosion modeling is used to schedule maintenance or predict structural life (Kelly *et al.*, 2003) and associated life-cycle costs.

Models have been developed for a variety of corrosion growth mechanisms including: pitting, for example, Pidaparti *et al.* (2004) and Turnbull (1993), crevice corrosion (Heppner *et al.*, 2002), intergranular corrosion (Huang *et al.*, 2006; Zhang *et al.*, 2003), and filiform corrosion (Williams and McMurray, 2003). All of the above types of corrosion occur in aluminum aircraft structure. Filiform corrosion on aircraft skins does not impact structural performance and is not a concern for structural health management. Crevice corrosion occurs in lap joints common to many airframes and can present a safety problem as well as a maintenance issue if multi-site damage is present (NTSB, 1988). Intergranular corrosion can lead to stress corrosion cracking (SCC) or exfoliation depending on the location of attack, part geometry and stresses. SCC is a significant issue for aging aircraft. However, the failure mode associated with SCC differs from failure modes associated with other corrosion types. The modeling efforts discussed in this chapter focus on corrosion that leads to thinning of structure and the associated loss of load bearing area. For example, exfoliation corrosion can lead to significant loss of load bearing area, or structural thinning, and is a common form of corrosion in legacy high-strength aluminum alloys used for wing spars and other critical structure.

6.1.2 Objectives of corrosion growth modeling for aircraft maintenance planning

The objective of the corrosion rates work discussed in this chapter was to determine the effects of alloy, temper and other parameters such as structural configuration (e.g., lap joints) on corrosion growth in aircraft aluminum. The corrosion growth rates effort was not focused on predicting when and where corrosion would initiate on a given piece of aircraft structure but on predicting growth of identified (such as by visual or nondestructive inspection) corrosion. The intended purpose for establishing corrosion growth rates was to provide corrosion thinning data to structural analysis tools such as the crack-growth analysis software AFGROW (Harter, 2004). Corrosion growth predictions for time periods of one or more depot cycles can help engineers and maintainers decide when the corrosion damage requires removal based on a structural integrity perspective.

The approach taken for corrosion prediction for planning aircraft structural maintenance is based on the likely realities that for a given part the level of detail known about material properties could consist only of product form (i.e., part made from plate, extrusion, etc.) and global grain direction (e.g., rolling direction). Consequently, the focus for IG/exfoliation rates was development of distributions for corrosion growth in both the L-T plane and the thickness, S, direction.

6.2 Material and environmental influences

6.2.1 Alloy/temper

Chapter 4 covers the affects of alloy and temper on corrosion susceptibility. The USAF funded corrosion modeling work discussed in this chapter focused on three bad-actor alloys found prevalently across legacy fleets: 2024-T3, 7075-T6, and 7178-T6. While 7178-T6 is used sparingly in current production aircraft (some will argue that it is no longer used, but the production J model of the C-130 aircraft is at least one example of continued use), the other two alloys are still quite common, with 7075-T6 used on internal structure of new commercial as well as military aircraft.

6.2.2 Effects of structural configuration and manufacturing processes

An airframe consists of a variety of structural configurations that may influence the local electrochemical environment. Lap joints and splice joints, for example, provide crevice environments that may accelerate corrosion compared with boldly exposed structure. Aluminum joints fastened with steel or titanium fasteners; and aluminum/steel and aluminum/composite joints are several examples of structure susceptible to bimetallic corrosion. Sealants, coatings and isolation layers are used as protective measures. Poor application methods, inferior products, and degradation of coatings with time all provide opportunities for bimetallic corrosion to initiate.

The original forging dies for legacy airframes are typically unavailable and too costly to replace. Current depot manufacturing processes utilize high-speed machining to produce replacement parts quickly. Parts that were originally forged are replaced with parts that are machined or 'hogged-out' of thick plate resulting in significant end-grain exposure at the part surface. For alloys like 7075-T6 that are susceptible to IG corrosion, relatively small flaws in the coatings protecting end-grain rich surfaces can lead to significant IG, exfoliation and/or stress corrosion cracking (SCC) over a relatively short time frame (1–5 years). It is not uncommon for the life-cycle of replacement 'hog-out' parts to be much shorter than the originally forged part made from 7075-T6. Material substitution should be considered whenever practical to stem the usage of 7075-T6.

6.2.3 Surface finishes and coatings

Aluminum aircraft structure undergoes one or more surface treatments to prevent the onset of corrosion. These treatments include: anodization, chromate conversion coating, and primer coating. Internal structure that is safety critical (e.g., wing box) or that is subjected to harsh environments (e.g., stone-spray from landing and take-offs) may receive a topcoat of glossy polyurethane in addition to a primer coating. The outer mold-line of military aircraft is typically coated with a low-gloss topcoat. These highly-filled topcoats are more prone to breakdown of barrier properties and do not provide the same degree of corrosion protection as do glossy topcoats used on commercial airliners.

The corrosion of the aluminum substrate will not occur until the protective coatings are compromised. Unfortunately, local defects in the coatings system may occur soon after depot repaint or field touch-up due to removal of access panels and other routine maintenance activities. Differences in modulus and thermal expansion coefficients between aluminum structure and steel or titanium fasteners are another cause of localized protective-coating failure. Thus, while the coating system on-the-whole may provide excellent barrier properties for decades (particularly for interior structure), many opportunities exist for localized mechanical or chemical degradation of the protective coatings.

6.2.4 Local versus global environment

Environmental factors such as time of wetness, distance from seacoast, temperature, chloride exposure, and sulfide exposure are known to affect corrosion rates in aluminum alloys. Internal aircraft structure is sheltered from the outside environment by the aircraft itself. Some structural areas, like wheel wells expose internal structure to the outside environment. Other internal structure, like structure beneath latrines, is exposed to corrosive environment from pooling of fluids inside the aircraft. Consequently, internal structure at various locations on the aircraft experience different local environments. Several studies by Battelle (Abbott and Kinzie, 2003) were done under USAF funding to evaluate local corrosion rates or local corrosion severity on various airframes including: C-141, C-130, F-15, F-16, and C/KC-135. The first study used the coupon racks, which had been used extensively in ground-based exposure evaluations. Mass loss data as a function of exposure time resulted from this work. The second type of study used resistive sensors to map the cumulative environmental exposure at many locations inside aircraft. Both types of studies showed that all internal areas, including the wheel wells, experience a more benign environment than the outside environment.

Thus, aircraft structure can experience a range in corrosion environments owing to local variation of the environment within the aircraft. This local variation combined with variation in the basing environment experienced by deployed aircraft suggest that a statistical approach to corrosion growth estimation is required for aluminum aircraft structure.

6.3 Data generation and correlation

The USAF funded Corrosion Effects on Structural Integrity (CESI) program (Ullett, 2007) generated corrosion rate data using three methods: laboratory accelerated testing, outdoor exposures, and analysis of real aircraft inspection and repair data. There were pros and cons to each method and each provided considerable insight into the relative influences of the factors discussed in section 6.2 above. The majority of the laboratory testing was performed at The Ohio State University under the supervision of Dr Gerald Frankel. Outdoor exposure evaluations were conducted both by Dr William Abbott of Battelle and by S&K Technologies. The analysis of real aircraft data and correlation among various data sets was also performed by S&K Technologies (SKT).

6.3.1 Laboratory evaluations

Accelerated testing in the laboratory provides corrosion data in months versus years for outdoor exposure testing and allows for careful control of test parameters such as local environment and stress. The laboratory environment allows control of the many parameters that influence corrosion growth rate such as humidity and temperature. Various alloys and tempers were included in the laboratory studies. Where possible, samples from aircraft structure were included in the test matrix. For example, sections of wing skin removed from C/KC-135 aircraft were made available to SKT and its subcontractors for outdoor exposure and laboratory analysis.

The conditions used to accelerate the corrosion may lead to 'un-natural' corrosion degradation morphologies and careful attention was paid to develop experiments that replicate corrosion morphologies observed in aircraft structure. The work conducted by Frankel and coworkers has been documented (Huang and Frankel, 2006, 2007a, and 2007b); Huang *et al.* (2002 and 2006); Zhao and Frankel (2006 and 2007); sample results are provided here.

Three experimental methods were utilized. The first method is referred to as foil penetration and is described in references (Huang *et al.*, 2002) and (Frankel and Rokhlin, 2003). The foil penetration method involves thin slices or foils removed from a sample through a particular material plane. Typical foil thickness is of the order of 0.10 mm to 0.80 mm. The sample thickness is varied in order to generate a range in IG growth depths.

The second method is applied to slices of the order of 0.40 to 8.0 mm encased in epoxy. Corrosion is initiated using an applied potential for 7 h and then the samples are removed from the electrochemical environment and aged in a controlled humidity environment. Intergranular fissure lengths are measured periodically using radiography. This controlled humidity method produces sharp intergranular fissures similar to those observed in actual aircraft structure.

Results for both the standard foil method and the controlled humidity method were fitted to an equation of the form:

$$d = A \times t^{\mathrm{B}} \qquad [6.1]$$

where d is depth or length of fissure (mm), A and B are constants, and t is time (or days depending on type of experiment).

The graph in Fig. 6.1 (Huang *et al.*, 2002) shows an example of raw data and curve fits for the experimental results for foils removed from the L, T, and S material planes. Each material direction data set is generated from multiple foil samples removed from the same sample of alloy. Depending on the grain aspect ratio, results may vary from one piece of alloy to the next.

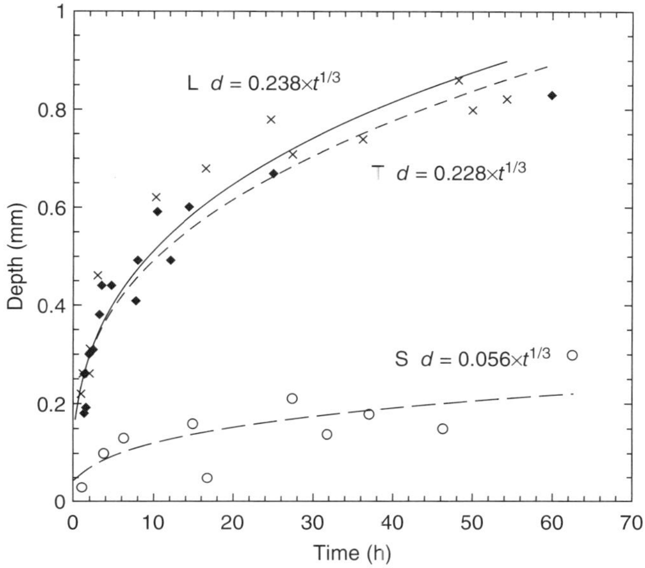

6.1 Results from foil penetration studies of AA7178-T6 material.

For the modeling efforts described in section 6.4, the laboratory measurements were used to determine values for '*B*' in equation 6.1 for IG/exfoliation corrosion for each alloy of interest (2024-T3, 7075-T6, and 7178-T6). Where available, corrosion damage data from USAF aircraft were used to develop a distribution for 'A' in Equation 6.1. Data generated from outdoor exposure coupons were used to generate distributions for '*A*' when real aircraft data were not available in sufficient quantity. When sets of data were available for different Air Force Base (AFB) locations, the effect of basing location was analyzed.

In other experiments with 7178-T6 samples, Frankel showed that the IG or exfoliation growth rate is very much dependent on the alloy microstructure and, in particular, the grain alignment and aspect ratio (Huang and Frankel, 2006). Outdoor exposure testing of 7178-T6 wing skin material removed from a retired KC-135 aircraft confirmed that grain aspect ratio has a significant effect on exfoliation corrosion rates.

The center photo in Fig. 6.2 (Battelle, 2005) shows an example of an outdoor-exposed wing coupon. The top plate on the left side was of different thickness than the plate on the right side of the splice. Observations of surface exfoliation and subsurface IG attack show the left plate to be more severely corroded near the fasteners than the right plate. This difference was observed in all such coupons that were exposed in a variety of outdoor locations. Since the coupons had been removed from an aircraft wing and

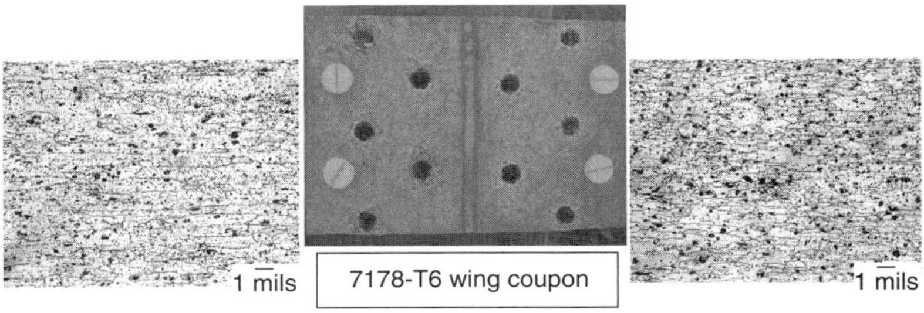

6.2 Section of 7178-T6 wing joint with steel fasteners (center photo) exposed to seacoast environment and micrographs of sections taken from left and right upper plates of coupon.

paint removed from both sides of the splice using the same process, the reason for the disparity in corrosion after one year of exposure was unclear. Metallographic analysis showed that the grain geometry was significantly different between the two plates. The plates that corroded more severely had longer grains than the plates that showed lesser corrosion.

This research helps to explain why there is so much variation in corrosion growth rates from one section of a wing to the next or from one aircraft to the next and why a distribution of rates versus a simple rate constant is required to describe corrosion growth for a population of structural components made from the same alloy/temper and product form.

6.3.2 Outdoor exposure experiments

Outdoor exposure testing was done to evaluate the effects of time and geographic location on general thinning of boldly exposed sheet and plate (both 2024-T3 and 7075-T6), and thinning resulting from crevice corrosion in lap joints constructed of 2024-T3. While the surface corrosion morphology observed on a micrometer scale is pitting, general or average thinning is of interest to aircraft maintainers. A severely corroded area that is smaller in size than the area of a fastener is repaired by drilling the corroded area out and placing a fastener in the new hole. Thus, predicting the depth of an individual pit is not particularly relevant to maintenance of aluminum aircraft structure. Average thinning, which impacts the load carrying area and crack-growth calculations was of prime importance for the experiments discussed here.

In addition to the coupons discussed above, 7178-T6 wing skin coupons with steel fasteners were exposed at a variety of locations. For these coupons, the maximum extent of IG/exfoliation corrosion in both the radial (radiating from fastener in the LT plane) and depth (plate thickness S) directions was measured from metallographic sections removed from the LS and TS

Table 6.1 Summary of Battelle exfoliation geometry data for 7178-T6 wing skin sections exposed on racks for one year

Location	Maximum depth in S direction, mm (in.)
Charleston (SC)	1.10 (0.0432)
Corpus Christi (TX)	0.10 (0.0040)
Daytona Beach (FL)	0.64 (0.025)
Eareckson AFB (AK)	0.78 (0.0309)
Hickam AFB (HA)	0.18 (0.0072)
Homestead AFB (FL)	0.44 (0.0174)
MacDill AFB (FL)	0.30 (0.012)
Mildenhall AFB (UK)	0.30 (0.012)
Minneapolis St Paul (MN)	None recorded*
Oklahoma City (OK)	None found
Patrick AFB (FL)	0.80 (0.0308)
Tyndall AFB (FL)	0.26 (0.0102)
West Jefferson (OH)	None recorded*

*No values smaller than 0.002 in. were recorded by Battelle. It is assumed that exfoliation damage considerably shallower than this value was discounted as insignificant.

planes. Table 6.1 provides example exfoliation depth data from coupons exposed at various locations for one year (Battelle, 2005).

Coupons located in mild inland locations had little to no measurable exfoliation or IG attack after one year. Coupons exposed outdoors in locations considered to have moderate to severe environmental severity had measurable corrosion. A large number (hundreds) of coupons were exposed from one to seven years. About 20 locations had several sets of coupons exposed at different time periods. For the same location, no statistical difference could be found for corrosion rates based on different sets of coupons. Also, no statistically significant difference could be found among data sets from different but similarly severe locations. Corrosion growth was statistically different among data sets from locations of grossly different severity (e.g., mild versus severe). Thus, the data in Table 6.1 are representative of the much larger data sets in that there is significant scatter in corrosion growth among coupons weathered at moderate to severe sites, but a measurable difference between coupons located at mild (e.g., Ohio) versus severe (Florida) sites. Similar results were found for other coupon types (e.g., simple sheet, lap joints) and other alloys evaluated.

6.3.3 Real aircraft data

The basis for the IG/exfoliation growth data discussed in this section was obtained from repair data for USAF transport and tanker aircraft. The repair data sets have been validated as appropriate by evaluating comparisons with corrosion and repair data from other data sources, which will be reviewed in the following sections.

Damage tracking on C/KC-135 upper wing skins was started by the C/KC-135 System Program Office in the late 1990s. This data source provides measurements of repairs, i.e., grindouts as opposed to the exfoliation damage that was removed. Because Technical Orders (TOs) recommend surface blending to avoid stress raisers, the lateral grind-out dimensions will most likely be larger than the actual corrosion damage. The depth data are assumed to be less conservative because technicians typically grind just past visible damage. A strength of this data source is the large amount (thousands of data sets for repairs accomplished over a ten-year period) of data available, orders of magnitude more than any other data source available. A second strength is that, although repair measurements are recorded versus actual damage, the data pertain to real aircraft structure not coupons. Thus, the structure saw the chemical and mechanical environments experienced by operational aircraft.

Aircraft basing records were reviewed to find aircraft that had been stationed exclusively in either moderate to severe environments or in mild to moderate environments. Environmental severity index (ESI) data for bases can be found in TO 1-1-691 (USAF, 2006) and were used to define a base as mild, moderate or severe. Four tails were identified that had spent the majority of their on-station time at Hickam (HA) AFB. The data from these tails was pooled and referred to as moderate to severe pooled data. The pooled data contained over 4000 records of grindout damage. Each data set includes the specific location on the wing that the damage was recorded, the grindout length in the LT plane (rolling plane) and the grindout depth. Four other aircraft were identified that had been stationed at a variety of mild to moderate ESI bases and that had collectively a large number of grindouts. The pooled mild-moderate data contained over 800 records. The pooled data from the two sets are presented in Fig. 6.3.

The pooled data sets were curve-fitted using lognormal statistics, which provided the best fit compared with Weibull, Normal, Gumball and other distribution statistics. Figure 6.4 shows the cumulative probability fits with 95% confidence bounds for the two pooled record sets of grindout depths. The data represents corrosion growth over a depot cycle, which is five years for the C/KC-135 aircraft.

Data for IG/exfoliation growth in the LT plane was also fit with lognormal statistics. Unlike the depth distributions, the distributions for lateral or radial (e.g., radiating from a steel fastener) growth crossed at higher cumulative probabilities (above 80%). The distribution for the moderate–severe pooled data was tighter with a mean value of 0.516 in. for five years. The distribution for the pooled mild-moderate data was more spread and had

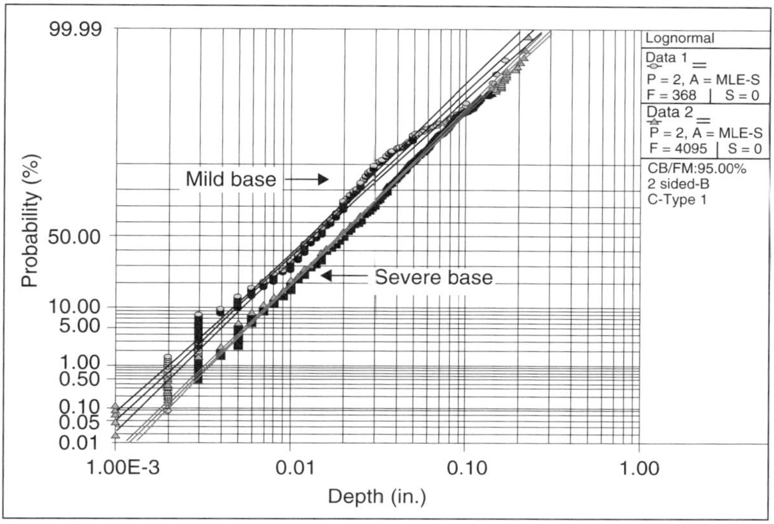

6.3 KC-135 grindout depth data (symbols) plotted with best-fit lognormal distributions.

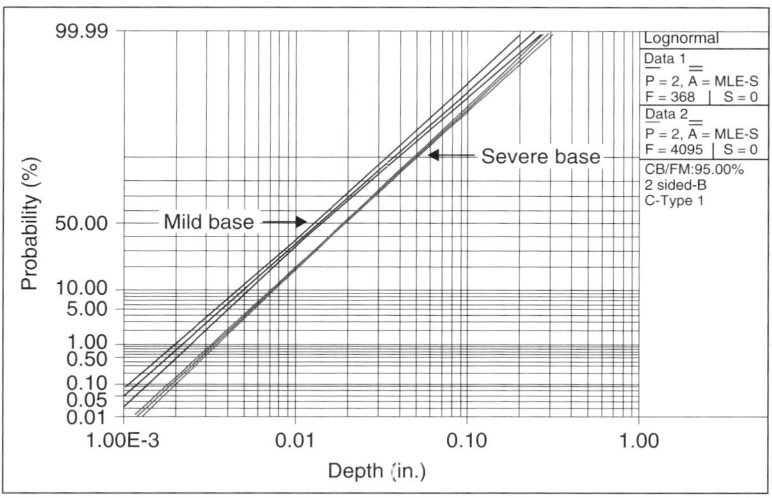

6.4 Best-fit lognormal distributions for depth grindout data with 95% confidence limits.

a mean value of 0.407 in. Looking at the data provides some insight into the differences in the high-end of the distributions.

Data for each lateral/radial distribution were grouped for values greater than or equal to 1.0 in. The pooled moderate–severe data had 100 datum points with values of 1.0 in. compared with 64 datum points for the mild–moderate pooled data. The moderate–severe data had 25 datum points greater than 1 in. but less than 1.5 in. – values were recorded to the nearest 1/8 in. The mild–moderate data had no datum points between 1 in. and 1.5 in. – grindouts greater than 1 in. appeared to be recorded to the nearest 1/2 in. The mild–moderate data had 25 datum points with values greater than 1.5 in. compared with only 3 datum points exceeding 1.5 in. for the moderate–severe pooled data.

In summary, the four tails that were primarily based at Hickam AFB had a total of 128 grindouts greater than or equal to 1 in., whereas the four tails with home bases having milder corrosion severity had a total of 89 grindouts greater than or equal to 1 in. Thus, the more severe environment resulted in more total large (≥1 in.) grindouts. Using the two distributions for corrosion prediction would result in a greater percentage of large grindouts predicted for the mild–moderate distribution compared with the moderate–severe distribution, which is contrary to observation. Consequently, all lateral grindout data was pooled to result in the Combined Mild–Severe distribution shown in Fig. 6.5.

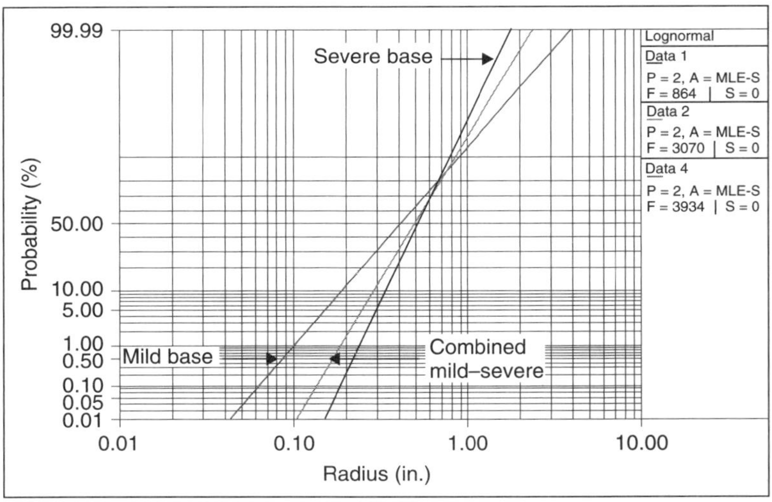

6.5 Comparison of best-fit lognormal distributions for surface radial grindout data for aircraft stationed at primarily environmentally mild bases or environmentally severe bases. The middle distribution is for pooled mild–severe data.

Given that grindout repairs blend into undamaged surrounding material so as to avoid sharp transitions in thickness, one may be concerned that the distribution based on grindout data is overly conservative in the lateral direction or that the distribution itself is not representative of actual damage distributions. The following discussion addresses these concerns.

Data were generated based on a set of photographs taken in 1965 during programmed depot maintenance (PDM) of KC-135 aircraft. The set of 8″ × 10″ photographs, made available to S&K Technologies by the Air Force Corrosion Prevention and Control Office, captures corrosion damage around fasteners located in the fuselage and horizontal stabilizer structure. The aircraft had been based at Ramey AFB in Puerto Rico after production. The damage is located on five separate tail numbers. This group of aircraft did not have a protective coating system (organic) on the outer surface, possibly because it was thought that the cladding on the aluminum would provide adequate protection against the environment. Also, these aircraft were stationed only at one location, Ramey AFB, before depot maintenance. The damage shown in the set of pictures is significant because it is actual corrosion damage data for the fuselage and stabilizer skin material as a result of the environment at Ramey AFB. A simple measure of the furthest extent of corrosion for each damage site provides an indication of the progress of corrosion for the period of 1960–1965. Examples of the measurement of this damage are shown in Fig. 6.6.

The maximum visible extent of corrosion was measured for damage around 57 fasteners in 7178-T6 skin material. The values were fitted to a lognormal distribution and compared with the 'combined' distribution developed from pooled grindout data from the grindout database as discussed above. The comparison is shown graphically in Fig. 6.7. The distributions are almost parallel indicating the grindout distribution shape is

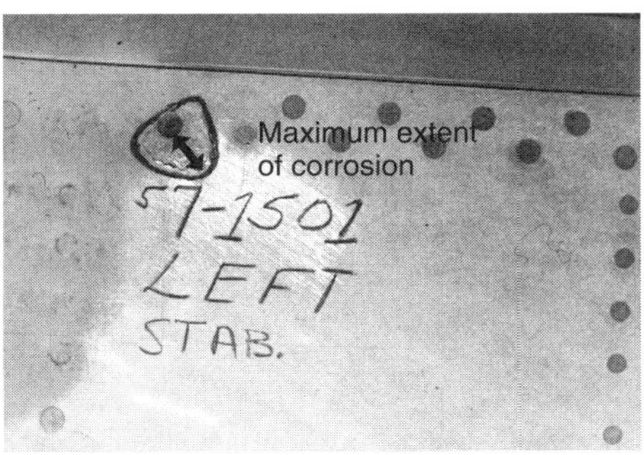

6.6 Photograph of visible exfoliation damage on KC-135 exterior skin with arrow denoting maximum visible damage extending from a fastener.

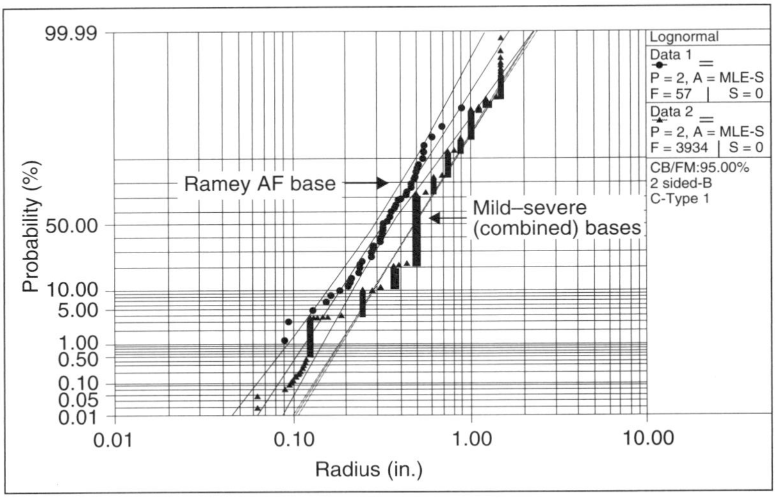

6.7 Lateral corrosion data (ovals) estimated from photographs plotted against lateral grindout data (triangles) with corresponding distributions (lines).

Table 6.2 Comparison of lateral-growth distribution data for pooled grindout repairs and for corrosion damage measured from photographs of KC-135 structure

Distribution (%)	50	60	70	80	90	95	99
Grindout (in.)	0.490	0.545	0.611	0.698	0.839	0.977	1.300
Ramey (in.)	0.327	0.365	0.412	0.474	0.576	0.677	0.915

representative of actual corrosion damage distribution. As expected, the grindout values are conservative compared with the damage measured from the photographs. Two factors contribute to the grindout data bounding the measured data: first, only corrosion damage visible in the photographs was captured – minor IG/exfoliation probably extended past the visible damage; and second, as discussed previously, mechanics grindout past the damage to blend the surface profile. Table 6.2 provides a comparison of discrete values from the two distributions.

The grindout distribution provides a good boundary for measured corrosion growth. The predicted values along any point in the distribution are conservative, which is appropriate given that corrosion damage will be repaired by grinding out past the observed damage radius.

No corrosion depth data were available from real aircraft. Instead, data generated from sectioning coupons exposed outdoors were used to compare

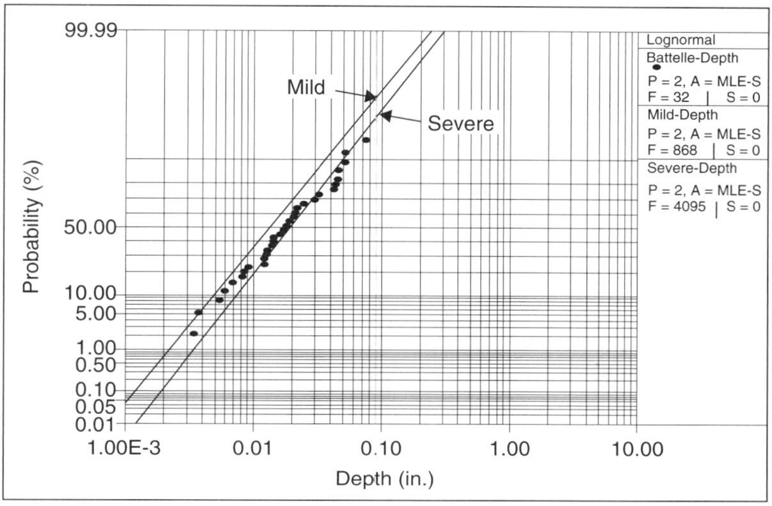

6.8 Battelle exfoliation depth data (ovals) plotted with lognormal distributions for grindout depth data.

with the distributions developed from grindout repair data. Figure 6.8 shows coupon data (symbols) plotted with the best-fit distributions for the grindout data sets. The distributions based on grindout repair data adequately encompass the majority of the coupon data.

6.4 Model development and implementation

The Structural Damage Management Tool, SDMT (S&K Technologies, 2006) is a software interface that facilitates aircraft structural failure analysis. The tool links the crack-growth software AFGROW with a Corrosion Prediction Module (CPM), developed by SKT, allowing the effects of corrosion thinning to be accounted for in fatigue crack growth calculations. Corrosion rate data were developed for three forms of corrosion: general thinning, crevice corrosion, and IG/exfoliation. Section 6.3 of this chapter focused on data development for IG/exfoliation corrosion of 7178-T6. Similar methods and types of data sets were used to develop corrosion rates for thinning in boldly exposed 2024-T3, thinning in lap joints made of 2024-T3, and IG/exfoliation in 7075-T6. Although IG corrosion can lead to SCC, when fatigue is the damage mode the corroded volume is the parameter of interest and is considered non-load bearing. Consequently, for all alloys, corrosion growth rate development focused on growth in the depth (S material direction) direction. Where data were available growth rate expressions were developed for the radial (from fasteners) or LT material direction also.

6.4.1 Assumptions

Results from both controlled laboratory experiments and outdoor exposure evaluations indicate that corrosion growth rate is influenced by a variety of parameters, many of which are not measured or tracked for operational aircraft. Consequently, simplifying assumptions were necessary to develop corrosion growth rate estimations for aircraft structure among these are:

- growth rate is independent of alloy microstructure,
- growth rate is independent of temperature,
- growth rate is independent of local structural environment,
- growth rate can be modeled using a power law expression
- for a given alloy/temper, the power constant is fixed, and
- for a given alloy/temper the constant *A* can be modeled as a distribution.

Providing a distribution for *A* in equation 6.1 allows the SDMT user to chose a conservative value (e.g., a 99% value from the cumulative distribution) when the environmental severity (local or global) is unknown. Where the user has knowledge from documented corrosion data, he/she may select a less conservative value for *A*. SDMT gives the user control for selecting values for *A*. Example data for IG/exfoliation are provided in Table 6.3.

Equation 6.1 expresses grindout depth *d* as a function of time. Research by Frankel and coworkers (Frankel and Rokhlin, 2003; Huang *et al.*, 2002) indicates that the power constant *B* has a value of approximately 1/3 for IG attack in 7178-T6 wing material. By setting *t* in equation 6.1 to five years, values for *A* can be calculated for points along the cumulative distribution curves (Fig. 6.4 and 6.5). Values calculated for *A* (units of in./$yr^{1/3}$) are given in Table 6.3. Once the *A* values are known, equation 6.1 can be used to estimate corrosion growth for any time period.

Corrosion growth for IG/exfoliation is predicted in both the LT plane and S or through-thickness direction, but only the through-thickness estimation has a bearing on the crack-growth calculations. It is assumed that the corrosion in the LT plane will extend from one fastener to the next adjacent fastener for the time-periods simulated in crack growth calculations. The growth of corrosion in the lateral direction is estimated and displayed in SDMT even though it is not used in the AFGROW calculations. The estimated corrosion depth is used to calculate effective load bearing thickness. The user can select either the mild or severe cumulative probability curves and can then select a cumulative probability value from the chosen distribution. Estimated IG/exfoliation growth values for an exposure period of ten years are shown in Table 6.3 for lateral and depth damage growth.

Table 6.3 Data corresponding to distributions fitted to pooled repair data for aircraft based at locations with mild and severe corrosion severity environments

	Probability (%)										
7178-T6	**10**	**20**	**30**	**40**	**50**	**60**	**70**	**80**	**90**	**95**	**99**
Depth-mild	0.0049	0.0070	0.0089	0.0110	0.0134	0.0163	0.0201	0.0257	0.0362	0.0481	0.0817
Depth-severe	0.0073	0.0102	0.0129	0.0158	0.0190	0.0230	0.0281	0.0356	0.0494	0.0647	0.1075
A – mild	0.0029	0.0041	0.0052	0.0064	0.0078	0.0095	0.0118	0.0150	0.0212	0.0281	0.0478
A – severe	0.0043	0.0060	0.0076	0.0092	0.0111	0.0135	0.0164	0.0208	0.0289	0.0379	0.0629
Predicted growth for 10 yr, mild	0.0062	0.0088	0.0112	0.0139	0.0169	0.0205	0.0253	0.0324	0.0456	0.0606	0.1029
Predicted growth for 10 yr, severe	0.0092	0.0128	0.0162	0.0199	0.0239	0.0290	0.0354	0.0448	0.0622	0.0815	0.1354

The planned uses for the Corrosion Prediction Model (CPM) include: evaluating the impact of basing rotation on corrosion growth; scheduling skin replacements; and estimating the impact of deferred corrosion maintenance. Corrosion growth predictions tied with crack-growth analysis allow the engineer to determine the structural impact of corrosion.

6.5 Summary

Corrosion prediction modeling is of increasing interest not only for military aircraft but for commercial aircraft and in a variety of industries. The varied microstructure of high-strength aluminum alloys adds considerably to the uncertainty of corrosion growth predictions for aircraft structure. Laboratory analysis was used to determine the time-dependency of corrosion growth. A power law model was chosen to capture the change in growth rates with time.

While aircraft basing environment has an effect on corrosion growth rate, statistically significant differences were only observed among locations with dramatically different environmental severity. This finding was derived from analysis of both outdoor exposure coupon sets and from corrosion repair data for aircraft with known basing histories. Consequently, data was pooled from various exposure sites of similar environmental severity and distributions fitted to the pooled data. The lognormal statistic provided the best-fit for data described here. Known exposure times were used to calculate the constant A in the power law rate expression for discrete values of each distribution.

The rate data and model constants for three legacy alloys have been incorporated in a Corrosion Prediction Model that can be used independently to predict corrosion growth or can be used in conjunction with a crack-growth model to predict the effect of corrosion growth on fatigue life.

6.6 Future trends

As described in section 6.3, the micro-environment local to a particular structural area may have a measurable effect on the local corrosion rate. The USAF is currently evaluating the use of sensors to capture local environmental parameters for key structure (Abbott and Kinzie, 2003). Within an aircraft structure, some areas may experience very benign environments while others experience more aggressive environments due to localized condensation, material spillage, and other factors. Protective coatings may also experience variations in damage from wear, mechanical damage and localized exposure to aggressive fluids. Sensors have the potential to capture local conditions and reduce the uncertainty in corrosion growth modeling.

Nondestructive inspection (NDI) is used to find corrosion damage on hidden surfaces and within structure that cannot be visually examined. Often the NDI results can only be used to assess the presence of corrosion but cannot reliably 'size' the corrosion. A subject of current interest is the use of multi-frequency NDI and the use of data fusion to reduce dimensional uncertainty and allow for more accurate sizing of the existing corrosion state. Subsequent NDI evaluations can then be used to estimate corrosion growth without the need for disassembly of structure for visual verification.

6.7 Sources of further information and advice

Corrosion growth modeling is a subject covered by journals such as: Corrosion, Corrosion Engineering Science and Technology, Corrosion Science, and the Journal of the Electrochemistry Society. Corrosion and corrosion modeling for aircraft structure are covered regularly in the Aging Aircraft conference held yearly in the United States and sponsored jointly by NASA, the Federal Aviation Administration and the Department of Defense.

6.8 References

ABBOTT, W. H. and KINZIE, R. (2003), 'Corrosion Monitoring on Operational Aircraft Status of Recent Work,' in: *Proceedings of Joint NASA/FAA/DOD Conference on Aging Aircraft Conference*, New Orleans, LA. September, 2003.

BATTELLE (2005), 'Battelle Monthly Report under Subcontract F0965-00-D-0018-0001', February 2005.

FRANKEL, G. S. and ROKHLIN, S. I. (2003), 'Intergranular and Exfoliation Corrosion Rate Studies', Ohio State University Final Report, 23 February 2003, received by S&K Technologies as a deliverable under Subcontract F0965-00-D-0018-0001.

HARTER, J. (2004), 'AFGROW Users Guide and Technical Manual', AFRL-VA-WP-TR-2004-XXXX, Available from: www.afgrow.org.

HEPPNER, K. L., EVITTS, R. W. and POSTLEWAITE, J. (2002), 'Prediction of the Crevice Corrosion Incubation Period of Passive Metals at Elevated Temperatures: Part I – Mathematical Model', *Can J Chem Eng*, **80**, 849–856.

HUANG, T.-S. and FRANKEL, G. S. (2006), 'Influence of Grain Structure on Anisotropic Localized Corrosion Kinetics of AA7xxx-T6 Alloys,' *Corr Eng Sci Tech*, **41**(3), 192–199.

HUANG, T.-S. and FRANKEL, G. S. (2007a), 'Effects of Temper and Potential on Localized Corrosion Kinetics of AA7075,' received January 2006 by S&K Technologies as a deliverable under Subcontract F0965-00-D-0018-0001, accepted for publication in *Corrosion*, **63**, 731.

HUANG, T.-S. and FRANKEL, G. S. (2007b), 'Sharp Intergranular Corrosion Fissures in AA7178', *Corr Sci*, **49**, 858–876.

HUANG, T.-S., FRANKEL, G. S., FARAHBAKHSH, B. and PEELER, D. (2002), 'Localized Corrosion Growth Kinetics in Al Alloys' in: *Proceedings of Joint NASA/FAA/DOD Conference on Aging Aircraft Conference*, 16–19 September 2002.

HUANG, T.-S., FRANKEL, G. S. and WOLFE, D. A. (2006), 'A Statistical Model for Intergranular Corrosion of AA7178', received January 2006 by S&K Technologies as a deliverable under Subcontract F0965-00-D-0018-0001, *Corrosion*, in press.

KELLY, R. G., SCULLY, J. R., ALTYNOVA, M and PEELER, D. T. (2003), 'An Algorithm for Modeling the Effects of Corrosion Damage on the Structural Integrity of Lap Joints in Aerospace Structures,' in: J. R. Scully, D. W. Shoesmith, eds. *Lifetime Prediction Modeling of Corrosion Processes*, Houston, NACE.

MACDONALD, D. and ENGELHARDT, G. (2003), 'Prediction of Long-Term Corrosion Damage in High Level Nuclear Waste Disposal Systems'. Available from: http://www.virginia.edu/cese/doe/Prediction%20of%20Long%20Term%20Corrosion%20Damage.pdf/ (accessed 18 August 2007).

NATIONAL TRANSPORTATION SAFETY BOARD (1988), 'Aloha Airlines, Flight 243, Boeing 737-200, N73711, Near Maui, Hawaii April 28, 1988,' *Aircraft accident report, NTSB–AAR-89-03*, adopted on 6/14/1989.

PIDAPARTI, R. M., PALAKAL, M. J. and FANG, L. (2004), 'Cellular Automation Approach to Model Aircraft Corrosion Pit Damage Growth', *AIAA Journal*, **42**(12), 2562–2569.

S&K TECHNOLOGIES (2006), 'SDMT User's Manual, F09650-00-D-0001-5018'.

TURNBULL, A. (1993), 'Review of Modeling Pit Propagation Kinetics,' *Brit Corr J*, **28**(1), 297–308.

ULLETT, J. S. (2007), 'Prediction of Corrosion Growth Rates in Legacy Alloys', in: *2007 USAF Corrosion Conference Proceedings*, 6–8 March 2007 Macon, GA.

UNITED STATES AIR FORCE (2006), 'TO 1-1-691 Cleaning and Corrosion Prevention and Control, Aerospace and Non-Aerospace Equipment'. Available from: http://www.robins.af.mil/shared/media/document/AFD-070108-271.pdf (Accessed 29 August 2007).

WILLIAMS, G. and MCMURRAY, H. N. (2003), 'The Kinetics of Chloride-Induced Filiform Corrosion on Aluminum Alloy AA2024-T3', *J Electrochem Soc*, **150**(8), B380–B388.

ZHANG, W., RUAN S., WOLF, D. A. and FRANKEL G. S. (2003), 'Statistical model for intergranular corrosion growth kinetics', *Corr Sci*, **45**(2), 353–370.

ZHAO, X. and FRANKEL, G. S. (2006), 'Effects of RH, Temper and Stress on Exfoliation Corrosion Kinetics of AA7178', *Corrosion*, **62**, 256–266.

ZHAO, X. and FRANKEL, G. S. (2007), 'Quantitative Study of Exfoliation Corrosion: Exfoliation of Slices in Humidity Technique', *Corr Sci*, **49**, 920–938.

7
Integrated health and corrosion monitoring systems in the aerospace industry

A.-D. NGUYEN, Los Gatos Research, USA;
and V. GODINEZ, Physical Acoustic Corporation, USA

Abstract: Integrated health monitoring methods that can be applied to aging aerospace structures for monitoring fatigue, cracks and corrosion in real-time are discussed. An overview of established methods and recent developments is presented, including the traditional piezo-electric transducer and novel fiber-optic sensor systems for measurements of acoustic emission, strain, temperature, ultrasonic Lamb waves and their applications in structural health and corrosion monitoring.

Key words: structural health monitoring, aerospace corrosion, piezo-electric transducer, fiber optic sensors, acoustic emission, strain, ultrasonic Lamb waves, metal fatigue.

7.1 Introduction

Over the past decade, integrated health and corrosion monitoring has emerged as an important consideration for overall condition monitoring of civil and aerospace systems. There is a clear need to develop integrated sensing systems capable of monitoring damages, corrosion and fatigue in aging structures, leading to improved performance, reduced costs, and improved safety. Corrosion costs US industry and government agencies an estimated $278 billion/year, according to the Federal Highway Administration (Koch *et al.* 2002). Civil infrastructure systems are generally the most expensive investments in any country; in the United States the investment is estimated to be $20 trillion. These systems are weakening at an alarming rate due to material or system deterioration caused by overuse, overloading, aging, damage or failure caused by external loads such as natural or man-made hazards (Liu *et al.* 2003). The US spends more than $200 billion each year on the maintenance of plant facilities and equipment. In the aerospace industry, the average major airline spent 12% of its total budget on inspection – almost $9 billion a year (Jackson 2001). Traditional airframe maintenance is labor intensive, involving the dismantling of the airframe, inspecting components manually and reconstructing the airframe. Often no airframe damage is found, but nevertheless frequent inspections are a safety requirement, and the authorities demand evidence of safety by mandatory inspection records. An integrated health and corrosion monitoring system can eliminate much of the current labor-intensive process, enabling airframe maintenance to be performed specifically on areas where cracking, corrosion, or fatigue is likely to occur. Selected use of condition-based maintenance would significantly reduce the cost of inspection programs.

There is a wide variety of structural damage and corrosion detection measurement techniques using direct methods or indirect methods. The direct methods include nondestructive evaluation (NDE) techniques such as ultrasonics, radiography, thermography, and eddy current. Direct methods are used to supplement visual inspection at routine maintenance intervals. When the inspected area is physically accessible, visual tests are commonly used for periodic checks. Sometimes, tools such as magnifying glasses or boroscopes are employed for further evaluation or for less accessible areas, respectively. The inspection involves a visual search for cracks, change of color, change of texture, or bulges. Surface corrosion at its embryonic stage can be visually detected from localized indication such as discoloration, faint powder lines, pimples on the paint and paint damage. Concealed corrosion is very difficult to detect since, in most cases, the characteristics of the damage are not sufficient to trigger an indication in conventional NDE tests.

Existing direct NDE methods for detection of crack and corrosion have limited capabilities and sensitivity. Frequently, corrosion and structural damages are detected only after several subsequent inspection schedules, in which case the damages are fairly extensive and may require the replacement of the structural components involved. In this case, indirect methods can be used to identify the environmental and structural health condition of a unit that can experience severe corrosion or structural damage. Indirect methods which are used to provide real-time crack and corrosion monitoring include pH, ion concentration, flow rate, weight loss, humidity, temperature, load, acceleration, impedance, acoustic emission, and acousto-ultrasonic measurements.

Among the available options for on-board health and corrosion monitoring systems, load, acoustic emission and acousto-ultrasonic sensors have the advantage of being unobtrusive and readily integrated into structures. Such systems can provide real-time measurements while the structure is in use, and have the potential to provide immediate notification of loading excess or the presence and severity of crack and corrosion-induced damages. The present report focuses on recent progress and development from several research projects conducted by Physical Acoustic Corporation (PAC) and Los Gatos Research. These research projects aim to develop integrated health and corrosion monitoring sensor systems that can be applied on aging structures for monitoring structural damages including cracks and corrosion. The sensing techniques include traditional electrically based piezoelectric transducer and novel fiberoptic Bragg grating demodulation methods for measurements of strain, temperature, acoustic emission, and ultrasonic guided Lamb waves. Several comprehensive literature reviews of the research and development in structural health and corrosion monitoring over the last few decades are discussed by Sohn *et al.* (2003), Moran *et al.* (1986), Spanner *et al.* (1987), and Measures (1999).

7.2 Acoustic emission

The first observation of acoustic emission (AE) in metals was the audible emission produced by mechanical twinning of tin during plastic deformation. The first documented studies on acoustic emissions were carried out in the 1950s. Kaiser performed tensile tests of conventional engineering materials to determine what acoustic emissions are generated from within the specimens, the acoustic processes involved, the frequency levels found, and the relation between the stress–strain curve and the frequencies noted for various stresses to which specimens were subjected (Spanner *et al.* 1987). The potential use of acoustic emission was soon recognized. Currently, AE testing is used in many areas including nondestructive testing and evaluation of materials and structures, real-time monitoring and failure analysis.

There are various definitions of acoustic emission and related terms in the literature. The definitions of basic AE terms proposed by Acoustic Emission Working Group (AEWG) (Spanner 1974), are presented below:

- *Acoustic emission* is a transient stress wave generated by the rapid release of energy from localized sources within a material.
- *Acoustic emission event* is rapid physical change in a material that releases energy appearing as an acoustic emission.
- *Burst emission* is an individual emission event which has clearly observed waveform initiation and end. Crack opening in metals and fiber debonding in composites are two examples of burst emissions.
- *Continuous emission* is a qualitative term applied to an acoustic emission if individual bursts are not discernable. Leakage and corrosion are two examples of continuous emissions.
- *Hit-based (qualitative) analysis* is the evaluation of AE testing using parameters extracted from AE waveforms. The main parameters are maximum amplitude, AE count, duration, rise time and energy. Hit-based analysis is valid for burst emissions.
- *Waveform-based (quantitative) analysis* is the evaluation of AE testing using complete AE waveforms.

7.2.1 Acoustic emission instrumentation

AE testing is a passive nondestructive testing method which means that the detection of a flaw source depends on energy emitted by that flaw source itself. Figure 7.1 is a schematic of AE testing and the basic instrumentation used in this testing method. The radiation of elastic stress waves in a material occurs when a flaw initiates or propagates and leads to the initiation of a source-time function $S(t)$. The stress waves propagating in the material (i.e., acoustic emissions) reach the surfaces of the material and lead to mechanical disturbance. There, the mechanical disturbance, for example, the dynamic surface displacement $u(t)$, (i.e., input signal) is transformed into an electrical signal $V(t)$ (i.e., output signal) by an AE transducer. As the dynamic surface displacement may have low energy content depending on the flaw source type, pre-amplification is often required before converting the analog electrical signal to a digital signal. The recorded AE signal is evaluated using several methods. As the dashed line in Fig. 7.1 indicates, the signal evaluation locates and characterizes the flaw source as the propagating elastic stress waves carry information about the characteristics of the flaw source. The success of AE testing relies on the proper evaluation of recorded AE signal.

In general, acoustic emissions from civil engineering materials are broadband energy sources spanning frequencies from 20 kHz to 1 MHz in the shape of pulse and produce complex transient signals (Stephens *et al.* 1971). The origin of acoustic emission is divided into macroscopic and microscopic origins. Macroscopic origin refers to circumstances where a relatively large (whether in volume or in surface) part of the test material is contributing to the AE event. Examples of macroscopic origin AE events include yield-

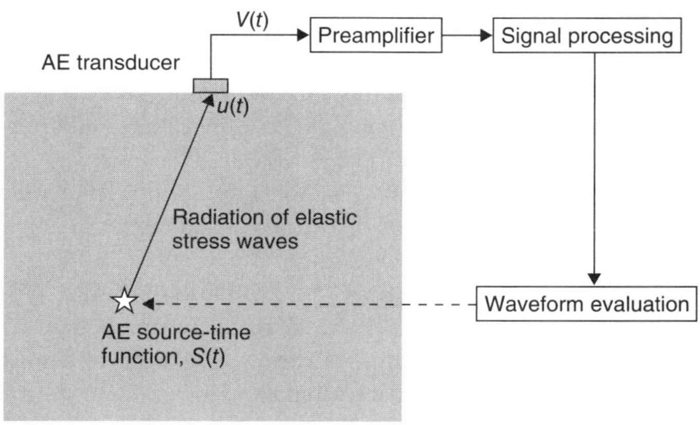

7.1 Schematic of AE testing and basic instrumentation.

ing, crack growth, corrosion for metals, fiber breakage and matrix debonding for composites. Microscopic origin refers to circumstances where the flaw source involves a small area or volume as compared with a macroscopic origin event. Examples of microscopic origin AE events include phase transformation, microcracks and dislocation movement. Each macroscopic origin AE event is composed of several microscopic origin AE events.

The frequency content and amplitude of acoustic emissions vary depending on the origin and type of flaw sources. For example, because the microscopic origin flaw sources involve a small area or volume as compared with macroscopic origin flaw sources, their source-time function amplitudes are relatively weak. The variation of amplitude and frequency content of different flaw sources requires a sensitive AE transducer and different frequency AE transducer depending on the origin and type of flaw sources. If the sensitivity (relationship between output signal and input signal) and frequency bandwidth of an AE transducer are not sufficient, the detection probability of different flaw sources is reduced.

7.2.2 Crack and corrosion detection using acoustic emission

Fatigue, cracks, and corrosion are responsible for the failure of most industrial components and structures. Acoustic emission has the potential to detect and monitor the initiation and propagation of cracks resulting from the different forms of corrosion. It is known that materials release characteristic AE transients during damage growth/propagation cycles (Pollock 1989). Stress corrosion cracks have been widely studied by AE techniques (Spanner 1974; Dunegan 1976). AE is a passive nondestructive testing method that can provide real time detection of damage mechanisms. When a damage site initiates or grows, the stress field changes around the damage that leads to the generation of elastic stress waves. The AE sensors mounted on the test structure convert this mechanical disturbance into electrical signals. AE methods can be used to detect corrosion growth on-line for periodically active corrosion or off-line for active corrosion during measurement. On-line monitoring may generate detection problems because of high background noise. Corrosion growth emission intensity is low compared with other damage mechanisms such as crack growth. To successfully detect active corrosion growth using AE methods, background noise should be minimized. AE methods can be used to detect active corrosion growing on metal components. Figure 7.2 illustrates possible AE sources on a metal structure during environmental exposure and stress (Yuyama *et al.* 1984). The evolution of hydrogen gas via cathodic reaction in acid solutions and the breakdown of thick, surface oxide films formed in high-temperature

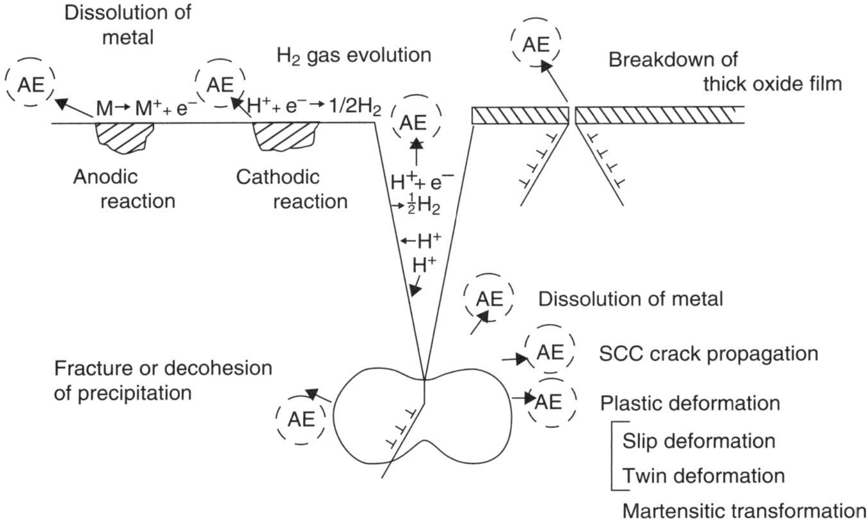

7.2 Possible AE sources during environmental corrosion, stress corrosion cracking and corrosion fatigue processes.

water are fundamental AE sources beside crack initiation and growth. The AE activity depends strongly on environmental factors, mechanical conditions and materials. The relation between AE total count and corrosion rates has been demonstrated by Rettig *et al.* (1976). It was found that a linear relationship exists between the total count and the amount of hydrogen collected.

7.3 Ultrasonic guided waves (or Lamb waves)

One of the emerging new powerful health and corrosion monitoring technique in recent years is the guided wave method. Guided wave technique has been used as an extremely effective health monitoring technique to evaluate high-strength, low-weight composites in advanced aircraft components. Applications for guided waves involve large areas of testing as well as cases where direct access to a specific component is not possible. Guided waves are ultrasonic stress waves that propagate in a different way from the more commonly used longitudinal and shear waves. Guided waves result from multiple reflections and interference of the pressure and shear waves over the specimen cross-section, such that a standing wave mode through the specimen thickness is obtained. From the measurement of the guided wave at a few points on the surface of the structure it is possible to detect defects in a large area with a fast and cost-effective method (Rose 1999).

Guided modes in structures with plate geometries have been theoretically predicted and extensively investigated (Rose 1999). The guided wave approach, however, can be extended to include any practical structure which has one dimension that is smaller than the other two. Examples of this include aircraft and spacecraft structures such as wing and fuselage skins, turbine blades, and combustion chambers, whose walls are usually thin. For example, Prosser et al. (1992) illustrated that guided modes were present in a thin-walled graphite/epoxy tube of a type similar to that proposed to be used in NASA's Space Station. Rose (2000) reported detection of cracks in helicopter blades with guided waves. Crack propagation in the transmission beam of an H-60 helicopter has been investigated using guided wave techniques (Fromme *et al.* 2002). Guided waves have been used to test bonding performance in lap splice joints, tear strap, and honeycomb structures commonly used in the aircraft industry (Rose 2000). The ultrasonic guided waves technique has been proposed for corrosion and fatigue crack growth monitoring at fastener and rivet holes in the fuselage (Rose 2000; Fromme *et al.* 2002). Comprehensive literature surveys on significant research work that has been carried out in guided ultrasonic wave mechanics are reviewed by Rose (2002a, 2002b), and Raghavan *et al.* (2007).

There are many techniques for generating and sensing guided waves. Guided waves can be generated by active probing or by crack-induced acoustic emission in thin plates (Gorman *et al.* 1991; Prosser *et al.* 1992). In the case of probing, the most common guided wave generation technique employs an angle beam transducer for the generation of guided waves by pulsing a piezoelectric element on the wedge placed on a test surface (Fig. 7.3). As a result of refraction at the interface between the wedge and the test specimen and by way of mode conversion and totally internal reflection, a discrete number of waves with distinct mode shapes can propagate in the structure. Different modes can be excited by changing the angle of incidence. Since these waves are confined inside the waveguide structure, they can travel at relatively large distances with very little amplitude loss. In two-dimensional structures such as thin plates and shells, guided waves penetrate the entire thickness of the plate and propagate parallel to the surface, allowing a large portion of the material to be interrogated from a single 'sender probe' transducer location. Fatigue cracks, corrosion, and defects along a waveguide can be detected and located by analyzing in the time domain the reflecting wave signals from a 'receiver probe' sensor placed near the 'sender' in a 'pulse-echo' mode, or by measuring the propagating wave intensity signals from a sensor placed on the far side of the waveguide in a 'pitch-catch' mode. Typically, the same sender transducer is used as a receiver sensor in the pulse-echo setup, and another angle beam transducer is used as a sensor in the pitch-catch setup.

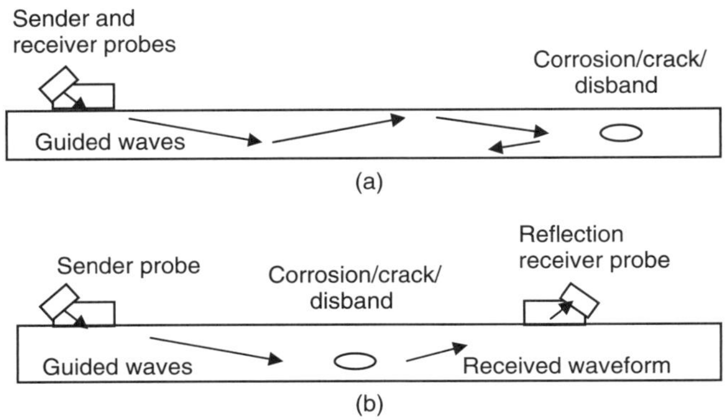

7.3 Basic ultrasonic guided wave pulse-echo and pitch-catch techniques. (a) Pulse-echo: a sending probe generates a guided wave mode propagating along the length of the plate. The wave then reflects from any defects and returns to the probe. (b) Pitch-catch: the guided wave mode propagates along the plate and interacts with a defect area. The depth, width, length, and shape of the defect will influence characteristics of the resulting waveform, which is then received at the receiving probe.

In a guided wave inspection of plate-like structures, the waveguide (Lamb) modes commonly excited and detected are the first symmetric (S_0) and antisymmetric (A_0) modes. The S_0 mode generally has large in-plane particle motion amplitudes near the middle of the layer, while the A_0 mode has more intense out-of-plane motion near the surface and is more dispersive than the S_0 mode. The A_0 mode is therefore more sensitive to cracks initiated on the surface of the structure, while the S_0 mode is more sensitive to a flaw located at the center of the material. Certain crack size and shapes can be inferred by comparing the reflection and transmission coefficients of the S_0 and A_0 modes (Rose 1999). One useful application of guided wave technique is the detection of crack and corrosion in high-strength composite materials. For anisotropic and heterogeneous media such as advanced composites, mode distribution and characteristics of Lamb waves can be used to study dispersion and scattering of the waves at and around inhomogeneities. In addition, wave propagation in these materials must be studied from the viewpoint of Lamb or plate waves, since advanced composites are typically used in the form of thin plates and traditional bulk wave characterization methods are not applicable (Henneke 1990). As a consequence, comprehensive, accurate detection, location, and characterization of damage, corrosion, or disbands in advanced structures, such as those employed in high-strength, light-weight aircraft components and systems, require sensors that can precisely detect both S_0 and A_0 Lamb modes with high sensitivity.

Commercially available ultrasonic wave sensors include piezoelectric, piezoceramic and piezocomposite transducers. Electrically based transducers are reliable, accurate, and easy to use; however, they suffer a few important drawbacks. For example, these sensors are generally sensitive to strong electromagnetic (EM) waves, their accuracy deteriorates in chemically corrosive environments over long periods of time, and they are difficult to multiplex for measurements at multiple locations. Most research work on acoustic emission and guided wave methods are based on conventional electrically based piezoelectric transducers. More recently, a number of optical fiber sensor techniques have been developed and proposed as potential viable alternatives to electrically based ultrasonic wave sensor technique. A major advantage of fiber optic sensors over electrical techniques is that they can be made environmentally rugged and unresponsive to EM interference. In addition, optical fiber sensors are generally smaller, lightweight, and can be easily integrated into structures and configured to create a distributed network for multi-use.

7.4 Strain monitoring

Strain gauges are common methods of inferring structural loads. Electrical, resistive foil strain gauges are currently the state of the art for strain evaluation purposes. For health monitoring of large and complex structures, such as bridges, ships, and airframes, the strain gauge sensor array has to cover a very large physical space. With the increase of sensing points and structure size, the amount of cabling, weight, power requirements, and cost for sensor hardware increases dramatically. The large amount of cabling also leads to the increase of installation and maintenance costs, reduction in reliability, deterioration in measurement resolution due to electromagnetic interference and increases of self-weight. Recent developments in micro electrical–mechanical systems (MEMS) sensors combined with wireless communication technologies provide alternative means for strain monitoring. Wireless MEMS strain sensors hold a distinct advantage over conventional strain gauges in that the overall sensor system (consisting of a strain sensor, signal conditioning and telemetry circuit, and antenna) does not require a wire connection. However, reliability and measurement accuracy are still problems that must be addressed for successful implementation of MEMS technologies to health monitoring. A comprehensive review of wireless sensors and their adoption in health monitoring has been reported by Lynch *et al.* (2006). A major disadvantage of electrically based strain sensor techniques is that sensor calibration is often a challenging task due to the device signal drift and the need to constantly monitor the changes in the sensor itself. Optical fiber sensors offer significant advantages over electrically based strain sensors. Optical sensor devices such as Bragg grating strain sensors require no powered line drivers, can be multiplexed and are of low weight and size. In addition, each Bragg grating sensor has an inherent self-referencing capability which is determined by the Bragg reflection wavelength associated with a particular strain level.

7.4.1 Fiber optic sensor for health monitoring

Optical fiber sensors can be incorporated into structures to provide health monitoring, providing strain, temperature, acoustic emission, and ultrasonic wave information indicating when crack and corrosion occur and where. There are various optical fiber sensor types being developed for health monitoring applications. The most common of these is fiber Bragg grating. Other types of fiber optic sensors include Fabry Perot, interferometric, microbend, and polarimetric sensors. All fiber optic sensing methods rely on either (1) the longitudinal deformation of a sensor fiber caused either directly by axial strain, or indirectly via the Poisson effect through lateral pressure on the fiber, or (2) the photoelastic effect which results in refractive index changes in the core and cladding materials of the fiber. For example, fiber optic interferometric sensors (FOIS) have been shown to possess the high sensitivity required to respond to the Ångström particle displacement amplitudes associated with in-plane and out-of-plane Lamb waves, and a wideband frequency response limited in many cases by detection electronics alone (Gachangan *et al.* 1995). The most attractive acousto-ultrasonic fiber-based sensor, however, is fiber Bragg grating (FBG). FBG sensors are ideally suitable for measuring static and dynamic fields such as temperature, strain, pressure, and acoustic waves. However, the principle advantage of FBG sensors is that the measurand information is wavelength-encoded, thereby making the sensors self-referencing, rendering them independent of fluctuating light levels and other optical noise sources. This wavelength-encoding property also offers convenient multiplexing, either time division multiplexing (TDM) or wavelength division multiplexing (WDM), along a single optical fiber for making highly localized strain, temperature, or stress wave measurements for condition-based monitoring over a distributed area. What makes FBGs most attractive for health monitoring is that they can be used for both strain-based load monitoring and acousto-ultrasonic damage detection. FBG sensors have been proposed and demonstrated as S_0 and A_0 Lamb mode detection and strain monitoring sensors (Perez *et al.* 2001), as well as acoustic emission sensors for aircraft and spacecraft applications (Betz *et al.* 2003). In the last few years, optical fiber sensors as effective means of detecting and quantifying ultrasonic waves have been reported by Perez *et al.* (2001), Betz *et al.* (2003), and Nguyen (2007). Extensive literature reviews on various optical fiber sensor technologies and their applications on health monitoring are discussed by Measures (1999) and Lopez-Higuera (2002).

7.5 Active corrosion detection

7.5.1 Active corrosion detection on a magnesium gearbox housing

The detection of active corrosion on aircraft components has been recently investigated by Physical Acoustic Corporation. In a collaboration research project with the US Navy, corrosion detection in magnesium helicopter transmission housing was performed at the Cherry Point Naval Air Station. To determine which AE sensor is the most sensitive to corrosion growth, the upper rim and side of the housing were tested using various AE sensors. The main difference among AE sensors is their operational frequency range. For example, R15I is a resonant type AE sensor having operational range as 100–300 kHz with 150 kHz peak frequency and integrated pre-amplifier. Mini 30 sensor is a high-frequency sensor operating in the range of 270–970 kHz. The corrosion growth was initiated using salty water and stopped using corrosion inhibitor.

Figure 7.4 shows the acoustic emission activities detected by R15I when salty water and corrosion inhibitor were applied. Absolute energy, presented at *y*-axis, is an important AE feature and calculated by the area underneath the AE waveform having a duration of 500 ms. Absolute energy can be calculated as either hit-based transient signals (when the AE signal

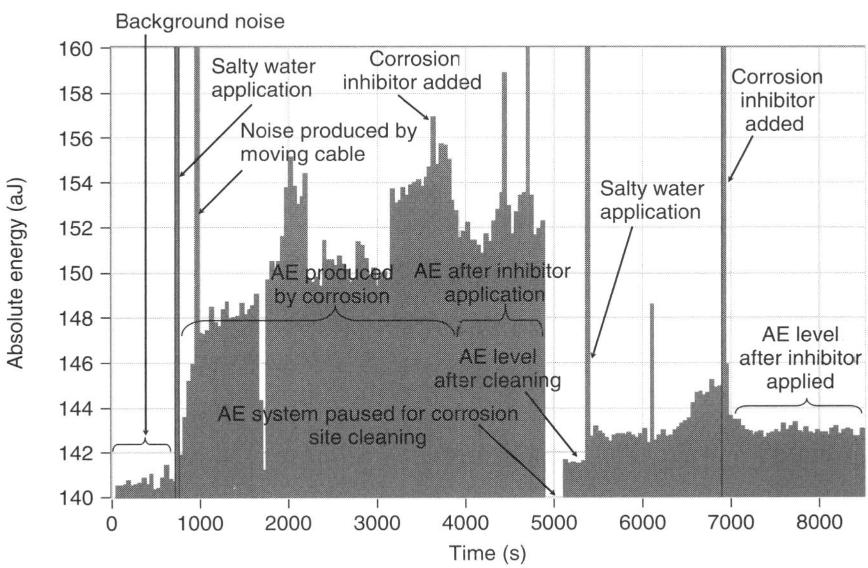

7.4 Acoustic emission due to corrosion growth detected by R15I sensor.

level at the sensor face exceeds the pre-defined threshold) or time-based (using certain duration of waveform and calculating absolute energy every time interval). The data shown in Fig. 7.4 is time-based absolute energy. As there was no background noise, the changes in the absolute energy level are caused by corrosion growth and the user interruption to add salty water and corrosion inhibitor. AE absolute energy level increases owing to corrosion growth when salty water is applied. There is some decrease in the AE level when the corrosion inhibitor is applied, but it is not known accurately whether the corrosion inhibitor stops the corrosion immediately or in a certain time period. When the corroded area is cleaned at the 5000th second of testing, the absolute energy level drops considerably and starts increasing when salty water is reapplied. The tests indicate that the AE sensors can detect active corrosion reliably if background noise level is sufficiently reduced.

7.5.2 Detection of active corrosion on storage tank floors

One of the most successful applications of acoustic emission is the detection of active corrosion in storage tank floors. TANKPAC™ was developed in cooperation with the oil and process industry over a 10-year period. More than 3000 tank floor tests have been carried out worldwide by PAC and licensees. The test consists of attaching sensors to the outside of the tank and monitoring for emission resulting from active corrosion of the floor. Noise sources are eliminated by sophisticated pattern recognition methods and the tank is graded from A to E depending on the amount of AE detected. The AE results have been confirmed independently by Yuyama *et al.* (2007) using other NDT techniques.

7.6 Integrated strain, temperature, and stress wave monitoring sensor

For the simultaneous measurements of strain, temperature, and ultrasonic stress waves, Nguyen (2007, 2008) has developed a laser-based demodulation technique to interrogate FBG sensors, either surface mounted or embedded inside monitoring structures. In this technique, a distributed feedback laser (DFB) is tuned to the mid-reflection wavelength of the Bragg grating. Light reflected back from the grating is detected by a photodetector, which converts the light signal to an electrical signal. The dc to low frequency ac signal corresponds to strain and temperature, and higher frequency ac signals correspond to ultrasonic stress waves. As the Bragg wavelength shifts due to the acoustic waves impinging on the grating, the light (and electrical) signal is modulated with an intensity proportional to the applied acoustic field, at the same frequency as the generated ultrasonic waves.

For strain measurements, a submicrostrain resolution requires a picometer wavelength detection sensitivity. Such a small wavelength shift detection is often obscured by environmental and system noise. To effectively remove the noise contribution, a lock-in amplifier is used to provide a feedback signal to the laser control electronics. This lock-in scheme enables the laser wavelength to be continuously locked to the stable point at the mid-reflection wavelength of the Bragg grating to produce the highest signal-to-noise, providing a direct strain measurement. Once the laser is locked, the dc strain signal remains stable, and the laser wavelength is highly resistant to environmental noise that tends to move the laser wavelength away from the stable point it is locked to, enabling high signal-to-noise strain measurements and reliable ac strain and stress wave detection.

Using the lock-in laser-based demodulation technique, highly accurate strain and temperature measurements with submicrostrain and milli-Kelvin resolution, respectively, can be routinely obtained while ultrasonic stress waves can be continuously monitored. As an example, Fig. 7.5 illustrates ambient temperature signals monitored as a function of time over 17 h by a FBG sensor (dark grey line) and a commercial temperature monitoring sensor (light grey line). For this measurement, the FBG sensor was bonded to a carbon fiber composite plate for simultaneous measurement of strain, temperature, and stress wave, and the temperature monitoring sensor was used to record room temperature. Light from a DFB laser was passed to the FBG sensor. The reflected light from the FBG was detected by an

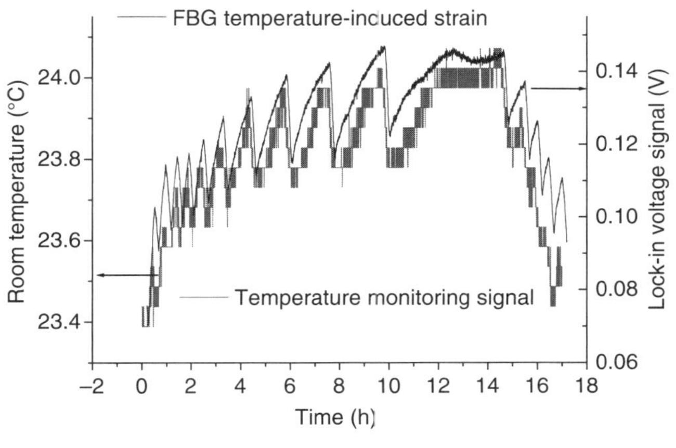

7.5 FBG thermal strain induced lock-in signal (dark grey) and commercial temperature sensor signal (light grey) recording room temperature fluctuation over 17 h.

InGaAs photodetector, while the DFB laser wavelength was continuously locked to the mid-reflection wavelength of the Bragg grating using a commercial lock-in amplifier. Over 17 h, both the FBG and the temperature sensor were exposed to room temperature fluctuation, and the FBG and temperature sensor data were recorded by a computer using a PCI-6111 National Instrument data acquisition board and LabView program. The FBG signal clearly demonstrated a superior signal-to-noise ratio compared with that of the commercial temperature sensor. The near periodic fluctuation in the recorded signals was due to the room-temperature fluctuation cycle with a period of 30 min to 2 h. From time $t = 0$ to $t = 17$ h, the temperature sensor recorded a temperature shift of 0.713 °C. This temperature shift induces a thermal strain in the FBG which corresponds to a wavelength shift of 6.4 pm. During this time, the photodetector registered a changed voltage signal $\Delta V(t) = 69.8$ mV, with a noise level of approximately 0.7 mV, corresponding to a signal to noise ratio (SNR) of 100:1 and a temperature-induced strain resolution of 0.06 $\mu\varepsilon$.

While the laser was still locked to the mid-reflection wavelength of the Bragg grating, ultrasonic Lamb waves at 200 kHz were launched into the carbon fiber composite plate using a PAC S-9208 PZT actuator and a PAC ARB-1410-150 pulser system with a 10-cycle sine burst signal input. The Lamb wave signals were collected by the same photodetector that was used to record the thermal strain signal. The digitized data were averaged 100 times, plotted by a LabView program, and saved in the desktop computer for further data analysis. Figures 7.6a and 7.6b show the detected Lamb wave signals at $t = 0$ and $t = 17$ h, respectively. The high SNR, reproducible Lamb wave signal after 17 h of laser locking in Fig. 7.6b demonstrates that robust, high-resolution strain and high sensitivity acoustic wave signals can be simultaneously measured by the same FBG sensor.

7.7 Crack detection

The ultrasonic Lamb wave method can be used to determine the presence and severity of crack- and corrosion-induced damages. Several experiments were performed to determine the damage monitoring capability of the FBG interrogation technique. Figure 7.7 displays a photograph picture of the lock-in based FBG interrogation prototype system at Los Gatos Research (Nguyen 2008). The FBG interrogation system is a multi-channel strain/ Lamb wave monitoring instrument that includes a PC/104 embedded computer system, a multi-channel data acquisition board, a laser driver controller, and a lock-in detection system. Figure 7.8 shows the picture of a commercial FBG sensor bonded at the center of a 60 cm × 60 cm × 0.17 cm aluminum plate. An APC 850 PZT actuator was bonded slightly off the center toward the right edge of the same aluminum plate, with a sensor–

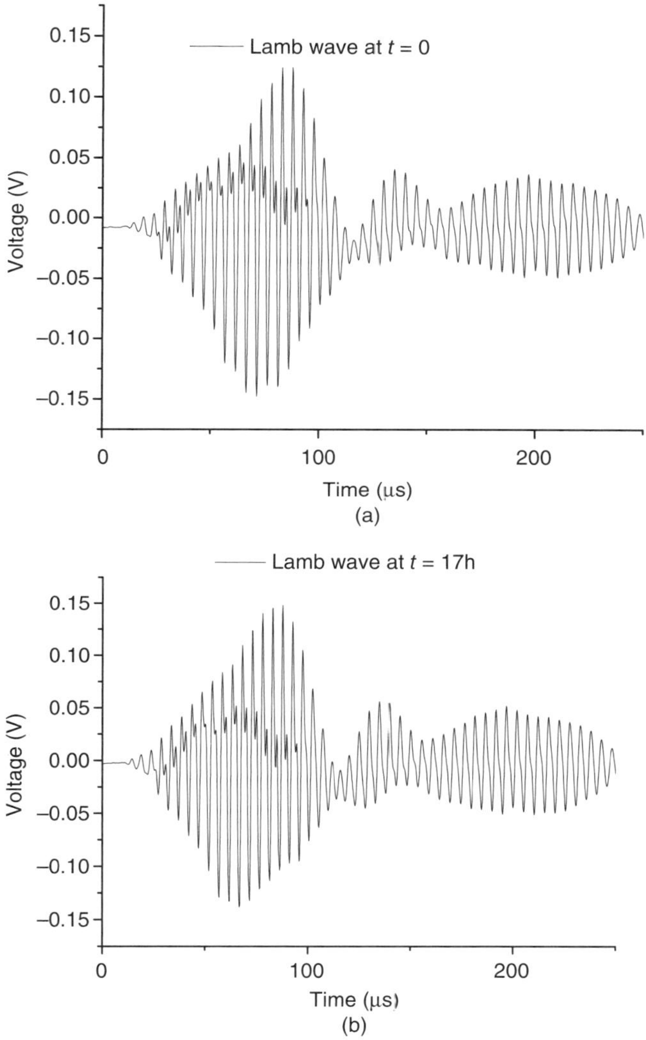

7.6 FBG sensor response to a 10-cycle sine tone burst Lamb waves on a carbon fiber composite plate at (a) $t = 0$ and (b) $t = 17$ h while the laser is locked to the FBG's mid-reflection wavelength and the temperature-induced strain data is continuously recorded.

actuator distance of approximately 4 cm. For performance comparison, another APC 850 PZT used as a sensor was bonded right next to the FBG (about 1 cm from the FBG). A 2 mm wide × 5 cm long × 0.5 mm deep trench simulating a damage was drilled on the surface of the plate at a location approximately 17 cm from the FBG sensor (appears as a straight line on

7.7 LGR's lock-in based FBG interrogation prototype system.

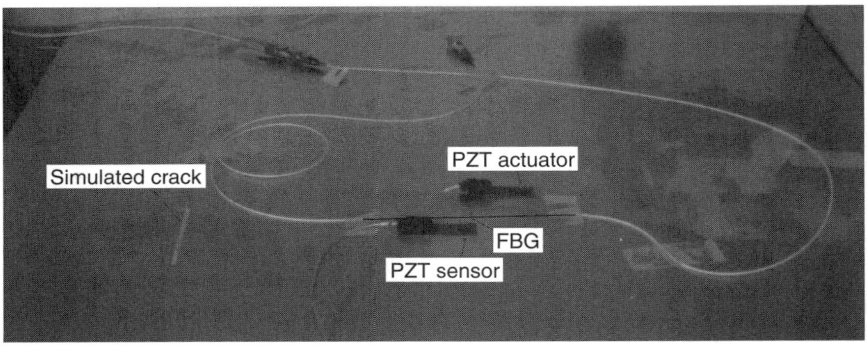

7.8 A FBG and PZTs bonded on the surface of an aluminum plate. A drilled crack line simulating a damage is approximately halfway between the FBG and the plate's left edge.

the left hand side of Fig. 7.8). The location of the crack, FBG sensor, and PZT actuator were arranged in a 'pulse-echo' configuration, and the orientation of the FBG was chosen in such a way that the FBG response to the crack-induced Lamb wave signal is highest while the direct Lamb wave signal remains low since the FBG sensor has highest response to stress waves presumably along the fiber axis and lowest along the direction normal to the fiber axis (Betz *et al.* 2003). The aluminum plate together with the bonded FBG and PZTs were located inside a fume hood. The FBG sensor was optically connected to the remote interrogator through a 500 ft long optical fiber. A 200 kHz, 2-cycle, sine tone burst generated by a PAC ARB-1410-150 pulser system was used to excite the actuator, which produces primarily the in-plane (S_0) Lamb wave mode. Light from a DFB laser located inside the FBG interrogator was locked to the mid-reflection wavelength of the Bragg grating throughout the entire experiment. The returned light signal detected by an InGaAs photodetector was high-pass filtered and recorded by the interrogator's data acquisition board. The digital signals were averaged a thousand times, and the entire data set was saved to the PC/104 computer and analyzed by a LabView program.

Figures 7.9a and 7.9b display the FBG's Lamb wave response to the PZT actuator before and after the crack was formed, respectively. For comparison, a PZT sensor's response to the same actuator signal after the crack was formed is shown in Fig. 7.10. The first wave packet arrived at

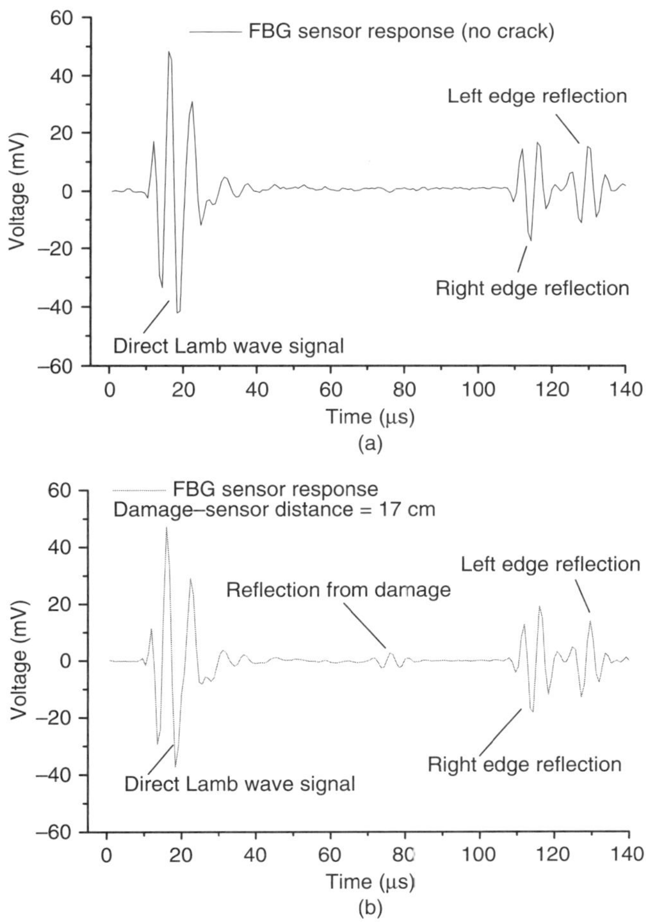

7.9 FBG sensor response to Lamb waves (a) before and (b) after a crack simulating damage was made.

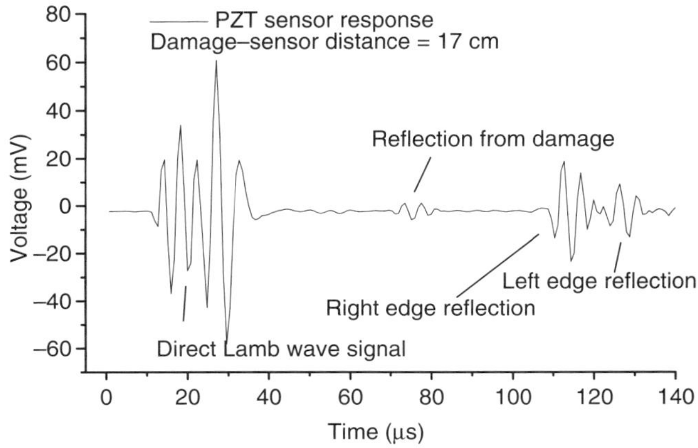

7.10 PZT response to Lamb waves after a crack simulating damage was made.

approximately $t = 10$ μs in Fig. 7.9a, 7.9b, and 7.10 are the direct Lamb wave signals corresponding to the S_0 mode. Subsequent wave packets are the reflection signals from this Lamb wave mode. The weak signals beginning around 70 μs in Figure 7.9b and 7.10 are the reflected signals from the crack, which are clearly absent in Fig. 7.9a plot data that correspond to a pre-crack condition. The two stronger signals near 108 and 123 μs in Fig. 7.9a, 7.9b, and 7.10 correspond to the reflections from the left and right edges of the plate, respectively. Note that the left edge reflection signals around 123 μs in the presence of a crack (Fig. 7.9b and 7.10) are slightly weaker than that in the absence of the crack (Fig. 7.9a) due to the attenuation of the wave passing through the crack. Comparing Figures 7.9a, 7.9b, and 7.10, it can be inferred that the FBG and PZT sensors have similar ultrasonic wave sensitivity. These observations clearly indicate that both FBG interrogation and PZT-based methods can be used as effective Lamb-wave based techniques to probe structural damages.

7.8 Corrosion monitoring

While the aluminum plate sample was inside the fume hood and the laser was locked to the mid-reflection wavelength of the FBG, an accelerated corrosion test was performed on the crack simulating damage by wet etching the damage area with an aluminum etchant. Cyantek Al-12S, an aluminum etchant with an etch rate of 1 μm/min, was continuously flooding the entire crack line trench for 2 days using an electronically controlled syringe pump method. During the wet etching process, crack-induced Lamb wave inten-

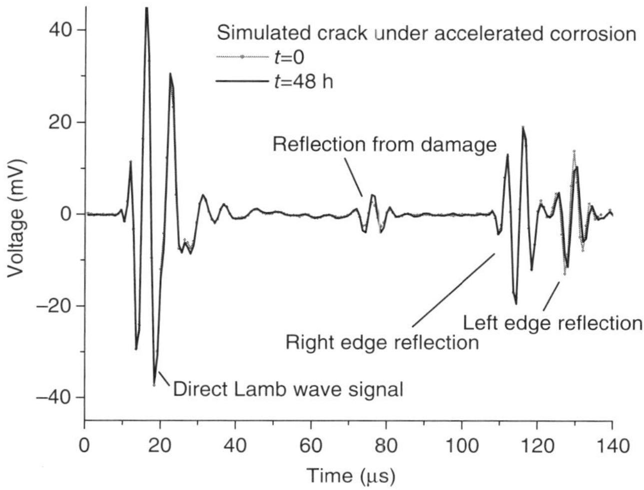

7.11 Lamb wave signal change due to corrosion.

sity was monitored using the FBG interrogator. Figure 7.11 compares the Lamb wave signals before the corrosion test begins and 48 h after the aluminum etching started. It is evident that the damage diffracted signal intensity (time $t \sim 70\ \mu s$, thick line) was enhanced after 48 h of aluminum etching. In addition, the left-edge reflected signal ($t \sim 123\ \mu s$, thick line) was reduced while the right-edge reflected signals ($t \sim 108\ \mu s$, thin and thick lines) remain the same during the same period, indicating that corrosion has altered the crack dimension, effectively enhancing the attenuation of the Lamb waves propagating through the damage. The slight time delay of the left edge reflection signal ($t \sim 123\ \mu s$, thick line) can be attributed to the dispersion of the ultrasonic waves traveling around the corroded damage.

7.9 Conclusions

The theory and development of several NDE methods has been reviewed including load, AE, and Lamb waves that can be used for health and corrosion monitoring applications. We also report several experimental results from electrically based PZT and optical fiber Bragg grating methods for the detection of crack and corrosion and for strain and temperature monitoring. The results demonstrate the potential of combining a multi-sensor platform with various sensing techniques for future on-board, integrated health and corrosion monitoring applications of aging structures.

7.10 Acknowledgements

Los Gatos Research gratefully acknowledges the support of the SBIR contract NND07AA04C, granted by NASA, and the SBIR contract OII-0539234, granted by the National Science Foundation.

7.11 References

BETZ D, THURSBY G, CULLSHAW B, STASZEWSKI W (2003), 'Acousto-ultrasonic sensing using fiber bragg gratings,' *Smart Mater. Struct.*, **12**, 122–128.

DUNEGAN H (1976), 'Round robin testing of the frequency spectra of acoustic emission signals,' *Proceedings of the 15th Acoustic Emission Working Group Meeting*.

FROMME P, SAYIR M (2002), 'Monitoring of fatigue crack growth at fastener holes using guided Lamb waves', *Rev. Quant. Nondestr. Eval.*, **21**, 247–252.

GACHANGAN A, PIERCE S, PHILP W, MCNAB A, HAYWARD G, CULSHAW B (1995), 'Detection of ultrasonic Lamb waves in composite plates using optical-fibers', *Proc. IEEE Ultrason. Symp.*, 803.

GORMAN M, PROSSER W (1991), 'AE source orientation by plate wave analysis,' *J. Acoustic Emission*, **9**, 283–288.

HENNEKE E (1990), 'Ultrasonic nondestructive evaluation of advanced composites,' in: J. Summerscales, ed., *Non-Destructive Testing of Fibre-Reinforced Plastics Composites*, 2, 55–159.

JACKSON P (2001), *Jane's all the world's aircraft*, London: Jane's Information Group, 102.

KOCH G H, BRONGERS P H, THOMPSON N G, VIRMANI Y P, PAYER J H (2002), 'Corrosion costs and prevention strategies in the United States,' *Federal Highway Administration*, FHWA-RD-01-156.

LIU S C, TOMIZUKA M (2003), 'Vision and strategy for sensors and smart structures technology research,' *Proceedings of the 4th International Workshop on Structural Health Monitoring*, 42–52.

LOPEZ-HIGUERA J (2002), *Handbook of optical fibre sensing technology*, West Sussex, England: John Wiley & Sons Ltd.

LYNCH J, LOH K (2006), 'A summary review of wireless sensors and sensor networks for structural health monitoring,' *Shock Vibr. Digest*, **38**(2), 91–128.

MEASURES R (1999), *Structural monitoring with fiber optic technology*, San Diego: Academic Press.

MORAN G, LABINE P (1986), 'Corrosion monitoring in industrial plants using nondestructive testing and electrochemical methods,' *STP 908*, Philadelphia: ASTM.

NGUYEN A-D (2007), 'High performance fiber optic strain and ultrasonic wave sensing,' *Proceedings of the 6th International Workshop on Structural Health Monitoring*, 1182–1190.

NGUYEN A-D (2008), 'Photonic sensor for nondestructive testing applications,' in T. Kundu, ed. *Proc. SPIE*, **6935**, 693525.

PEREZ I, CUI H, UDD E (2001), 'Acoustic emission detection using fiber Bragg gratings,' *Proc. SPIE*, **4328**, 209–215.

POLLOCK A (1989), 'Acoustic emission inspection,' *Technical Report, Physical Acoustics Corporation*, TR-103-96-12/89.

PROSSER W, GORMAN M, DORIGHI J (1992), 'Extensional and flexural waves in a thin-walled graphite/epoxy tube,' *J. Compos. Mater.*, **26**, 418–427.

RAGHAVAN A, CESNIK C (2007), 'Review of guided-wave structural health monitoring,' *Shock Vibr. Digest*, **39**(2), 91–114.

RETTIG T, FELSEN M (1976), 'Acoustic emission method for monitoring corrosion reactions,' *Corrosion*, **31**(4), 121–126.

ROSE J (1999), *Ultrasonic waves in solid media*, New York: Cambridge University Press.

ROSE J (2000), 'Ultrasonic guided waves for anomaly detection in aircraft components,' *Mater. Eval.*, **50**(9), 1080–1086.

ROSE J (2002a), 'A baseline and vision of ultrasonic guided wave inspection potential,' *J. Pressure Vessel Technol.* **124**, 273–282.

ROSE J (2002b), 'Standing on the shoulders of giants – an example of guided wave inspection,' *Mater. Eval.*, **60**, 53–59.

SOHN H, FARRAR C R, HEMEZ F M, SHUNK D D, STINEMATES D W, NADLER B R (2003), 'A review of structural health monitoring literature: 1996–2001,' *Technical Report, Los Alamos National Laboratory*, LA-13976-MS.

SPANNER J (1974), *Acoustic emission: techniques and applications*. Evanston: Intex Publishing Company.

SPANNER J C, BROWN A, HAY D R, MUSTAFA V, NOTVEST K, POLLOCK A (1987), 'Fundamentals of Acoustic Emission Testing,' In P. McIntire and R. K. Miller, eds. *Nondestructive testing handbook, second edition: volume 5, acoustic emission testing*, Columbus, Ohio: American Society for Nondestructive Testing, 11–44.

STEPHENS R, POLLOCK A (1971), 'Waveforms and frequency spectra of acoustic emission,' *J. Acous. Soc. Amer.*, **50**(3), 904–910.

YUYAMA S, YAMADA M, SEKINE K, KITSUKAWA S (2007), 'Verification of acoustic emission testing of floor conditions in above ground tanks by comparison of acoustic emission data and floor san testing,' *Mater. Eval.*, 929–934.

YUYAMA S, KISHI T, HISAMATSU Y (1984), 'Fundamental aspects of AE monitoring on corrosion fatigue processes in austenitic stainless steel,' *J. Mater. Eng. Sys.*, **4**(4), 212–221.

8
Corrosion and fatigue modeling of aircraft structures

N. C. BELLINGER and M. LIAO,
Institute for Aerospace Research, Canada

Abstract: The development is described of a new physics-based life management framework that takes into account both cyclic and environmental effects and their interaction. The current design paradigms and their limitations in predicting the life of corroded components is reviewed. The processes employed to include different corrosion degradation modes (pitting, exfoliation) into the new methodology are discussed. Examples are provided that demonstrate the advantage of including corrosion effects (not just corrosion fatigue crack growth rate curves) into the sustainment phase of an aircraft.

Key words: aircraft life management, corrosion risk assessment, metal fatigue, pitting, exfoliation, Holistic Structural Integrity Process (HOLSIP).

8.1 Introduction

Traditionally, corrosion has been treated as a durability issue in aircraft structures with the primary objective being its prevention through the proper use of design materials, coatings, assembly and maintenance. When aircraft are designed and built, most manufacturers make little allowance for corrosion since aircraft are not expected to remain in service beyond their design service life (about 20 years). However, in the mid-1980s, it became very apparent that commercial and military aircraft would continue to operate well beyond their original projected service life, which in turn raised concerns regarding the possible interaction between environmental/age degradation and fatigue. In this chapter, the existing corrosion control 'find-it and fix-it' policy is summarized and the new 'anticipate and manage' maintenance approach, which is being proposed by a number of operators,

Note: The authors would like to thank our sponsors, including the National Research Council Canada, and collaborators for their support over the years as well as all the technical officers and students who have helped to carry out the various experiments that have been performed during the various projects.

is outlined. The present life philosophies will also be presented along with a newly proposed physics-based total life management approach. Some examples will be presented to demonstrate the advantage of including the effect of corrosion on estimating the remaining life of a component. This effect is included in the new life management approach known as the Holistic Structural Integrity Process (HOLSIP).

8.1.1 Present and future life methodologies

To date, there are two main design paradigms that have been used to determine the life of aircraft structures; the safe-life and damage tolerance approaches. In fact, there is a third paradigm known as the fail-safe approach but for this section it is being considered as a derivative of the damage tolerance approach since the difference is the presence of multiple load paths in the fail-safe design. The safe-life paradigm is currently used to design safety critical components, such as landing gears, in order to prevent cracks from forming during the life of the component. To avoid the formation of cracks, a large safety factor is applied to the fatigue (or endurance) limit of the material in order to take into account the inherent material scatter. In addition, these types of components are designed such that the operating stress would be below the 'fatigue limit' of the material. A problem with this methodology is that it assumes the material is free from 'defects' or discontinuities. Therefore, if any type of damage (such as scratches or corrosion pitting) is found it must be immediately replaced.

The damage tolerance methodology, on the other hand, assumes that all fatigue critical components contain a growing crack and that failure can occur when actual conditions are different to those modelled. In this methodology, the crack growth is calculated using the appropriate material properties, which then drives directed in-service inspections allowing undamaged components to remain in service reducing the cost of maintaining aircraft. However, once again, the effect that unexpected damage may have on crack nucleation and/or crack growth is not taken into account. Therefore, once unexpected damage is found, such as corrosion, the component must be repaired or replaced.

This failure of the two life prediction methodologies to adequately take into account complex damage scenarios has led to the present 'find-it and fix-it' maintenance philosophy, which in turn has significantly increased the cost of maintaining aging aircraft. It has been estimated that the United States Department of Defense spends between $10 billion and $20 billion annually (Under Secretary, 2005) in corrosion-related activities. These methodologies cannot adequately predict the life of components since neither takes into account the inherent discontinuities present in materials nor the environmental and age degradation modes that can occur during the life of an aircraft. To address these concerns a new life management methodology, known as the Holistic Structural Integrity Process (HOLSIP), is being developed by an international consortium with representatives from Canada, USA, UK, Japan and Australia to augment and enhance the traditional methodologies and to provide a powerful design and sustainment tool. As a design tool, HOLSIP is capable of determining the life cycle of a structure from the early stages of damage formation to final failure by enclosing material selection, structural design, manufacturing procedures, and inspection and maintenance requirements. As a sustainment tool, HOLSIP can be applied at any stage during the service life of an aircraft when damage is detected or fleet and operational variations are encountered. HOLSIP includes both cyclic and environmental effects in estimating structural life and residual strength capability, accounting for time in operation as well as time at rest.

8.1.2 Holistic Structural Integrity Process

Both the safe-life and damage tolerance methodologies were developed based on a combination of basic scientific principles that were available at that time as well as practical experience. As when they were first developed, these methods still rely on the generation of empirical material properties, such as the fatigue properties used for the strain life method or crack growth rate properties to determine the number of cycles, or flight hours, to a pre-specified small crack length or the time required to grow a crack to a critical size. This life estimation is then monitored throughout the life of an aircraft using specified nondestructive inspection (NDI) techniques in order to reduce the risk of failure for a specific component. This reliance on both analytical and experimental techniques to predict the life of aircraft components that do not take into account environmental and age degradation modes should no longer be allowed.

The Holistic Structural Integrity Process contains physics-based models to predict failures before an aircraft enters service or throughout the life of an aircraft. These models have been developed so that the causes, or reasons, behind potential failures are well understood (physics-of-failure). This process is being developed to determine the total life of a component by including all four phases of life: nucleation, short/small crack growth, long crack growth and unstable fracture. Presently, HOLSIP contains processes that take into account both cyclic and age/environmental effects by predicting the progression of initial discontinuity states (IDS) through the life of a component (Hoeppner, 1981 and Brooks and Simpson, 1998) as shown in Fig. 8.1. This is accomplished by accounting for the time in the air as well as the time while moving or at rest on the ground. The term 'IDS' is used to describe the as-produced or as-manufactured state of the material, which

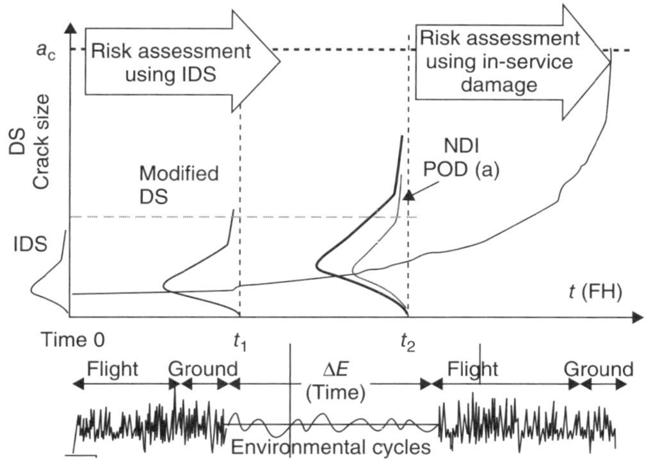

8.1 HOLSIP takes into account the progression of discontinuity states (cyclic operation spectra and time on ground spectra both with environmental time dependent effects).

involves the characterization of the state of a material. HOLSIP also includes the probability that damage is present and that it will be subsequently detected. The reliability of an airframe relating to its capability to perform its intended function is also included.

HOLSIP augments and enhances traditional safe-life and damage tolerance design methodologies to provide a powerful design and sustainment tool. The ultimate goal of this process is to ascertain the basic fatigue response of a structure from the as-manufactured state, including the early stages of damage formation. It is also being developed to assess the criticality of structures subjected to cyclic loading as well as age and environmental degradation, such as fretting, corrosion pitting, general corrosion and corrosion pillowing. It is the intent of this process that these assessments be applied to the design stage to optimize component design, predict future problem areas and minimize maintenance requirements.

8.2 Corrosion pitting and fatigue modeling

Corrosion pitting is a localized type of attack that leads to the formation of deep and narrow cavities.

8.2.1 Fundamentals of modeling

The effect of environmental degradation and fatigue, which can act not only concurrently but also independently, is characterized within HOLSIP by

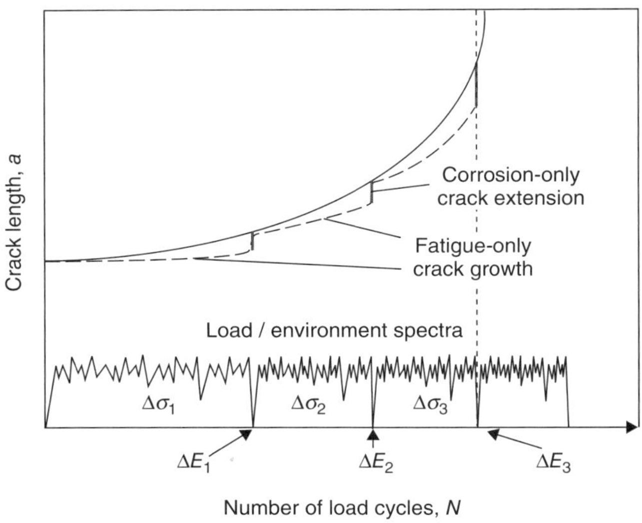

8.2 Corrosion fatigue analysis procedure in HOLSIP.

changes in the crack tip stress intensity. Fracture mechanics principles and special environmental parameters are used to formulate the required stress intensity factor solutions that are used to calculate the crack growth increment (Liao *et al.*, 2002). Over the past few years, the National Research Council Canada (NRC) in conjunction with other organizations has carried out extensive research to develop special parameters to represent the effects of corrosion (corrosion-related thickness loss and corrosion-pillowing stress) and corrosion fatigue interaction (local geometry stress risers, topography change and corrosion-induced sustained stress) on the stress intensity factor. In HOLSIP the total crack growth is determined through the summation of the cyclic damage and time-dependent damage as shown in Fig. 8.2, which is used to grow an initial discontinuity state to final instability (Brooks *et al.*, 1999).

8.2.2 Full-scale fatigue testing and corrosion finding

A full-scale CF-18 wing test was carried out at NRC under an International Follow-on Structural Test Project supported by the Canadian Forces and the Royal Australian Air Force. During the test, a long fatigue crack of 46 mm (1.81 in.) was discovered on the right hand Al 7149-T73511 upper outboard longeron (UOL) of the centre fuselage at 2932 simulated flight hours, as shown in Fig. 8.3. In this test, the fuselage was used as a transition structure and was in fact a retired United States Navy (USN) F-18 aircraft that was in-service for ten years (1984–1994) (Rutledge *et al.*, 1996). Exten-

8.3 Fatigue crack on the right upper outboard longeron of CF-18 centre fuselage.

sive corrosion was reported on the UOL by USN during its service life, which was repaired and then recorded (Rutledge *et al.*, 1996). During the full-scale wing test, the component was discovered to contain a crack after a fairing and sealant were removed to perform a scheduled life enhancement procedure. Corrosion pits were found in the area near the apparent crack nucleation site. Since this test was carried out in an environmentally controlled facility, it was reasonable to assume that these pits formed in-service and went undetected after the last maintenance cycle before the aircraft was retired. This cracking scenario will be used to demonstrate the capability of the three life prediction methodologies; safe-life, damage tolerance and HOLSIP.

8.2.3 Safe-life and durability analysis

A safe-life analysis was carried out on the UOL using the strain-life based 'crack initiation' program, CI89, which was tailored specifically for use on the CF-18 by L3 Communications Ltd. (formerly part of Bombardier Aerospace), who are responsible for maintaining the aircraft. In this analysis, the time to a crack length of 0.254 mm (0.01 in.) was calculated to be 6695 h, which is the data point mark A in Fig. 8.4. In addition, the crack growth data obtained from a full-scale CF-18 fuselage test that was carried out at L3 Communications Ltd., which did not contain any corrosion damage is marked B in Fig. 8.4.

The damage tolerance methodology was then used to carry out a durability analysis on the longeron. This type of analysis is used to obtain a quantitative measure of a structure's resistance to fatigue cracking under specified service conditions but does not include environmental degradation. In this analysis, an initial crack length of 0.254 mm (0.01 in.) was assumed to be present at the critical location, which was then grown using fracture mechanics principles until failure to determine the life of the component. To perform this analysis, the crack growth database for Al 7149-T73511 available in the fracture mechanics program, NASGRO, was used as input to the United States Air Force (USAF) crack growth program, AFGROW, to grow the 0.254 mm (0.01 in.) crack to failure. The result from this analysis is marked C in Fig. 8.4. As can be seen from the result, the durability analysis significantly underestimated the life (conservative) of the longeron when corrosion was not present (labelled B in Fig. 8.4) while it significantly overestimated the life (non-conservative) for the corroded upper outboard longeron (labelled D in Fig. 8.4).

To estimate the total life of this component, the time to reach a crack length of 0.254 mm obtained from the strain-life analysis was added to the long crack growth curve calculated from the durability analysis. The result from this analysis is labelled E in Fig. 8.4 and shows that although the total life of the longeron was overestimated (non-conservative), it was closer to the life of the undamaged component that was obtained from the full-scale centre fuselage test.

8.2.4 Damage tolerance analysis

To carry out a damage tolerance analysis on the longeron, an initial crack size of 1.27 mm (0.05 in.) was assumed to be present at the critical location,

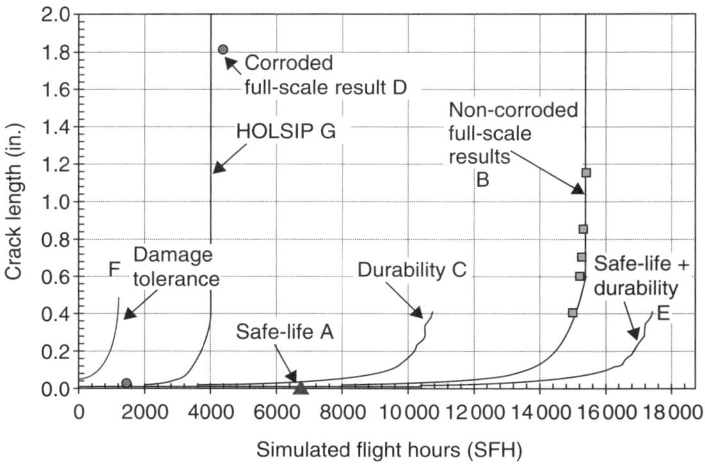

8.4 Life predictions for the upper outboard longeron using various methods.

which is the size of crack used to perform some damage tolerance analyses. The crack was then grown using AFGROW and the crack growth database obtained from NASGRO, the results of which are labelled F in Fig. 8.4. As can be seen from this figure, this analysis clearly underestimates the life of the component, as it does not include environmental effects.

8.2.5 Holistic structural integrity process analysis

To carry out a holistic structural integrity process (HOLSIP) analysis a number of parameters need to be determined for the particular material being studied: in this case 7149-T73511 aluminum extrusion as well as the crack geometry and crack growth rate data. It should be pointed out that research is continuing in order to expand the current IDS distributions (2024-T3, 7075-T6 and 7049-T6) and short (2024-T3 and 7050-T6) and long crack growth databases to include additional materials into the HOLSIP framework. To determine the cracking scenario, a replica was taken of the crack, which verified that it started from corrosion pits that were present at the round corner, Fig. 8.5. A corner crack model was then added to the HOLSIP framework to simulate this cracking scenario. The combined length of the pits along the thickness direction was approximately 0.63 mm (0.0248 in.) and the pit depth was 0.44 mm (0.0172 in.), as shown in Fig. 8.5.

The corner stone behind the HOLSIP framework is the determination of the initial discontinuity state (IDS) distribution for the particular material. To accomplish this, standard metallurgical procedures are carried out to obtain the size and shape of the intrinsic discontinuities present in the material. In addition, a limited number of cyclic tests are carried out to determine the fatigue critical discontinuities that result in the formation of cracks. However, owing to the unavailability of pristine material, it was decided to use existing fatigue life data to estimate the IDS value for this alloy at the critical location. Crack growth data from the full-scale CF-18 fuselage tests carried out at Bombardier Aerospace were used to estimate

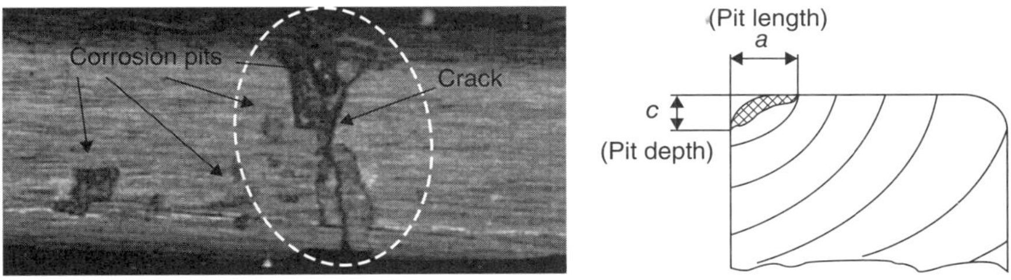

8.5 Corrosion pits and cracking scenario.

the IDS value in which a fatigue crack nucleated at the same location on the UOL, but no corrosion was present in this case. Revised crack growth rate data for this alloy, which were originally taken from the NASGRO material database and modified to include the small crack growth regime (Brooks *et al.*, 2001) to estimate an IDS value. This revised crack growth rate data was also used in the HOLSIP framework and the geometry for the IDS on the corner of the UOL was assumed to have a quarter circular shape. Using this crack growth data in conjunction with the spectrum loading that was used for the full-scale test, an IDS value was estimated and then employed to analyze the corrosion fatigue scenario discovered in the full-scale wing test.

To predict the corrosion growth rates, the cube root power law was included in the HOLSIP framework, which has been shown to effectively describe the pit growth in aluminum alloys (Godard, 1960), (Hoeppner, 1979), (Harlow and Wei, 1998). Based on the pit depth and the assumption that the corrosion had developed during the ten years of service in the USN, the corrosion pit growth curve was established. In addition, using data from the full-scale wing test, the loading spectrum in the area of the crack was determined and used in the analysis.

Using the estimated IDS, the analysis was carried out from the start of the service life (1984) to when the crack was found during the wing test. The results from the corrosion and fatigue analysis are labelled G in Fig. 8.4. As can be seen, the physics-based HOLSIP prediction gave closer results to the full-scale results than any other life prediction model did.

8.3 Exfoliation corrosion and modeling

In heavily rolled or extruded material, such as aluminum 7075-T6 and 2024-T3, where the grains are flattened and elongated in the direction of working, the presence of corrosion can lead to layering and flaking, producing a delamination effect with surface grains being pushed out by the underlying corrosion products (Wallace and Hoeppner, 1985). This is known as exfoliation corrosion, which is essentially a severe form of environmentally assisted cracking (intergranular corrosion) that occurs in the direction of grain flow and is normally associated with machined edges in thick sheets such as at rivet holes in wing skins.

8.3.1 Exfoliation and fatigue tests

The use of nondestructive inspection (NDI) techniques and search shot peening has offered increased detection capability for exfoliation. Since the current corrosion maintenance philosophy requires that even the smallest exfoliation damage be removed, the capabilities of different NDI techniques need to be determined. To accomplish this, various service-exposed 7178-T6 upper wing skins were inspected using different NDI techniques, including ultrasonics, x-ray, pulsed eddy current and thermography. Afterwards, sections were cut from these skins, mounted and progressively polished to characterize the exfoliation damage present; this was then compared with the NDI results. Neither x-ray nor pulsed eddy current techniques were able to detect the visible exfoliation that was present. However, both the ultrasonics and thermography techniques were capable of detecting not only exfoliation but also small levels of intergranular corrosion. The disadvantage to using the ultrasonic technique is the fact that due to the rough surface caused by the exfoliation, inspections have to be carried out from the undamaged (opposite) surface.

To determine the effect that exfoliation corrosion has on the fatigue life of upper wing skins, a specimen that was loosely based on the ASTM E647 standard was designed so that it could be machined from service-exposed Boeing 707 exfoliated 7178-T6 upper wing skins. The stiffener, which was located along the back-face of the upper wing skins, was removed leaving only a small portion of the flange around the fasteners to form a washer around the nut. Those holes that were not affected by the exfoliation were cold expanded and an interference fit steel plug was inserted into the hole to prevent premature failure.

All the tests were carried out under constant amplitude compression dominated cyclic loading, with a maximum gross section stress of 82.7 MPa (12 ksi) and a minimum gross section stress of 137 MPa (20 ksi). All the specimens were tested to failure in laboratory air at a frequency of 10 Hz and the results are presented in Fig. 8.6. The maximum depth of the

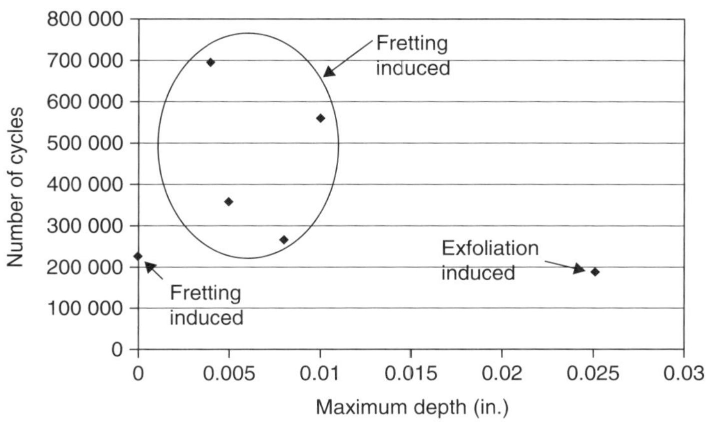

8.6 Results from fatigue tests carried out on coupons containing various levels of exfoliation.

exfoliation was measured using ultrasonic inspection techniques from the back face of the specimen. As can be seen from this figure the majority of the specimens failed due to fretting fatigue. Although exfoliation was present around the holes where failures occurred, the primary nucleation site for the majority of the specimens was a small pit, or pits, that were present along the countersunk hole (Bellinger *et al.*, 2002, 2003). The results from these tests suggested that given the level of damage examined, exfoliation did not have a detrimental effect on the fatigue life of this material. Since the type of loading used in these tests was not representative of the load applied to an upper wing skin, additional tests were carried out using a truncated upper wing skin spectrum. The specimen chosen contained a maximum exfoliation damage of 29%, which successfully completed 20 passes of the truncated spectrum representing two lifetimes with no observable cracking. After completion of this test, the specimen underwent a simulated repair, in which the exfoliation was ground out in accordance with repair guidelines and then re-tested. Once again, the specimen successfully completed 20 passes of the spectrum (two lifetimes) without any evidence of cracking. Thus, this specimen successfully completed four lifetimes without cracking, which raises some questions into the reasoning behind the requirement to grind-out low levels of exfoliation when a high level of exfoliation did not cause failure. It may be more cost effective to apply corrosion inhibitor compounds to slow or stop the growth of the exfoliation damage, thus delaying the maintenance procedure.

8.3.2 Exfoliation modeling

To simulate exfoliation damage, a 'soft inclusion' technique, which was previously developed and verified to simulate compressions after impact in composite structures (Xiong and Poon, 1992; Bellinger *et al.*, 1992), was applied to simulate the exfoliation present in a specimen. In this technique, the exfoliation damage is simulated by reducing the Young's Modulus of those elements in the affected area. To simplify the model generation, a technique was developed to automatically generate the 3D geometry of the soft inclusion (damage zone) from the NDI input using the PATRAN Command Language, Fig. 8.7. In addition, the non-test holes were merged with the steel plugs and the fastener in the test hole was merged with the flange. To simulate the contact between the various bodies, the Coulomb friction model with a coefficient of friction of 0.2 was used, as shown in Fig. 8.8. Models were created to simulate different exfoliated specimens and analyzed to determine the stress increase caused by the damage that was present. The results showed that the increase in the maximum principal stress was less than 5% for even the most severe exfoliation.

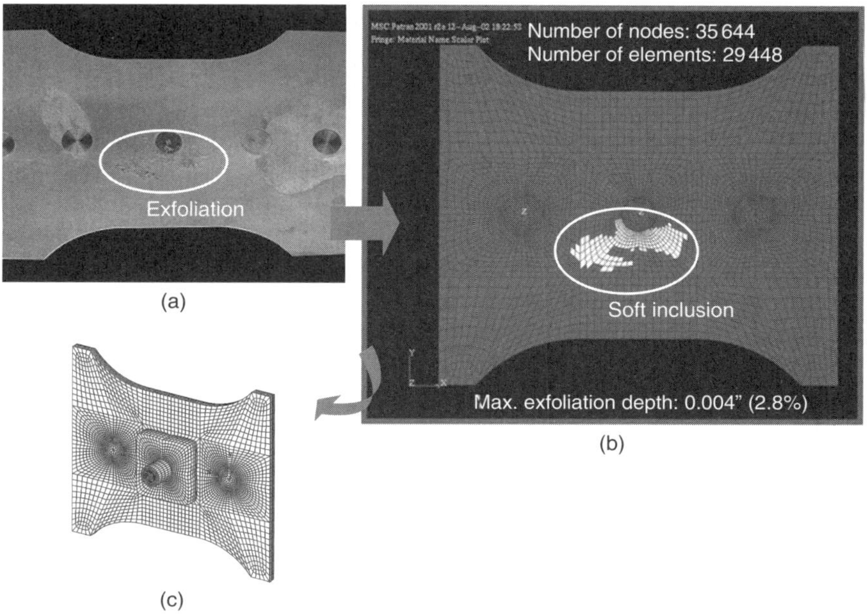

8.7 Typical three-dimensional finite element mesh with soft inclusion.

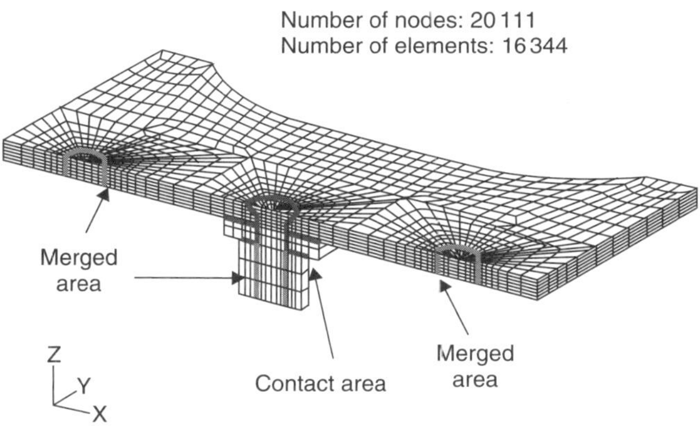

8.8 Boundary conditions for finite element analysis.

8.4 Residual fatigue life analysis

Fractographic analyses that were carried out on exfoliated specimens revealed several mechanisms, such as pitting and fretting fatigue, had caused cracks to form, which grew to failure. As mentioned earlier, the HOLSIP framework already contains models to take into account the effect that pitting has on the life of a component. However, owing to a lack of information such as service loading and service environment, these models could not be used to predict the time required to form a pit-like discontinuity in the upper wing skins. Therefore, only a crack growth analysis was carried out to predict the life of a specimen using the 3D finite element (FE) analysis results and the crack growth rate program, AFGROW (Liao *et al.*, 2003a). Since the current version of AFGROW did not contain a crack model at a countersink hole filled with a fastener, the crack model at a straight open hole associated with corrected beta factors was used for this analysis. To estimate these beta factors, 2D FE analyses were carried out to obtain the stress distribution along the assumed crack path for both a straight open hole and the countersunk hole with a fastener. Then the normalized stress values were calculated by dividing the ratio of the stress distribution from the countersunk hole to the straight hole model at the crack origin to the ratios along the assumed crack path. The results indicated that the predicted life was more than 50% less than the test results, which may be attributed to the fact that the nucleation phase was not included.

In another simplified fatigue model, the exfoliation damage was assumed to be a surface crack with a depth that was determined by NDI or a grindout database (Liao *et al.*, 2007). In one simulation, the exfoliation depth was estimated from a grindout database with an assumed lognormal distribution. Using this along with an effective surface crack aspect ratio, a Monte Carlo AFGROW analysis was carried out to calculate the life distributions, the results of which are shown in Fig. 8.9. If required, safety factors or risk analyses could be carried out using these life predictions, which are not available from the traditional fatigue or damage tolerance analysis without corrosion considerations.

8.5 Risk assessment of corrosion maintenance actions

A NRC in-house program, ProDTA (Probabilistic Damage Tolerance Analysis), has been developed to calculate the probability of failure (POF) of aircraft structures with both fatigue damage and environment-related time-dependent degradation modes, Fig. 8.10. The methodologies used by ProDTA were based on the HOLSIP framework and enhanced probabilistic techniques. The enhanced probabilistic techniques, which include prob-

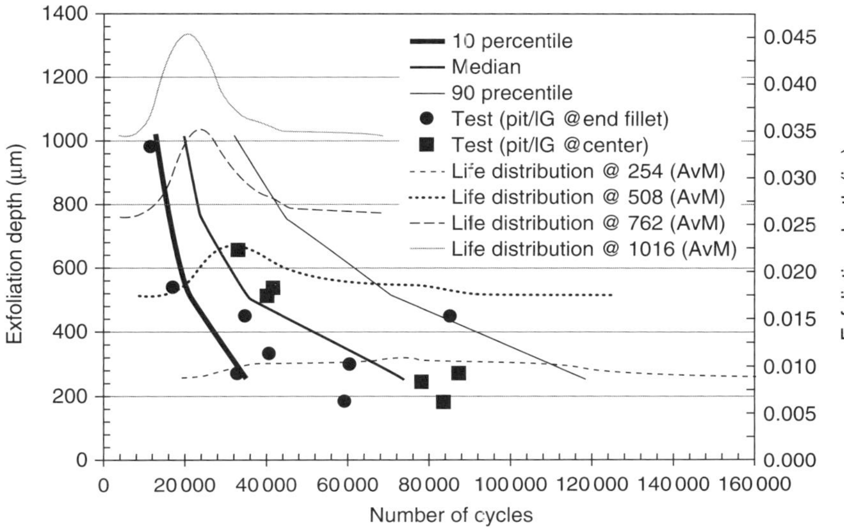

8.9 Probabilistic modeling based on grindout database.

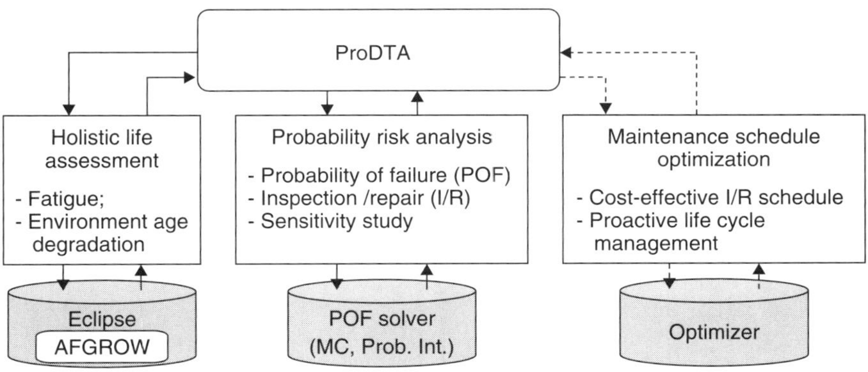

8.10 Major elements of the NRC in-house program, ProDTA.

ability integration and Monte Carlo simulation, were developed to incorporate environment-related random variables, such as corrosion growth rate (CGR) and pit depth (PD), in the POF calculation. In addition, nondestructive inspection (NDI) results for corrosion were incorporated into ProDTA in order to quantify the effect of NDI on the probability of failure. ProDTA was first evaluated and verified with existing codes in several risk analysis examples on non-corroded structures. ProDTA was then applied to risk analyses on a typical Boeing lap joint containing fatigue and environmental damage. The effects of corrosion maintenance actions, such as the corrosion NDI, corrosion prevention compound (CPC) application, and grindout, on the damage state and failure risk of the structure were quantified using ProDTA.

8.5.1 Risk assessment methods and tools

ProDTA is included in the HOSLIP framework, in which fatigue and age degradation can act not only sequentially but also concurrently. In brief, HOLSIP applies fracture mechanics principles to determine the total crack growth through the summation of cyclic damage and age degradation (time-dependent), which are used to grow an IDS to final instability. For each crack growth integration interval, beta correction factors are determined or updated in the framework as a function of crack length, time, thickness loss, pit depth, surface topography and corrosion pillowing stress. These beta correction factors are passed between the different models to calculate the crack growth for the next integration interval.

Figure 8.11 schematically illustrates the probabilistic analysis procedures employed in ProDTA. The analysis starts from the IDS distribution for the as-produced and as-manufactured material. Two methods were developed in ProDTA to grow the IDS distribution to the next inspection/repair time. The first method is to project the crack size distribution based on a single Master crack growth curve, which is the same as the method used in the

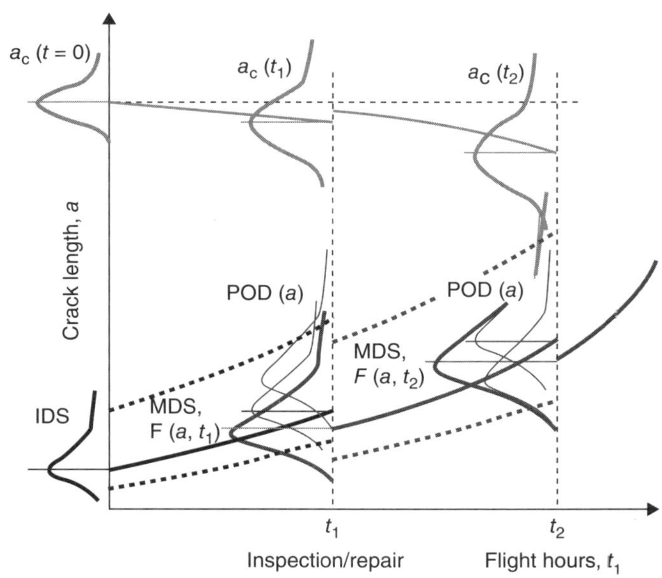

8.11 Analysis procedures employed in ProDTA.

USAF risk assessment program, PROF. The second method is to grow the crack size distribution using Monte Carlo analysis, which allows ProDTA to use more random variables especially age degradation parameters, such as corrosion growth rate, thickness loss, pit depth, and corrosion protection breakdown time.

The critical crack size distribution, on the other hand, is governed by all failure criteria, such as critical fracture toughness Kc, net section yield strength or critical residual strength, and function impairment limit. The POF at any given time is basically determined using the time-dependent stress-strength interference model. When inspection/repair actions take place at a given time, the crack size distribution would be modified based on a probability of detection function and a repaired crack size distribution. The modified crack size distribution is grown to the next inspection/repair time, which is repeated until the end of the life cycle.

8.5.2 Risk assessment of corrosion maintenance actions

Corrosion risk analyses were carried out based on an in-service aircraft, which undergoes Planned Depot Maintenance (PDM) every five years. During this time significant teardown, inspections and repairs are carried out. This fleet of aircraft is operated at a number of bases around the world with widely varying environments that can cause different degrees of corrosion damage to an aircraft. For these analyses, a fuselage lap joint was used which has two layers of Al 2024-T3 sheet, 7075-T6 stringer with three rows of flush-head rivets. Figure 8.12 presents the probabilistic damage

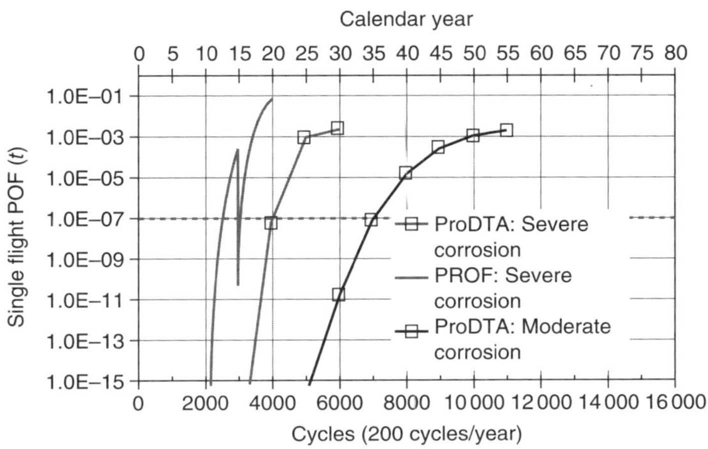

8.12 Single flight probability of failure showing effect of corrosion severity.

tolerance and risk analysis results for moderate and severe corrosion cases, which are similar to corrosive environments at Atlantic City Airport and Hickman Air Force base Hawaii, respectively. The results indicate that the single flight POF(t) would reach the critical level, 10^{-7}, at the time just after year 20 for severe corrosion, and year 35 for moderate corrosion. The POF results from PROF for the severe corrosion case are also presented in this figure. As expected, PROF's results were higher than those of ProDTA at the same number of cycles, for the same corrosion condition, due to the larger scatter in the crack size distribution generated by the Master curve method.

More simulations were carried out to demonstrate the capability of ProDTA to quantify the effects of CPC and corrosion NDI on the damage state, risk level, and the resulting maintenance actions. Four types of corrosion maintenance action were assumed (in order of increasing severity and cost); 1) no repair, 2) CPC application, 3) grindout and CPC application, and 4) replacement. The CPC performance would be affected by the method of application such as wicking, and the condition of the lap joint such as wet or dry, pristine or corroded.

Figure 8.13 shows the effect that the different CPC performances had on the risk analysis. Figure 8.13a assumed that a CPC was first applied at year 20 (4000 cycles) in order to preserve the POF less than 10^{-7} at the next PDM. The results indicate that a medium performance CPC (50%) would not be able to maintain the acceptable POF at the next PDM. However, an excellent performance CPC (90%) would be able to keep the structure safe till the next PDM before costly corrosion maintenance has to be performed. Figure 8.13b assumed that a CPC was applied at each PDM and reduced the CGR by 25, 50, and 75%. The results indicate that a high performance CPC (75%) could extend the life by a factor of 2.12, a medium performance CPC (50%) by a factor of 1.37, and a low performance CPC (25%) by 1.16, while maintaining the POF less than 10^{-7}.

Figure 8.14 shows the effect that NDI uncertainty for corrosion has on the risk analysis. In this simulation, it was assumed that at year 30 a 16% average thickness loss was found in the first layer of the simulated lap joint by using two different NDI techniques. The uncertainty of corrosion NDI was modeled by a normal error distribution and the standard deviation for the first NDI technique was 0.0015″ and 0.00075″ for the second one. These two numbers were chosen because they are typical of commercial and laboratory NDI techniques (Liao *et al.*, 2003c). Based on the inputs provided by corrosion NDI, risk analyses were carried out using ProDTA, and the single flight POF results are presented in the left hand figure. The results indicate that the POF would reach 10^{-7} at about 47 years and 53 years by using the inputs from the first and second NDI techniques, respectively. In

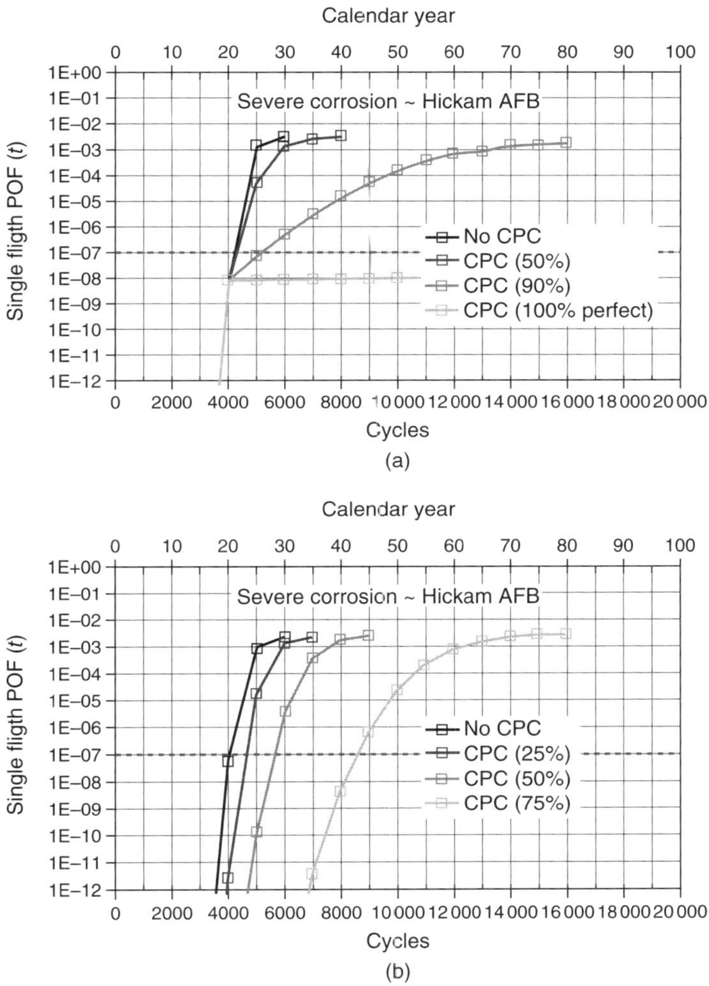

8.13 Effect of corrosion prevention compounds on single flight probability of failure (a) assuming CPC is applied at 20 years, and (b) assuming CPC applied after every PDM.

other words, given that two NDI techniques gave the same mean value, the one with a smaller error would result in a longer life extension.

8.6 Conclusions

The new physics-based life assessment framework, HOLSIP, has shown both the detrimental effects that corrosion can have on the life of a

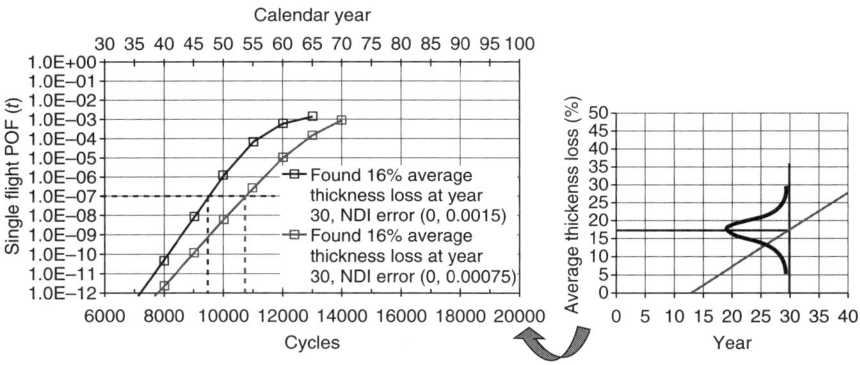

8.14 Effect of NDI uncertainty on the single flight probability of failure.

component as well as the benign effect that some types of corrosion may have. This contradiction in the effects that corrosion may have on the life of a component has raised concerns into the justifications that were used to determine when corrosion must be repaired. This new life management framework enables more accurate life predictions to be calculated resulting in a significant decrease in maintenance costs.

8.7 References

BELLINGER NC, FOLAND T and CARMODY D (2003), 'Structural integrity impacts of aircraft upper wing exfoliation corrosion and repair configurations', *Proceedings of the seventh joint DoD/FAA/NASA conference on aging aircraft*, September, New Orleans, Louisiana.

BELLINGER NC, KOMOROWSKI JP, LIAO M, CARMODY D, FOLAND T and PEELER D (2002), 'Preliminary study into the effect of exfoliation corrosion on aircraft structural integrity', *Proceedings of the 6th joint FAA/DoD/NASA conference on aging aircraft*, September, San Francisco, California.

BELLINGER NC, XIONG Y and POON C (1992), 'Determination of finite width correction factors for composite laminates containing impact damage' *Institute for Aerospace Research Report*, LTR-ST-1889.

BROOKS CL, PROST-DOMASKY S and HONEYCUTT K (1999), 'Correlation of life prediction methods with corrosion-related tests', *Proceedings of the 1999 USAF ASIP Conference*, San Antonio, Texas.

BROOKS CL, PROST-DOMASKY S and HONEYCUTT K (2001), 'Monitoring the robustness of corrosion and fatigue prediction models', *Proceedings of the 2001 USAF ASIP Conference*, San Antonio, Texas.

BROOKS CL and SIMPSON D (1998), 'Integrating real time age degradation into the structural integrity process', *Proceedings of NATO RTO workshop on fatigue in the presence of corrosion*, RTO-MP-18, Corfu, Greece.

GODARD HP (1960), 'The corrosion behavior of aluminum in natural waters', *Canadian Journal of Chemical Engineering*, **38**, 167–173.

HARLOW DG and WEI RP (1998), 'A probability model for the growth of corrosion pits in aluminum alloys induced by constituent particles', *Engineering Fracture Mechanics*, **59**(3), 305–325.

HOEPPNER DW (1979), 'Model for prediction of fatigue lives based upon a pitting corrosion fatigue process', *Proceedings of an ASTM-NBS-NSF Symposium, Fatigue Mechanisms*, JT Fong, Ed., ASTM STP 675, American Society for Testing and Materials, 1979, pp. 841–870.

HOEPPNER DW (1981), 'Estimation of component life by application of fatigue crack growth threshold knowledge', *Fatigue and Creep of Pressure Vessels for Elevated Temperature Service, Proceedings of Winter Annual Meeting of ASME*, edited by C. W. Lawton, R. and R. R. Seeley, ASME, Washington, DC, pp. 1–85.

LIAO M, BELLINGER NC and KOMOROWSKI JP (2003a), 'Modeling the effects of prior exfoliation corrosion on fatigue life of aircraft wing skins', *International Journal of Fatigue*, **25**(9–11), 1059–1067.

LIAO M, BELLINGER NC, KOMOROWSKI JP, RUTLEDGE R and HISCOCKS R (2002), 'Corrosion Fatigue Prediction Using Holistic Life Assessment Methodology', *Proceedings of the 8th International Fatigue Congress, FATIGUE 2002*, Stockholm, Sweden, June 2002, Vol. 1, pp. 691–700.

LIAO M, FORYSTH DS, KOMOROWSKI JP, SAFIZADEH M, LIU Z and BELLINGER NC (2003c), 'Risk analysis of corrosion maintenance actions in aircraft structures', *Proceedings of the 22nd Symposium of the International Committee on Aeronautical Fatigue (ICAF 2003)*, Lucerne, Switzerland, 5–9 May 2003.

LIAO M, RENAUD G and BELLINGER NC (2007), 'Fatigue modeling for aircraft structures containing natural exfoliation corrosion', *International Journal of Fatigue*, **29**(4), 677–686.

RUTLEDGE RS, SULLIVAN AB and LAFLEUR LJ (1996), 'FT245 fatigue test investigation of SN 162829 for use as a transition structure', *Memorandum* LM-ST-780, National Research Council Canada.

UNDER SECRETARY OF DEFENSE (ACQUISITION, TECHNOLOGY and LOGISTICS) (2005), 'Status Update on Efforts to Reduce Corrosion and the Effects of Corrosion on the Military Equipment and Infrastructure of the Department of Defense', www.dodcorrosionexchange.org.

WALLACE W and HOEPPNER DW (1985), *AGARD Corrosion Handbook Volume 1 Aircraft Corrosion: Causes and Case Histories*, AGARD-AG-278 Volume 1.

XIONG Y and POON C (1992), 'Failure prediction of composite laminates containing impact damage', LTR-ST-1898, Institute for Aerospace Research, National Research Council Canada.

Part III

Corrosion protection and prevention

9
Corrosion control in space launch vehicles

L. M. CALLE, NASA, Kennedy Space Center, USA

Abstract: The John F. Kennedy Space Center on the east coast of central Florida in the United States is home to NASA's Launch Operations Center. The natural marine environment at the Kennedy Space Center (KSC) is one of the most corrosive in the continental United States. Corrosion control at KSC involves the flight hardware, ground support equipment, and facilities. A description is given of the space launch vehicles' environment and the process of materials selection for this environment. The corrosion control necessary for the Space Shuttle Orbiter and its ground support equipment is described. There is also a brief history of orbiter corrosion.

Key words: corrosion control, space shuttle orbiter, ground support equipment, orbiter corrosion, NASA's Launch Operations Center.

9.1 Introduction

The east coast of central Florida in the United States is home to NASA's Launch Operations Center. Since its establishment in July 1962, the spaceport has served as the departure gate for every American manned mission and hundreds of advanced scientific spacecraft. The center was renamed the John F. Kennedy Space Center in late 1963 to honor the president who put America on the path to the moon. In 1966, during the Gemini/Apollo Programs, NASA established a beachside atmospheric exposure facility near the launch pads at the Kennedy Space Center (KSC) to perform corrosion studies. The site was originally used for the evaluation of long-term corrosion protection performance of carbon steel coatings. In the years that followed, numerous studies at the site identified materials, coatings, and maintenance procedures for launch pad structures (hardware), ground support equipment, and facilities exposed to the highly corrosive environment at KSC. In 1968, KSC's Materials Science Laboratories began testing and evaluation of the corrosion behavior and corrosion protective properties of different materials in the launch environment. The Corrosion Laboratory was established in 1985 to conduct applied research and perform materials characterization in different corrosive environments. More recently, the lab became the NASA Corrosion Technology Laboratory (CTL).[1] The NASA CTL is a network of capabilities, people, equipment, and facilities that provide technical innovations and engineering services in all areas of corrosion for NASA and external customers.

Corrosion control at KSC involves the flight hardware, ground support equipment, and facilities. This chapter begins with a description of the space launch vehicle environment and materials selection for this environment, followed by a description of corrosion control for the orbiter and its ground support equipment. In conclusion, a brief history of Space Shuttle Orbiter corrosion is presented.

9.2 Space launch vehicles environment

The natural marine environment at KSC is one of the most corrosive in the continental United States. The two Space Shuttle launch pads, 39 A and 39 B, are located less than 1.6 km (a mile) from the Atlantic Ocean and are exposed daily to salt spray and high humidity. The hydrochloric acid (HCl) and intense heat, generated during a launch, exacerbate the natural corrosive conditions and have a detrimental effect on the protective coatings, structures, and machinery at the pads. In this section, the natural marine environment at KSC, the launch pad environment, and the various environments to which space vehicles are exposed during their processing at KSC are described.[2]

9.2.1 Corrosivity of the natural marine environment at the Kennedy Space Center

The Kennedy Space Center (KSC) is located on a barrier island and covers an area about 55 km long and varies in width from 8 to 16 km. The total land and water area is just over 56 656 hectare, however, only 2428 hectare are actually used for Space Shuttle operations. KSC is also a national wildlife refuge.[3] A wildlife refuge and a space center can share the same property, since both require a lot of land and minimal urban development.

The natural marine environment at KSC has been documented by the American Society for Metals (ASM) as having the highest corrosion rate of any site in the continental United States (Table 9.1).[4] Structures and ground support equipment at KSC weather or degrade at rates far in excess of those experienced at other locations around the country, or even the world, due to their exposure to the naturally occurring oceanfront environment combined with the intense Florida heat, ultraviolet radiation, rainfall, and humidity. Figure 9.1 shows a variation in the corrosion rate (measured as weight loss) and salt content of the atmosphere as a function of distance from the Atlantic Ocean. The data points on the figure are labeled from

Table 9.1 Corrosion rates of carbon steel calibrating specimens at various locations

Location	Type of environment	Corrosion rate* ($\mu m\ yr^{-1}$)	Corrosion rate* ($mil\ yr^{-1}$)
Esquimalt, Vancouver Island, BC, Canada	Rural marine	13	0.5
Pittsburgh, PA	Industrial	30	1.2
Cleveland, OH	Industrial	38	1.5
Limon Bay, Panama, CZ†	Tropical marine	61	2.4
East Chicago, IL	Industrial	84	3.3
Brazos River, TX	Industrial marine	94	3.7
Daytona Beach, FL	Marine	295	11.6
Point Reyes, CA	Marine	500	19.7
Kure Beach, NC (24 m from ocean)	Marine	533	21.0
Galeta Point Beach, Panama CZ†	Marine	686	27.0
Kennedy Space Center, FL (beach)	**Marine**	**1070**	**42.0**

* Two-year average.
† Canal Zone.

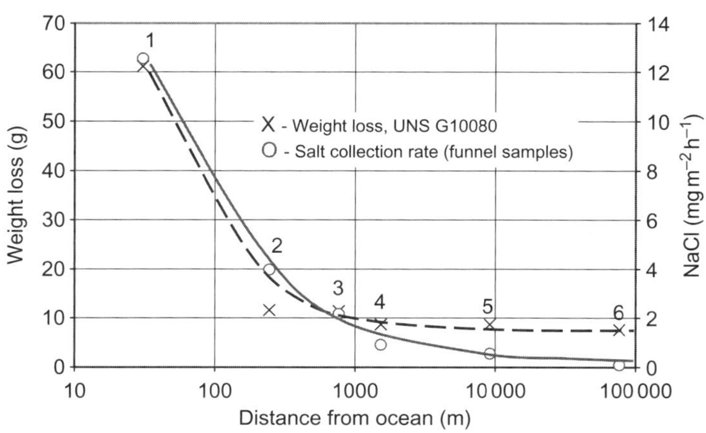

9.1 Corrosion rate (measured as weight loss) and salt content of the atmosphere as a function of distance from the Atlantic Ocean. The KSC atmospheric exposure site is at point 1. Launch complex 41 is at point 2. Complex 39 A (Space Shuttle Launch Site) is at point 3. Points 4 and 5 represent other locations at KSC. Point 6 corresponds to an urban area of Orlando, FL.

one (KSC Beachside atmospheric exposure facility) to six (Orlando, Florida, urban area). Points two to five correspond to the entrance to launch complex 41 at KSC, perimeter fence at launch complex 39 A, a pump station near the launch complex 39 A, and the KSC urban area.

9.2.2 Corrosivity of the environment at the Space Shuttle launch pads

With the introduction of the Space Shuttle in 1981, the already highly corrosive natural conditions at the launch pads were rendered even more severe by the acidic exhaust from the solid rocket boosters (SRBs). The two SRBs emit more than 91 tonne (100 tons) of exhaust during the first 10 s of launch, including 28 000 kg (61 729 lb) of aluminum oxide and 17 000 kg (37 478 lb) of HCl. During each lift off, a cloud of HCl and powdery aluminum oxide dust falls over a 2.6 km (1 square mile) area surrounding the pad. Without regular maintenance that includes high-performance coatings, corrosion-resistant metals and non-metals, the Shuttle launch pads could become highly unreliable or unsafe within several years. Serious structural damage requiring partial pad replacement could occur within 5 years if left unprotected.

9.2.3 Orbiter flight and ground environment

The Space Shuttle experiences corrosion environments from benign to severe as it progresses from flight to flight. The relative severity of each environment is illustrated in Fig. 9.2. At KSC, the orbiters are stored in the Orbiter Processing Facility (OPF) under temperature and humidity control. OPF temperature is maintained at 21.1 ± 2.8 °C. Relative humidity is 50% maximum. The OPF is the prime facility for orbiter processing: preflight

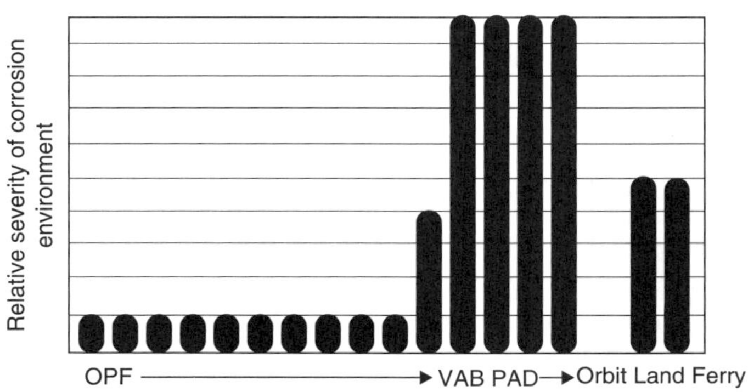

9.2 Variation of orbiter corrosion environment during typical flow.

and post landing. Access is provided to all external and limited internal surfaces of the orbiter to perform the following operations: draining and purging all fuel systems, ordinance removal, repair and replacement of damaged components, inspection and refurbishment of the Thermal Protection System (TPS), inspection and testing of orbiter systems (landing gear, main and auxiliary propulsion, power units, flight instrumentation, communications and orbiter hydraulics), payload bay (configuration and testing), and payload installation, connection, and removal. The orbiter is towed a short distance from the OPF to the adjacent Vehicle Assembly Building (VAB) where it is mated to the external tank (ET), which is already mated with the SRBs on the mobile launcher platform (MLP). Unlike the OPF, the VAB is not under temperature and humidity control but it offers protection from wind, rain, salt spray, and sunlight. When moved to the launch pad, the orbiter is subjected to an almost constant salt spray from the nearby Atlantic Ocean. The high humidity allows the formation of condensation on all surfaces open to the atmosphere. Once the orbiter reaches low earth orbit, any water that may have collected during earlier exposure evaporates in the vacuum of space. Corrosion is not a concern in space. However, the environment of space can have detrimental effects on materials due to the presence of high energy protons and electrons, ultraviolet radiation, atomic oxygen, high and low temperature extremes, high vacuum, galactic cosmic radiation, micro meteors, and man-made debris. These effects must be considered in selecting materials and corrosion protection finishes. As part of NASA's Materials International Space Station Experiment (MISSE), hundreds of samples representative of a variety of materials, including coatings, have been attached to the exterior of the International Space Station for long-term periods of exposure and brought back to earth for evaluation.[5] On landing, the orbiter is then exposed to the arbitrary environment of the landing site. If the orbiter lands in California, it must be ferried across the country atop the Shuttle Carrier Aircraft (SCA) to KSC. Until the orbiter returns to the OPF, this is additional exposure to an uncontrolled environment. All these factors were taken into account when the overall corrosion protection scheme for the orbiter was developed.

The Space Transportation System (STS) consists of the orbiter spacecraft, two SRBs for launch, and an ET for the liquid hydrogen and oxygen fuels which feed the three main engines housed in the orbiter aft fuselage. STS processing includes a scheduled period at the pad for payload installation, final servicing, and checkout before launch. The average pad stay is approximately 31 days. On occasion, problems with the flight hardware have extended the time normally scheduled at the pad. The longest time an STS has spent at the pad was 166 days before Space Shuttle Columbia's tenth flight. Extended stays are of concern because the pad is located a few hundred meters from the ocean. The coastal exposure is very severe due the heat, high humidity, salt air, and the daily condensation of dew deposited onto the orbiter structure.

9.3 Materials selection

The Space Shuttle Orbiter is the world's first reusable spacecraft as well as the first spacecraft that can carry large satellites to and from earth orbit. The Space Shuttle Orbiter launches like a rocket, maneuvers in earth orbit like a spacecraft and lands like an airplane. The casings of the SRBs are recovered from the Atlantic Ocean, refurbished, and reused. The ET is dropped on the Atlantic Ocean and is not recovered.

9.3.1 Space Shuttle Orbiter

Most of the orbiter structure is aluminum. The TPS protects the vehicle from the effects of aerodynamic re-entry heating by maintaining a maximum skin temperature of 350 °F. The TPS is reusable and was designed to last 100 missions without replacement. The three main engines in the aft are attached to a boron/epoxy composite reinforced titanium thrust structure which distributes thrust loads to the aluminum structure. The payload bay doors are constructed of graphite epoxy. With the exception of the TPS, developed for the Space Shuttle Program, the orbiter is essentially a state-of-the-art vehicle that combines a variety of aircraft and spacecraft technology.

9.3.2 Launch pad and ground support equipment

Since the late 1960s, Pads 39 A and 39 B at KSC's Launch Complex 39 have served as backdrops for America's most significant manned space flight endeavors – Apollo, Skylab, Apollo–Soyuz and Space Shuttle. The pads were originally built for the huge Apollo/Saturn V rockets that launched American astronauts on their historic journeys to the moon and back. Following the joint US–Soviet Apollo–Soyuz Test Project mission of July 1975, the pads were modified to support Space Shuttle operations.[6]

Both pads were designed to support the concept of mobile launch operations, in which space vehicles are assembled and checked out in the protected environment of the VAB and transported by large tracked vehicles to the launch pad for final processing and launch. During the Apollo era, key pad service structures were mobile. For the Space Shuttle, two permanent service structures were installed at each pad for the first time.

On April 12, 1981, Space Shuttle operations commenced at Pad 39 A with the launch of Columbia on STS-1 (one is the number of the mission). After

23 more successful launches from 39 A, the first Space Shuttle to lift off from Pad 39 B was the ill-fated Challenger in January 1986. Pad 39 B was designated for the resumption of Space Shuttle flights in September 1988, followed by the reactivation of Pad 39 A in January 1990.

Major features of pads 39 A and 39 B

Both pads are octagonally shaped and share identical features. Each pad covers about 0.6 km^2 (0.25 square mile) of land. Launches are conducted from atop a concrete hardstand 190 m by 99 m, located at the center of the pad area. The Pad 39 A and Pad 39 B hardstands are 15 m and 17 m above sea level, respectively.

Fixed service structure

The fixed service structure (FSS) is the pad's most prominent feature, standing 106 m from ground level to the tip of the lightning mast. The structure is made of carbon steel. The lightning mast is 24 m tall and made of fiberglass. The FSS is equipped with three swing arms which provide services or access to a Shuttle on the pad. They are retracted when not in use. There are 12 floors on the FSS positioned at 6-m intervals. The first is located 8 m above the pad surface.

Rotating service structure

The rotating service structure (RSS) provides protected access to the orbiter for installation and servicing of payloads at the pad, as well as servicing access to certain systems on the orbiter. The majority of payloads are installed in the vertical position at the pad, partly because of their design and partly because it allows payload processing to take place further along in the launch processing schedule. This structure is constructed of carbon steel.

The RSS is 31 m long, 15 m wide, and 40 m high. It is supported by a rotating bridge that pivots about a vertical axis on the west side of the pad's flame trench. The RSS rotates through 120° on a radius of 49 m. Its hinged column rests on the pad surface and is braced against the FSS. The RSS is retracted for launch.

The major feature of the RSS is the Payload Change out Room, an enclosed, environmentally controlled area that supports payload delivery and servicing at the pad and mates to the orbiter cargo bay for vertical payload installation. Clean-air purges help ensure that payloads being transferred from the payload canister into the Payload Change out Room are not exposed to the open air. The payload is removed from the canister, and later installed inside the orbiter cargo bay using the Payload Ground Handling Mechanism (PGHM). Five platforms are positioned at five levels to provide access to the payload when it is installed on the PGHM. Each platform has extendible planks that can be configured to accommodate a particular payload.

Flame trench and deflector system

The flame trench, built with concrete and refractory brick, bisects the pad at ground level. It is 149 m long, 18 m wide and 13 m deep. The flame deflector system includes an inverted, V-shaped steel structure covered with a high-temperature concrete material 12.7 cm thick that extends across the center of the flame trench. One side of the V receives and deflects the flames from the orbiter main engines, the opposite side the flames from the SRBs. There are also two movable deflectors at the top of the trench to provide additional protection to Shuttle hardware from the SRB flames.

Liquid oxygen and liquid hydrogen storage

Liquid oxygen (LOX), used as an oxidizer by the orbiter main engines, is stored in a 3.4 million liter tank on the pad's northwest corner, while the liquid hydrogen (LH2), used as a fuel, is kept in a 3.2 million liter tank on the northeast corner. The propellants are transferred from the storage tanks in vacuum-jacketed lines that feed into the orbiter and ET via the tail service masts on the MLP. These lines were originally made of 300 series stainless steel and had to be replaced in the late 1980s owing to numerous failures caused by pitting.

Shuttle era pad modifications

Since the startup of Space Shuttle operations, Pads 39 A and 39 B have undergone additional modification. Many of these changes were made during the stand-down in launch operations after the Challenger accident in 1986. Others have been made since the resumption of flights in September 1988.

New weather protection structures were installed and tested at Pad 39 B in 1986. The additions supplement the RSS, which shields most of the vehicle when it is on the pad. Without shielding, the orbiter's fragile heat protection tiles are subject to damage from hail and wind-blown debris, as well as from heavy rains which can erode the waterproofing on the tiles.

The protective hardware includes rolling doors which can be deployed between the orbiter's belly and the ET to shield the lower portions of the orbiter. The doors are supported by a structural framework on both the RSS and the FSS. They replaced a fabric drape system. Other weather protection modifications were designed to provide a roof seal and to enclose the space between the RSS and the ET. A fully assembled Space Shuttle was rolled out to Pad 39 B in the fall of 1986 to test the various additions and changes. The same weather protection modifications and additions were later made at Pad 39 A.

While no facilities or ground support equipment (GSE) failures were identified in the Challenger accident investigation, NASA did conduct an extensive in-depth system design review which resulted in a number of modifications considered mandatory for return to flight. Launch pad changes were aimed at assuring processing timelines and improving margins of safety, and eliminating critical single failure points in the system wherever possible.

In June of 1991, Pad 39 B was taken offline to undergo needed repairs and refurbishment. It had served as the takeoff point for 12 of the 16 Shuttle launches conducted in the September 1988 to June 1991 timeframe. About 50 modifications and repairs were made to pad structures and processing equipment and the environmental control system. The upgrades also included work on the RSS, hardware and electrical system improvements, and structural corrosion control. Improvements to the safety and overall efficiency of the pad were also made. The upgrades took about six months to complete. Pad 39 B was scheduled to again support launches starting with the maiden flight of the newest orbiter, Endeavour, which launched on STS-49 in the spring of 1992.

9.4 Corrosion control

Corrosion control of flight hardware, launch pad structures, GSE and facilities has been, and continues to be, a challenge in the highly corrosive marine and launch environment at KSC. For this reason, KSC is a major source of worldwide corrosion expertise with over 40 years of technical information on the long-term corrosion performance of many materials in its highly corrosive environment. This knowledge, which was not available when the launch pads were constructed, is critical for the design of future flight hardware and spaceports so that they can be operated more safely and require less maintenance between launches. The fact that many of the current corrosion control methods rely on materials that are no longer available, due to health considerations and new environmental regulations, is an important driver in the search for new materials and methods for corrosion control. The following sections describe the current methods used for corrosion control of flight hardware and GSE.

9.5 Corrosion control in the Space Shuttle Orbiter

Corrosion protection of the orbiter was based on the objective of having no structural failure due to corrosion within a 10-year or 100-mission life. A detailed material control plan addressing every material that would be utilized on the orbiter was implemented. The plan required that all metals meet what was termed an 'A' rating (materials that are not susceptible or highly resistant to corrosion and/or stress corrosion under normal operating conditions). Metallic materials were evaluated either by test or engineering judgment to meet NASA requirements. Metals were required to meet MSFC-SPEC-250,[7] class II requirements. The MSFC-SPEC-250 is the NASA general specification that describes the requirements for the protective finishes for space vehicle structures and associated flight equipment. The specification classifies the levels of protective finishing into three classes: class I includes finishes used for protection against severe corrosive environments such as exposure to seawater; class II includes finishes used for protection against moderately corrosive environments such as extended exposure to sea coast or industrial environments; class III includes finishes used for the protection against mildly corrosive environments such as inland, non-industrial environments. Metals not listed in the specification were subjected to a 1500-hour salt spray test. Metallic materials that were proposed for use but not 'A' rated by MSFC-SPEC-250 were evaluated by their use, location, or protection schemes and upgraded to 'A' status if possible.

MSFC-SPEC-522[8] is a guideline for determining the rating of material for stress corrosion cracking (SCC). For example, the requirement for aluminum alloys is freedom from cracking after 30 days alternating exposure in 1-h cycles of 10 min in a 3.5% sodium chloride solution followed by drying for 50 min, while stressed to 75% of the material's yield strength. However, where essential materials were used that did not meet the 'A' rating, the designer would consider, as a minimum, the following actions to reduce the probability of stress corrosion:

1. Selecting less susceptible alloys or tempers. In general, all aluminum used in structural applications is required to be conversion coated (MIL-C-5541)[9] or anodized (MIL-A-8625)[10] and coated with one coat of chromated epoxy primer. Primer coats are 0.01524 to 0.02286 mm thick.[11]
2. Reducing sustained stress levels on the part below stress corrosion threshold levels.
3. Protecting the part from the detrimental environment by hermetically sealing or coating the part or by inhibiting the environment.
4. Avoiding or reducing residual stresses in parts or assemblies by stress relieving, by avoiding interference fits, or by shimming assemblies.

5. Avoiding galvanic couples, which may tend to accelerate the stress corrosion.
6. Providing for regular inspection of parts to determine surface flaws and cracking during the life cycle of the part.
7. Improving the surface quality of the part by reducing surface roughness or increasing surface compressive stresses.

Additionally, the Standard Design Manual restricted the use of galvanically dissimilar metals by requiring that they not be used in contact unless suitably protected against electrolytic corrosion. Faying surfaces of dissimilar metals must be sealed against water intrusion or separated with a layer of corrosion-inhibiting epoxy or room-temperature vulcanized (RTV) silicone rubber. Dissimilar metals were considered compatible if they were in the same grouping as specified in MSFC-SPEC-250, or if the difference in potential was less than 0.25 V. Also imposed was the requirement that all fasteners be installed wet with epoxy. The epoxy of choice was a chromated primer known under the brand name Super Koropon. To address specific corrosion problems associated with aluminum alloys, additional restrictions were applied. Alloys susceptible to exfoliation were eliminated from design consideration as were all forms of alloys that exhibit stress corrosion thresholds of less than 172 MPa (25 ksi).

The orbiter TPS, bonded to the structural exterior, provides adequate corrosion protection under most circumstances. Exterior assemblies on the orbiter that are not covered by the TPS, rudder speed brake (RSB), wing leading edge (WLE) spar, and ET doors are primed with a chromated epoxy primer (Koropon) and, in some cases, also painted. Condensation can occur on these surfaces, and this may initiate corrosion activity when the primer/paint barrier is violated. These areas have shown repetitive corrosion problems.

Super Koropon primer (MB0125-055) plays a significant role in the corrosion protection of areas throughout the orbiter. One area where little understanding of the Koropon primer existed was in the level of risk associated with age related degradation. Recently, efforts were undertaken to better understand the useful life of the Koropon primer and to gain some insight into the aging process of this coating.[12] In this study, an aluminum access panel from the Orbiter Enterprise (the first Space Shuttle Orbiter) was used to investigate the performance of the old Koropon film. A control panel was also used to study the performance of newly applied Koropon coating. Preliminary investigations into the performance of aged Super Koropon primer indicated a significant decrease in corrosion protection. After only 500 h of salt fog testing (ASTM B117), corrosion was observed in the scribe line of the access panel from the Orbiter Enterprise. Corrosion was also observed away from the scribe after 1000 h of salt fog testing, while the freshly applied Koropon coating easily passed the 1500 h salt fog test with no corrosion.[13] As a result, subsequent testing was performed to determine the cause of the corrosion on the panel from the Orbiter Enterprise after salt fog testing. Further analysis of both the Enterprise Koropon and the freshly applied Koropon was performed to determine the distribution of chromium within the matrix of the film.

The control Koropon sample and the Enterprise Koropon sample were mounted on scanning electron microscopy (SEM) sample studs using silver paint. In an effort to minimize the possibility of charging during SEM analysis, both samples were further sputter coated with a gold/palladium film. A Quanta 3D Dual Beam focused ion beam/scanning electron microscope (FIB/SEM) was used to mill a cross section of the Koropon sample from the Enterprise and the freshly applied Koropon sample. SEM-x-ray energy dispersive spectrometry (XEDS) data was also gathered.

The back-scattered electron (BSE)-SEM images of the surface of the Koropon sample from the Orbiter Enterprise and the control Koropon sample are shown in Fig. 9.3. Even though the SEM image of the surface of the Enterprise sample is at a slightly higher magnification, the differences in the two surfaces are readily apparent. The surface of the Enterprise sample is very rough and 'flakey', whereas the surface of the control sample appears to be much smoother. The chromium is very visible on both surfaces. The chromium appears as light 'dots' on the surface of both samples. A higher magnification FIB image of the surface of the control Koropon sample is shown in Fig. 9.4a. The smooth surface can be seen in much greater detail in this image. A region containing chromium is also labeled

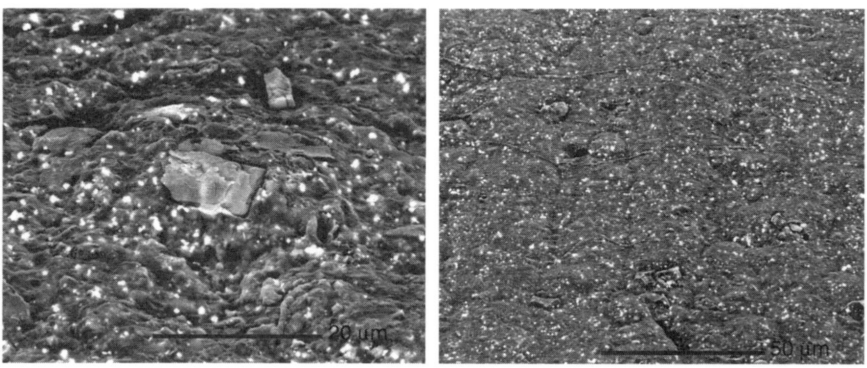

(a) (b)

9.3 (a) BSE-SEM image of the surface of the Koropon sample from the Orbiter Enterprise. (b) BSE-SEM image of the surface of the Koropon from the control sample.

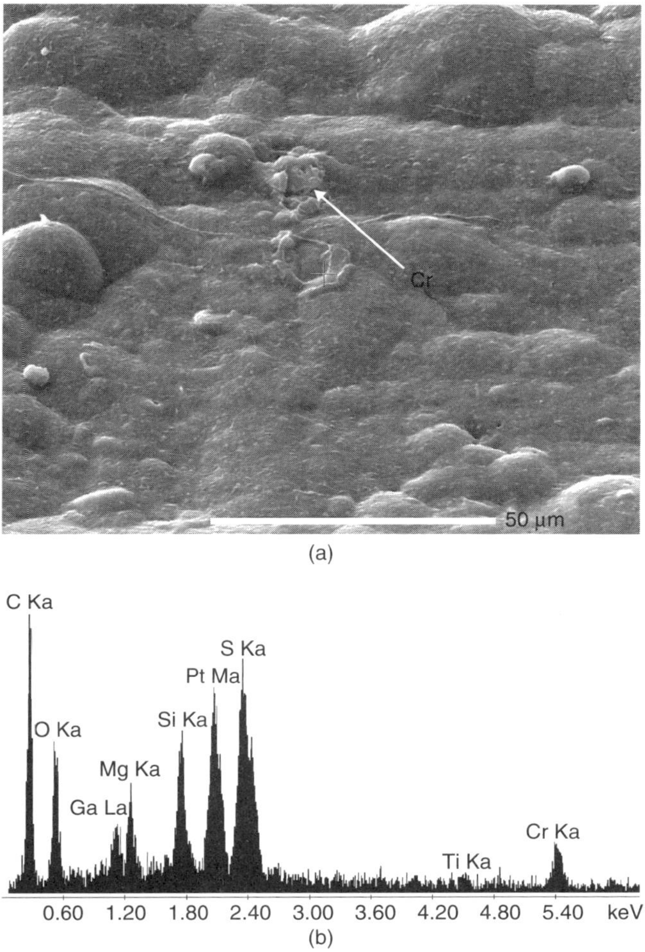

9.4 (a) FIB image of a region of the surface on the control Koropon sample. (b) SEM/XEDS of the chromium region of the surface of the same sample.

in the image and the SEM/XEDS spectrum obtained from this region is shown in Fig. 9.4b. The platinum in the spectrum may be attributed to the platinum that was deposited on the surface to protect the region of interest from damage during the milling process. The gallium may be attributed to the gallium ion beam used to image in FIB mode and also used to mill the SEM trench cut. The titanium and oxygen in the spectrum may be attributed to the paint used to mount the sample. The sulfur, silicon, magnesium, and carbon may be attributed to the epoxy resin of the Koropon matrix.

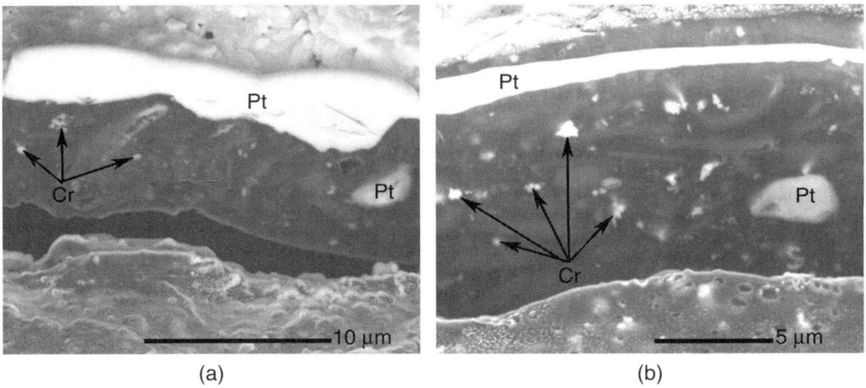

9.5 (a) BSE-SEM image of the Enterprise Koropon sample showing the chromium distribution throughout the matrix in this region. (b) BSE-SEM image of the control Koropon sample showing the chromium distribution throughout the matrix in this region.

The cross-section BSE-SEM images for the Enterprise Koropon and the control Koropon films are shown in Fig. 9.5. The platinum pad that was deposited to protect the surface of the region of interest from the gallium ion beam during milling is labeled in both images. There was also a void (labeled Pt in the bulk matrix) that filled with redeposited platinum during milling. Again, the chromium appears as light 'dots' and/or very light regions in both images. There are some regions of chromium just beneath the surface that appear as light gray areas. The overall particle size distribution varies widely, but some of the chromium particles were larger than expected. This may be due to some particle agglomeration.

The poor performance of the 30-year-old Koropon in a corrosive environment appeared to be directly related not only to a reduced level of chromium but to the uneven distribution of the chromium within the matrix of the film. In some regions of the film, the chromium level was minimal while in other areas an increased chromium level was observed. When comparing the Enterprise Koropon film to the freshly applied film, it appeared that the old Koropon film had a reduced level of chromium overall. There appeared to be a higher level of chromium in the control Koropon film that was more evenly distributed. The inconsistency in the thickness of the coating of the Enterprise Koropon and the specified Koropon thickness may be due to degradation of the polymer as a result of aging and/or the degradation as a result of chemical interaction between the polymer and stripper used to remove the coating from the access panel.

9.5.1 Water intrusion/entrapment design features

Drain system

The drain system of the orbiter consists of passive 'through hole' and ground 'vacuum line' systems. Drain holes were designed into the orbiter to prevent water from accumulating within the open orbiter structure. The drains were placed in selected areas of the structure so that drainage will occur in both horizontal and vertical orientations. Where passive gravity drainage cannot be accomplished, the active vacuum line system allows for water to be removed by ground-based vacuum pumps. Water accumulation is unacceptable from a corrosion standpoint, undesirable launch weight, and as a contaminant in space.

Purge system

The purge system consists of a series of onboard ducts that permit purging of un-pressurized compartments by ground-supplied air or nitrogen. Purges maintain positive pressure within the vehicle aiding in preventing the ingress of moisture. A dry nitrogen gas system was implemented to purge the interior spaces of the orbiter vehicle. The purpose of the purge is to maintain a dry environment by preventing condensation. The nitrogen purge is continuously operating while the vehicle is in the VAB and again, once the vehicle has been mounted on the MLP. To date, there have been no significant corrosion issues in any areas of the vehicle maintained with this purge.

9.5.2 Maintenance

Long-term planning for the maintenance and inspection of the orbiters was not given full consideration until after the orbiters were already constructed. As a result, many areas of the structure are difficult to access for corrosion inspection. Subsystem components, such as wire bundles, tubing, ducting, tanks, thrusters and line replacement units (LRUs) obstruct structure, some of which is significant. Borescopes are used extensively to inspect in locations where access is limited. A concern for inspecting the orbiter external surface for corrosion is that most of it is covered with TPS, which includes tiles, flexible insulation blankets, and reusable felt surface insulation. In many cases, the interior surface is accessible for inspections which allow for detection of through corrosion conditions. Tile removal, necessary for TPS servicing as well as a sampling plan to ensure TPS integrity, provides random access to some of the structure. However, the TPS limits the amount of structure available for inspection. The interior structure of the orbiter is often lined with thermal control system (TCS) blankets, which also obstruct access to the structure for the performance of corrosion inspections. Although these blankets can sometimes be removed or lifted to provide access in some areas, blanket removal would require destroying them and reinstallation would be impossible without structural disassembly. An example is the cavity between the forward fuselage and the Crew Module (CM) pressurized cabin. This area varies from approximately 0.9 m (3 ft) to only a few centimeters and encompasses the entire area surrounding the CM (excluding windows and airlock hatches). Another passive TCS obstruction to inspections is RTV heat-sink material, which is applied to the payload bay floor and the forward reaction control system cavity. Sampling inspection techniques in which strips of RTV are removed to inspect the skin underneath are employed in these cases. There are also cavity interiors, such as those within the RSB, that require inspection for corrosion but are only accessible with a borescope. Although there are some indications of minor corrosion forming within these cavities, access for further investigation and corrosion control is limited without cutting into or disassembling the structure.

The Space Shuttle orbiter has well exceeded its original design life. The original orbiter fleet was designed to be maintenance free for 10 years or 100 flights. Using the orbiter well past its design life is complicated by the fact that the vehicles utilize a wide variety of materials used over a wide range of operating temperatures and pressures. Owing to its unique design and operational requirements, the orbiter is subjected to some unusually harsh environments. These environments include those introduced by the aggressive fluids systems used by orbiter sub-systems, the sea-coast exposure seen during launch pad stays and ferry flight, and the vacuum of space.

The structural design of the orbiter is very similar to what is considered normal for airframe design. Examples include minimization of galvanic couples, sealing of faying surfaces, wet installation of fasteners and finish specifications. However, in some cases weight was often more important than corrosion resistance. Specific examples include less than adequate galvanic barriers and lack of corrosion protection in electrical bonding scenarios. Logistical support has also become an increasingly difficult challenge because vendors are going out of business or no longer will support the program because of the extra expense involved in the unique hardware design.[14]

The orbiter project office (OPO) has addressed the aging fleet on several occasions. From a Materials and Processes (M&P) perspective, this includes the formation of a Corrosion Control Review Board (CCRB), an age life assessment of non-metallic materials used in critical systems, and the recent formation of an Aging Orbiter Working Group (AOWG).

In 1993, the OPO chartered the CCRB. The board was formed as an advisory panel with the goal of assuring that all orbiter corrosion issues would be properly addressed and that solutions be recommended. The CCRB draws its core membership from the M&P discipline and Safety and Mission Assurance (S&MA) Engineers from both NASA and the prime contractors. The CCRB also receives regular support from structures engineering and from specialists from various NASA organizations in fields such as chemistry, materials science, and nondestructive evaluation.

The objectives of the CCRB include: Assessment of the extent and causes of corrosion, providing long-term and short-term corrective actions, generation and maintenance of a historical corrosion database, development and implementation of methods for detecting corrosion, and development and implementation of corrosion training and certification programs.

The CCRB has published three orbiter corrosion history reports.[2,15,16] A database was created in the mid 1990s and then subsequently updated and improved in 2004. Reviews of inspection, reporting and training requirements have been completed. Numerous fleet wide and select unique corrosion issues have been reviewed and corrective actions implemented. The CCRB has initiated several proactive measures to prevent corrosion such as galvanic barriers, corrosion preventative compounds (CPCs), design changes, washing of exposed surfaces and depainting/repainting. Additionally, the CCRB has been involved in the review and disposition of a number of subsystem corrosion issues.

Recently, an extensive corrosion program was completed. The program was part of the overall Aging Vehicle Assessment (AVA) program. This program provided a complete review of the orbiter's corrosion control program and provided the CCRB with an extensive list of products for the remainder of the Space Shuttle Program. The major products of the AVA program included: baseline, prioritization, prevention and detection, reaction and mitigation, trending, reports, documentation and process definition.

In 2006, using the tools developed during the AVA program, the CCRB developed a project plan. This plan assumes a Space Shuttle Program end date of September 30, 2010. The project plan was divided into three categories: near term (approximately one year), mid-term (approximately three years), and continuous (end of program). The near-term project goals include creating a CCRB website, finalizing recommendations for the implementation of non-chromated primer, and performing life cycle testing of CPCs. The mid-term goals were defined as completing the development of any nondestructive evaluation (NDE) for corrosion under the thermal protection system, and to finalize the recommendation for the development of laser de-painting. The continuous goals were defined as documenting lessons learned, maintaining the database, revising the Fair

Wear and Tear document, updating the CPC specification, networking, and benchmarking.

The AVA program was part of a certification assessment and verification for return-to-flight. The objective of certification verification is to assess the integrity of hardware certification relative to actual vehicle operational and processing environments. The goals were to ensure that actual ground processing and operational practices over time had not exceeded the engineering bases of certification or had introduced any unknown risks. The certification verification assessed the adequacy of hardware inspection requirements for critical hardware.

To assist the subsystem engineers in their assessments, Boeing M&P Engineering performed an age life assessment for the purpose of age life extension. The goal of this study was to extend material age life from 20 to 40 years. This study included 75 families of materials with approximately 1000 individual materials and covered approximately 500 critical parts. Age life conclusions were based on independently available data on material performance and on program data on a material's environment and historical performance. This program found approximately 20% of the materials to be good for 40 years. For approximately 70% of the materials, analysis of the data found no reason to suspect that the materials were degrading, but not enough data to extend the life out to the end of the program. For each material, corrective actions were recommended. For the remaining 10% of materials, analysis found the age life to be limited and corrective actions were recommended.

In 2004, the OPO started the AOWG. The AOWG was designed to provide the OPO oversight for aging vehicle issues such as corrosion, non-destructive evaluation, non-metallics, wiring, and subsystems.

9.6 Corrosion control in the launch pad and ground support equipment

Fuel, oxidizer, high pressure gas, electrical, and pneumatic lines connect the Shuttle vehicle with ground support equipment and are routed through the FSS, RSS and MLP. Pads 39 A and 39-B are virtually identical and roughly octagonal in shape. The distance between pads is 2657 m. The pad base contains 52 000 m^3 of concrete. Corrosion of metal structures are endemic to the oceanside environment. The metal pad structures are stripped and repainted on a recurring basis. The launch pads are removed from service every three to five years for maintenance and modifications. This 'Mod Period' lasts six to nine months. Pad 39 A underwent a renovation between June and September 1993 in which some 49 210 l of paint were used on two coats and 1 tonne of sand was used in the sandblasting operation.

The established requirements for the corrosion maintenance of the launch structure are outlined in the NASA Technical Standard NASA STD-5008 entitled Protective Coating of Carbon Steel, Stainless Steel, and Aluminum on Launch Structures, Facilities, and GSE.[17] This standard was developed to establish uniform engineering practices and methods and to ensure the inclusion of essential criteria in the coating of GSE and facilities used by or for NASA. This standard is applicable to GSE and facilities that support space vehicles, payload programs or projects and to critical facilities at all NASA locations worldwide. The standard is for the design of nonflight hardware used to support the operations of receiving, transportation, handling, assembly, inspection, test, checkout, service, and launch of space vehicles and payloads at NASA launch, landing, or retrieval sites. These criteria and practices may be used for items used at the manufacturing, development, and test sites located away from the launch, landing, or retrieval sites. The information provided in the standard is to be used for the preparation of written individual coating specifications for specific projects for the prevention of corrosion through the use of protective coatings on facilities, space vehicle launch structures, and GSE in all environments.

The launch pad structure is divided into zones of exposure that have been established to define coating system requirements for surfaces located in specific environments with regard to direct/indirect rocket engine exhaust impingement, acid deposition, and temperature. Corrosion protective coatings for launch structures are chosen from the NASA STD-5008 Qualified Products List (QPL). The carbon steel launch structure is coated with an inorganic zinc primer. Above the 30–34 m level, most of the zinc primer is coated with an inorganic topcoat. In the high heat areas, at the 30 m level, an ablative topcoat is used. An epoxy polyurethane coating is used on the wing covers.

9.6.1 Inorganic zinc coatings

All coatings and coating systems must meet the following minimum requirements to be listed in the NASA STD-5008 QPL:

1. self-curing, two package;
2. dry-temperature resistance to 400 °C (750 °F) for 24 h;
3. minimum shelf life of 12 months;
4. minimum of 83% zinc by weight in the applied dry film;
5. asbestos free, lead free, cadmium free, and chromate free (less than 0.01% by weight of mixed coating);

6. zinc dust pigment to be Type II in accordance with ASTM D520;
7. attain a numerical rating of not less than 9 in accordance with ASTM D610 and ASTM D1654 and 9F in accordance with ASTM D714, when applied to composite test panels and exposed at the KSC beachside atmospheric exposure test site (BAETS).

The coatings are evaluated for initial acceptance following an exposure period of 18 months. The coatings must continue to provide acceptable protection and performance for a period of five years to remain on the QPL. Application characteristics must be judged acceptable before atmospheric exposure testing.

9.6.2 NASA beachside atmospheric exposure test site

The NASA beachside atmospheric exposure test site (BAETS) has been documented by the American Society of Materials (ASM) as the most corrosive of any long-term atmospheric exposure site in North America.[4] The BAETS is located at approximately 2 km south from launch complex 39 A and is approximately 30 m from the high tide line of the Atlantic Ocean. As part of the facility, a fully instrumented weather station is enclosed within the site and provides continuous information on air temperature, humidity, wind direction and speed, rainfall, total incident solar radiation, and incident ultraviolet B radiation levels. The site has approximately 183 m of front row exposure for atmospheric corrosion specimens. Many types of test samples can be accommodated, including standard size test coupons 10.2 cm × 15.2 cm, stress corrosion cracking specimens, and full-scale test articles. These experiments can be performed in either a boldly exposed or sheltered configuration. Both power and data connections are available within the site to power test articles and record onboard data instrumentation outputs. The site has recently been outfitted with network connectivity for data acquisition through the Internet. The site has provided over 40 years of technical information on the long-term performance of many materials and continues to be upgraded with state-of-the-art capabilities to meet the current and future needs for corrosion protection of NASA and external customers.

9.6.3 Coating evaluation and development at the Kennedy Space Center

The evaluation of protective coatings for carbon steel, stainless steel, and aluminum has been an ongoing process for many years at KSC. In 1969, a study was initiated to identify coatings for the long-term corrosion protection of carbon steel exposed to the seacoast launch environment.[19] Both organic and inorganic zinc-rich coatings were applied to test panels exposed at the BAETS. These panels were evaluated for corrosion after 18 months, three years, five years, and ten years. The results of that study were that inorganic zinc-rich primers (ZRPs) were the best choice to provide long-term corrosion protection of launch equipment and ground support structures. The inorganic ZRPs outperformed organic zinc in the KSC seacoast environment. In general, organic topcoats were found to be detrimental to the long-term performance of inorganic ZRPs.

Un-topcoated inorganic ZRPs were used for many years at KSC for the long-term protection of carbon steel.[18] Several of the original panels exposed in 1969, that were painted with a single coat of ZRP without a topcoat, are still showing complete corrosion protection of the carbon steel at the BAETS.

In 1981, the Space Shuttle introduced a more aggressive environment to the launch pads at KSC. Exhaust from the SRBs resulted in deposits of small particles of alumina (Al_2O_3) with HCl adsorbed onto their surface. The impingement of this exhaust resulted in the failure of the carbon steel corrosion protection provided by the unprotected ZRPs, despite the fact that a pressure wash down was carried out as soon as possible after the launch. In response to the SRB exhaust problem, KSC initiated a study of new coatings to resist this new, more aggressive environment.

Tests were conducted in 1982 and 1986 to identify topcoat materials to enhance the chemical resistance of the coating systems in use at KSC. The 1982 study determined that two-component coatings were far superior to single-component types, epoxy/urethane topcoats provided some protection to the ZRPs, and that repair techniques, other than abrasive blasting, were ineffective in the launch environment.[20] The 1986 study focused on higher-built topcoat products to improve chemical resistance. As a result of this study, 10 topcoat systems were approved for use in the Space Shuttle launch environment.[21]

The coating systems selected as a result of the aforementioned studies were all solvent-based inorganic ZRPs topcoated with a variety of systems. In general, the topcoat systems that were successful in the 1986 study were epoxy mid-coats followed by polyurethane topcoats. All topcoat systems were solvent thinned. The results of these programs provided valuable data and resulted in the selection of appropriate coatings for the protection of KSC structures in their uniquely aggressive marine and launch environment. However, the Clean Air legislation and environmental regulations began to restrict the use of solvents in paints and coatings. These regulatory developments indicated that all inorganic ZRPs and topcoat systems approved for use at KSC would eventually become unavailable for use.

To address this challenge, studies were undertaken in 1990, and continue to this day, to identify inorganic ZRPs and topcoat systems that would provide superior protection while complying with the anticipated strengthening of environmental quality standards.[22,23] Many of the coating systems tested started with water-based inorganic ZRPs followed by water-based acrylic topcoats that could result in protective coating systems with essentially zero volatile organic compound (VOC). This approach would not only allow compliance with air quality regulations, but would also significantly reduce the use of flammable solvents and associated hazardous waste. In addition to liquid applied coatings, several powder-coating materials were evaluated for corrosion protection performance.

An evaluation program was initiated in 1994 to identify alternative inorganic topcoat coating materials for use at KSC and to study the performance of a new high-gloss polysiloxane topcoat for inorganic zinc-rich primers. Evaluation at the 18-month exposure period provided information for the revision of approved coating systems at KSC. This revision approved the use of several inorganic topcoat systems for general use with ZRPs at KSC's launch pads.

In an effort to reduce the time spent refurbishing facilities between launches, sprayable silicone ablative coatings were investigated as a replacement for ceramic-filled epoxy coatings. A 1994 study determined that sprayable silicone ablative coatings provided excellent heat and blast protection for launch structures.[24] Previous ablative materials were ceramic-filled epoxies developed in the 1960s for the manned space flight programs. The sprayable silicone ablative coatings were developed in response to concerns about damage to the protective tiles used on the Space Shuttle. The potential for damage resulted from the tendency of the ceramic-filled epoxy ablatives to spall when subjected to the thermal, impact, and pressure stresses involved in the exhaust plume of SRBs. In addition to their performance characteristics, sprayable silicone ablatives could be applied by plural component spray over an inorganic ZRP. This results in a significantly higher production rate than possible with the ceramic-filled epoxies. Ceramic-filled epoxy application requires labor-intensive mixing of a three-component system and manual application to a substrate primed with the epoxy components (without the ceramic filler). The use of sprayable silicone ablative coatings decreased the time required to refurbish the umbilical tower and other affected areas in preparation for follow up launches. The ablative material was applied at Launch Complex 39 B in 1994 on the entire 29 m level and on camera and communication boxes.

In the mid 1980s, researchers at KSC became interested in polyanilines (PAN) as protective coatings for metallic surfaces. As mentioned earlier, during the previous 20 years, extensive coating testing at KSC had led to the conclusion that inorganic ZRPs significantly outperformed organic zinc-rich type primers in the marine atmosphere of Florida. This was partially attributed to the increased conductivity of the inorganic ZRP coating film. The materials typically used to produce the organic zinc-rich films (e.g., epoxies, vinyls, etc.) produced an undesirable insulating effect on the zinc particles. This effect resulted in decreased galvanic activity of the zinc particles for protection of the carbon steel substrate. On the other hand, the organic zinc-rich primers had one advantage in that they allowed for less than perfect surface preparation on steel to achieve performance. The organic polymers provided better adhesion to marginally prepared substrates than the inorganic materials. This result led researchers at KSC to consider the use of conductive organic materials to formulate these zinc coatings to obtain the best of both types of ZRPs. The intent is to achieve a conductive organic vehicle that will provide the increased conductivity needed for superior galvanic protection of the steel substrate and better adhesion with less than perfect surface preparation. Other ways of increasing the conductivity of the organic vehicle are currently under investigation in the CTL.

9.6.4 Corrosion performance of alloys in the Space Shuttle launch environment

Testing and evaluation of the corrosion behavior and corrosion protective properties of different materials in the Space Shuttle launch environment have been conducted at KSC since 1968. In 1987, a study was begun to find a replacement alloy for 300 series stainless steel in the metal flex hoses used in various supply lines that service the orbiter at the launch pad. These convoluted flexible hoses, which were originally made out of UNS S30403 stainless steel, had failed by pitting. In the case of vacuum jacketed cryogenic lines, pinhole leaks caused by failure of the flex hose produced a loss of vacuum and subsequent loss of insulation. Nineteen alloys were investigated and evaluated using a variety of techniques that included exposure at the BSAETS, electrochemical characterizations, salt fog chamber exposure, and ferric chloride immersion. As a result of that investigation, a nickel–chromium–molybdenum–tungsten substitute alloy (UNS N06022) for UNS S30403 stainless steel was identified. Flex hoses of this alloy are now performing without failures at the launch pad.[25–28]

The standard specifications at the launch pad call for construction and replacement of tubing that used 300 series stainless steels. While the 300 series stainless steel is usually the material of choice for modern high-performance piping systems, there can be problems in high-chloride environments such as pitting and localized corrosion at welds and crevices. These concerns led to an investigation of a replacement-tubing alloy that would reduce the occurrence of corrosion problems. If a more corrosion-resistant alloy could be utilized, it would contribute to decreased risk of injury to personnel, reduced maintenance costs, reduced downtime, and better reliability.

In the year 2000, NASA began an extended atmospheric exposure test of various metal tubing samples in the oceanfront launch environment at the NASA BAETS. The objective of the project was to examine various types of corrosion-resistant tubing alloys for replacement of the existing tubing at the KSC Space Shuttle launch sites. NASA selected eleven different types of tubing alloys for testing. The tubing alloys tested were low-carbon, molybdenum-containing austenitic, super-austenitic, duplex, and superferritic stainless steels, along with nickel–chromium–molybdenum alloy, nickel–molybdenum–chromium–iron–tungsten alloy, and an austenitic nickel-base super-alloy.

Samples were fabricated into tubing assemblies containing flairs, bends, and welds connected with fittings typically used for handling gases and other pressurized fluids. The tubing assemblies were then pressurized with gaseous nitrogen before being placed at the BAETS. A drop in pressure would be indicative of pitting. The test matrix consisted of four separate conditions that are experienced at the launch facilities. They are summarized as follows: normal seacoast unsheltered, normal seacoast sheltered, acid/unsheltered environment, and acid/sheltered environment. A total of 98 tubes were mounted on four stands. Two of the four racks were sprayed once every two weeks with an HCl/alumina slurry to accelerate the corrosive effects. The HCl/alumina slurry was used to simulate SRB deposition. Two of the four racks, one with HCl/alumina slurry exposure and one without, have a protective roof (cover) to simulate partial shelter. The racks and samples remain at the BAETS.

Electrochemical testing of the corrosion properties of the alloys was also conducted. The workability, welding, and bending characteristics were assessed. Metallography and SEM were performed on the failed tubing samples to determine the extent of the corrosion damage. In addition, X-ray photoelectron spectroscopy (XPS) and EDS were used to examine the welds and surface oxide of the chosen material.

AL6XN tubing alloy was selected to replace the existing 300 series tubes. At KSC this alloy demonstrated increased corrosion resistance and resulted in a saving of cost compared with the seamless 300 series stainless steel. In addition, results of the testing showed that the procured tubing should be specified to be furnace annealed, pickled and passivated and meet the NASA specification KSC-SPEC-Z-007E.[30–37]

9.7 Corrosion control and treatment program

NASA's corrosion control and treatment program is described in NASA technical memorandum, TM-584C Revision C (November 1, 1994). This manual provides guidelines for the control of corrosion of materials in facilities, systems, and equipment at KSC and is available to download from the NASA CTL Website (http://corrosion.ksc.nasa.gov/publications.htm). The manual gives a thorough description of corrosion control, types of corrosion and their causes, corrosion removal and treatment of aluminum and aluminum alloys, carbon steel and low alloy steel, and stainless steel. It also includes descriptions of how to treat typical areas such as communication, electronics and associated equipment.

9.8 Space Shuttle Orbiter corrosion history

The first Space Shuttle Orbiter, the Enterprise, never flew in space but was used for approach and landing tests at the Dryden Flight Research Center and several launch pad studies in the late 1970s. Enterprise is now the centerpiece of the McDonnell Space Hangar at the National Air and Space Museum's Steven F. Udvar-Hazy Center in Virginia (USA). The Enterprise was followed by Columbia, which was the first Space Shuttle Orbiter to be delivered to NASA's KSC in March 1979.

The first Space Shuttle Orbiter flight was launched on April 12, 1981. Columbia and the STS-107 crew were lost February 1, 2003, during re-entry. The Columbia Accident Investigation Board (CAIB)[38] ruled out corrosion of the orbiter as a possible cause of the accident but made recommendations for long-term corrosion detection measures. The Orbiter Challenger was delivered to KSC in July 1982 and was destroyed in an explosion during ascent in January 1986. Each of the three Space Shuttle Orbiters now in operation – Discovery, Atlantis and Endeavour – were designed to last 10 years and fly at least 100 missions. Discovery was delivered in November 1983. Atlantis was delivered in April 1985. Endeavour was built as a replacement following the Challenger accident and was delivered to Florida in May 1991. There have been 123 shuttle launches as of May 2008. Currently, NASA plans to continue flying the Space Shuttle until as late as the year 2010. Although the orbiter flight and ground environment is often compared with that of commercial and military aircraft, there are significant differences. This is made evident by the relatively small number of orbiter corrosion problems experienced to date. However, trends indicate a gradual increase in corrosion reports in recent years, which is not unexpected as the orbiters age.

There are four major different types of orbiter corrosion mechanisms or causes: galvanic, concentration cell, uniform, and mechanical. These corrosion mechanisms can result in pitting, intergranular corrosion, cracking, filiform, erosion, fretting, and tarnishing.[37] Maintenance inspection requirements may continue to increase as the orbiters age.

9.9 Summary

Results from coating evaluation studies at KSC have shown that: inorganic ZRPs outperform organic ZRPs in the KSC seacoast environment; inorganic ZRPs are the best choice to provide long-term protection of launch equipment and ground support structures; in general, organic topcoats are detrimental to the long-term performance of inorganic ZRPs; inorganic topcoats perform well when used with inorganic ZRPs.

Studies performed to find a replacement for 300 series stainless steel in flex hoses at the launch pads identified a nickel–chromium–molybdenum–tungsten substitute alloy (UNS N06022) as the best material for this application. In another study to find a replacement for 300 series stainless steel tubing, alloy AL6XN was selected as the best material for tubing at the launch pads.

9.10 Future trends

NASA is planning to retire the Space Shuttle by the year 2010 and is now engaged in the Constellation Program. This program will create a new generation of spacecraft for human spaceflight, consisting primarily of the Ares I and Ares V launch vehicles, the Orion crew capsule, the Earth Departure stage and the Lunar access module. These spacecraft will be capable of performing a variety of missions, from Space Station resupply to lunar landings. The ambitious new endeavor calls for NASA to return human explorers to the moon and then venture even further, to Mars and beyond. As the nation's premier spaceport, KSC will play a critical role in this new chapter in exploration, particularly in the conversion of the launch facilities to accommodate the new launch vehicles. To prepare for this endeavor, the launch site and facilities for the next generation of crew and cargo vehicles must be redesigned, assembled and tested. One critical factor that is being carefully considered during the renovation is protecting the new facilities and structures from corrosion and deterioration.

NASA has relied on corrosion protection coatings that are not likely to be available for the design of the Constellation vehicles and launch pads. One example is the hexavalent chromium compounds (HCC) that are widely used for corrosion protection of aluminum alloys, as a conversion coating, and as paint pigments. Effective February 28, 2006, the Occupational Safety and Health Administration (OSHA) permissible exposure limit (PEL) for hexavalent chromium was reduced significantly from 100 $\mu g\ m^{-3}$ to 5 $\mu g\ m^{-3}$ (with the exception for aerospace applications, that the PEL of 25 $\mu g\ m^{-3}$, with reduction to 5 $\mu g\ m^{-3}$ be achieved by personal protection equipment). It is anticipated that many facilities will experience difficulties in ensuring that worker exposure meets these more stringent limits while continuing maintenance practices in using chromate primers.[39]

HCC are likely to be banned world-wide in the near future due to their significant environmental and human hazards and costs. Currently, there is no identified replacement product available. In addition, many of the coatings and coating systems currently approved for use in the NASA Standard 5008A and its associated appendices for the QPL are no longer available from the manufacturer or are harmful to the environment. The search continues to find a replacement for HCC, for lowering VOC in coatings, and for new depainting techniques that minimize the risk to workers and the environment.

Smart coatings are a recent development in corrosion control. The intelligence of the so-called 'smart coatings' relies on their capabilities to respond to physical, chemical or mechanical stimuli by developing readable signals which may often actuate, in addition to simple sensing, corrective action such as self-mending or healing.[40] Different smart coatings are under intensive studies for various applications including coatings for corrosion detection and control and self healing coatings that are able to repair mechanical damage.

The Constellation program has identified corrosion control as a top priority for its ground operations. The NASA CTL is engaged in developing smart coating technology for corrosion detection and control. As NASA prepares to move forward, the CTL at KSC will evolve to provide a better understanding of the corrosion processes affecting NASA's redesigned launch sites, structures, facilities and launch vehicles.

9.11 Sources of further information

Many of the documents containing information on corrosion control in space launch vehicles and GSE are available on the web. The following websites are recommended:

http://usago1.ksc.nasa.gov/usago/orgs/Engproc/MPE/AVA/ava_main.html (Aging Vehicle Assessment – FPR23000: Orbiter Corrosion Control Program)

http://corrosion.ksc.nasa.gov (NASA Corrosion Technology Laboratory Website)

http://ntrs.nasa.gov (NASA technical reports server for orbiter corrosion history)

http://www.agingaircraftconference.org (9th Joint FAA/DoD/NASA conference on Aging Aircraft, March 6–9, 2006, Atlanta, GA)

9.12 References

1 NASA, 2007. *Corrosion Technology Laboratory Website* (Online). NASA. Available at: http://corrosion.ksc.nasa.gov/ (accessed 4 December 2007).

2 NASA, 1995. Technical Memorandum 104810 (1995), *Space Shuttle Orbiter Corrosion History, 1981–1993: A Review and Analysis of Issues Involving Structures and Subsystems*, A report of the Orbiter Corrosion Control Review Board. Available at: http://ston.jsc.nasa.gov/collections/TRS/_techrep/TM-1995-104810.pdf (accessed 4 December 2007).

3 NASA, 2007. *Alligators and Rocketships*. Available at: http://www.nasa.gov/centers/kennedy/shuttleoperations/alligators/kscovr.html (accessed 4 December 2007).

4 COBURN S, 1978. 'Atmospheric Corrosion', in American Society for Metals, *Metals Handbook, Properties and Selection, Carbon Steels*, Metals Park, Ohio, 9th ed., Vol. 1, p. 720.

5 NASA, 2007. Material International Space Station Experiment (MISSE). Available at: http://missel.larc.nasa.gov/ (accessed 4 December 2007).

6 NASA, 1992. *Launch Complex 39, Pads A and B*, KSC Release No. 252-85. Available at: http://www-pao.ksc.nasa.gov/kscpao/release/1985/252-85r.htm (accessed 4 December 2007).

7 NASA, 1977. MSFC-SPEC-250A (1977), *General Specifications for Protective Finishes for Space Vehicle Structures and Associated Flight Equipment*. Available at: https://repository.msfc.nasa.gov/docs/multiprogram/MSFC-SPEC-250.pdf (accessed 4 December 2007).

8 NASA, 1987. MSFC-SPEC-522 (1987), *Design Criteria for Controlling Stress Corrosion Cracking*. Available at: (http://ntrs.nasa.gov/archive/nasa/casi.ntrs.nasa.gov/20000031371_1998106059.pdf) (accessed 4 December 2007).

9 US MILITARY SPECIFICATION, 1990. MIL-C-5541 (1990), *Chemical Conversion Coatings on Aluminum and Aluminum Alloys*.

10 US MILITARY SPECIFICATION, 2003. MIL-A-8625 (2003), *Anodic Coatings for Aluminum and Aluminum Alloys*.

11 SMITH C C, 1978. Corrosion Prevention for the Space Shuttle Orbiter, *SAMPE Journal*, 1978, March–April, 4–8.

12 LOMNESS J K and CALLE L M, 2006. *Comparison of the Chromium Distribution in New Super Koropon Primer to 30 Year Old Super Koropon using Focused Ion Beam/Scanning Electron Microscopy*, KSC Corrosion Technology Laboratory, Publication No. 268-86a (Online). Available at: (http://corrosion.ksc.nasa.gov/KoroponPrimer.htm) (accessed 4 December 2007).

13 EICHINGER E, 2006. Corrosion Performance of a 30 Year Old Chromated Primer, *9th Joint FAA/DoD/NASA Conference on Aging Aircraft*, Atlanta, GA, USA, 6–9 March 2006. Available at: http://www.agingaircraftconference.org/all_files/37/37d/37d-doc.pdf (accessed 4 December 2007).

14 RICHARD W R, 2007. An Overview of the Space Shuttle Orbiter's Aging Aircraft Program. Available at: http://ntrs.nasa.gov/archive/nasa/casi.ntrs.nasa.gov/20070018803_2007018406.pdf (accessed 4 December 2007).

15 BATSON K W and HESS J C, 2006. Space Shuttle Orbiter Corrosion History. Available at: http://www.agingaircraftconference.org/all_files/24/24a/58_doc.pdf (accessed 4 December 2007).

16 BOEING REPORT, in progress. Space Shuttle Orbiter Corrosion History, 1997–2004.

17 NASA STANDARD, 2004. NASA-STD-5008A, *Protective Coating of Carbon Steel, Stainless Steel, and Aluminum on Launch Structures, Facilities, and Ground Support Equipment*. Available at: http://corrosion.ksc.nasa.gov/pubs/nasa-std-5008a.pdf (accessed 4 December 2007).
18 NASA REPORT, 1972. MAB 3221-69, *Study of Corrosion Protection Methods for GSC Applications at Kennedy Space Center*. KSC Corrosion Technology Laboratory, Publication No. 3221-69. Available at: http://corrosion.ksc.nasa.gov/3221-69.htm (accessed 4 December 2007).
19 NASA TECHNICAL NOTE, 1973. NASA TN D-7336, *Performance Characteristics of Zinc-Rich Coatings Applied to Carbon Steel*. Available at: http://ntrs.nasa.gov (accessed 4 December 2007).
20 NASA TECHNICAL MEMORANDUM, 1984. NASA-TM-103503, *Evaluation of Carbon Steel, Aluminum Alloy, and Stainless Steel Protective Coating Systems After 18 Months of Seacoast Exposure*. Available at: http://ntrs.nasa.gov (accessed 4 December 2007).
21 NASA REPORT, 1988. MTB-268-86B, *Evaluation of Protective Coating Systems for Carbon Steel Exposed to Simulated SRB Effluent after 18 months of Seacoast Exposure*. Available at: http://ntrs.nasa.gov (accessed 4 December 2007).
22 NASA REPORT, 1993. FAM-93-2004, *Volatile Organic Content (VOC) Compliant Coating Systems for Carbon Steel Exposed to the STS Launch Environment – Application, Laboratory and 18 Month Exposure Results*. Available at: http://corrosion.ksc.nasa.gov/93-2004.htm (accessed 4 December 2007).
23 MACDOWELL L G, 1993. Testing VOC-Compliant Coating Systems at Kennedy Space Center, *Materials Performance*, **32**, 26–33.
24 MACDOWELL L G and DIVELY R W, 1995. Evaluating Silicone Ablatives on Launch Structures, *Journal of Protective Coatings and Linings*, **12**, 25–32.
25 ONTIVEROS C and MACDOWELL L G, 1990. Localized Corrosion of High-Performance Metal Alloys in An Acidic/Salt Environment. In NACE (National Association of Corrosion Engineers) *Corrosion 90*, Las Vegas, USA 23–27 April 1990. NACE: Houston, USA, Paper No. 94.
26 CALLE L M and MACDOWELL L G, 1994. Application of Electrochemical Impedance Measurements to Corrosion Prediction in the Space Transportation System Launch Environment. In NACE (National Association of Corrosion Engineers) *Corrosion 94*. Baltimore, USA 28 February–4 March 1994. NACE: Houston, USA, Paper No. 320.
27 CALLE L M and MACDOWELL L G, 1993. Electrochemical Impedance Spectroscopy of Metal Alloys, *NASA Tech Briefs*, **17**, 66.
28 NASA REPORT, 1990. MTB-610-89A, *Evaluation of High Performance Metal Alloys in the STS Launch Environment Using Electrochemical Impedance Spectroscopy*. Available at: http://corrosion.ksc.nasa.gov/610-89a.htm (accessed 4 December 2007).
29 CALLE L M, KOLODY M R, and VINJE R D, 2005. Electrochemical Investigation of Corrosion in the Space Shuttle Launch Environment. In EFC (European Federation of Corrosion) *Eurocorr 2005, European Corrosion Congress*. Lisbon, Portugal 4–8 September 2005.
30 CALLE L M, KOLODY M R, and VINJE R D, 2004. Corrosion Behavior of Stainless Steels in Neutral and Acidified Sodium Chloride Solutions by Electrochemical Impedance Spectroscopy. In ECS (The Electrochemical Society) *Abstracts of the*

206th International Meeting of the Electrochemical Society, Honolulu, USA 3–8 October 2004, The Electrochemical Society, Pennington, USA.

31 CALLE L M, KOLODY M, and VINJE R D, 2004. Electrochemical Impedance Spectroscopy of Alloys in a Simulated Space Shuttle Launch Environment. In *EIS 2004 (The 6th International Symposium on Electrochemical Impedance Spectroscopy)*, Cocoa Beach, USA 16–21 May 2004. Available at: http://corrosion.ksc.nasa.gov/Electrochemicalimpedance.htm (accessed 4 December 2007).

32 CALLE L M, VINJE R D, and MACDOWELL L G, 2004. Electrochemical Evaluation of Stainless Steels in Acidified Sodium Chloride Solutions. In NACE (National Association of Corrosion Engineers) *Corrosion 2004*. New Orleans, USA 28 March–1 April 2004. NACE: Houston, USA, Paper No. 04303.

33 CALLE L M, VINJE R D, and MACDOWELL L G, 2003. Corrosion Resistant Alloys in the Space Shuttle Launch Environment. In Australasian Corrosion Association Inc and Australian Institute for Non-Destructive Testing *Corrosion Control and NDT*, Melbourne, Australia, 23–26 November 2003. Paper No. 104.

34 BARILE R G, MACDOWELL L G, CURRAN J, CALLE L M, and HODGE T, 2002. Corrosion of Stainless-Steel Tubing in a Spacecraft Launch Environment. In NACE (National Association of Corrosion Engineers) *Corrosion 2002*. Denver, USA 7–12 April 2002. NACE: Houston, USA, Paper No. 02152.

35 BARILE R G, CALLE L M, CURRAN J, VINJE R D, MACDOWELL L G, and HODGE T, 2002. Corrosion Behavior of Tubing Alloys in a Seacoast Atmospheric Launch Environment. In ICC (International Corrosion Council) *15th International Corrosion Congress: Frontiers in Corrosion Science and Technology*, Granada, Spain 22–27 September 2002. Paper No. 370.

36 NASA SPECIFICATION, 1994. KSC-SPEC-Z-007E, *Tubing Steel, Corrosion Resistant, Types 304 and 316, Seamless, Annealed, Specifications For*. Available at: http://www-lib.ksc.nasa.gov/lib/s-s/specz0007.pdf (accessed 4 December 2007).

37 NASA REPORT, 2003. *Columbia Accident Investigation Board Report*. pp. 221 and 222). Available at: http://caib.nasa.gov/news/report/default.html (accessed 4 December 2007).

38 CURTIS C, 2006. Space Shuttle Corrosion Control Program, *9th Joint FAA/DoD/NASA Aging Aircraft Conference*. Atlanta, USA 6–9 March 2006. Available at: http://www.agingaircraftconference.org/all_files/24/24c/176_doc.pdf.

39 OSHA STANDARDS, US DEPARTMENT OF LABOR, 2007. *Hexavalent Chromium*. Available at: http://www.osha.gov/SLTC/hexavalentchromium/standards.html (accessed 4 December 2007).

40 XANTHOS M, 2003. Functional Additives as Sensors in Intelligent Polymer Coatings, *Smart Coatings II, European Coatings Conference*, Berlin, Germany, 6–7 February 2003.

10

Coating removal techniques in the aerospace industry

D. L. MONETTE, Canada

Abstract: In the aerospace industry, protective coatings – organic, inorganic, chemical conversion or metallic – are the first line of defence against corrosion. However, because of ageing, many of these coatings eventually require removal and reapplication. A flawed coating removal process can have disastrous results, it is, therefore, imperative that correct processing and coating removal risks be known. In this chapter the current state-of-the-art coating removal technologies in the aerospace industry are summarized. It contains a brief review of various coating removal techniques and the operational advantages and disadvantages. Coating removal effects on aircraft materials and worker health are also discussed.

Key words: aerospace coatings, coating removal, composite stripping, selective stripping, health and safety.

10.1 Introduction

One of the most important and yet most underestimated support activities for corrosion control is protective coating removal. Coating removal methods can actually degrade the substrate if they are not suited for the surfaces or are not performed correctly. This could have potential catastrophic consequences. An important role of organic coatings is to protect substrates from the severe environment in which they operate. Aerospace coatings are typically exposed to extreme temperature ranges that range from –40 °C to +40 °C. In addition, they are exposed to damaging elements such as sand, aircraft chemicals and mechanical damage caused by hail and maintenance activities (tool marks and ground handling vehicle impacts). Once damaged, the coating cannot protect the metallic or composite material from localized environmental degradation (corrosion). Another purpose of coatings is aesthetics: the appearance of an aircraft is important from a corporate or organizational point of view as it provides a positive or negative image depending on the condition of the coating system. Cracking, peeling and blistering paint may suggest that other systems on the aircraft that are not visible are also degrading. Typical aerospace coatings have a useful life of five to six years; on the other hand, aircraft are expected to have a service life of 20 or more years. In this chapter, a brief overview of the issues and the risks associated with coating removal methods is given and the different coating removal methods are reviewed.

10.2 Background

10.2.1 What is a coating?

There are several different types of coating systems that can be found on aircraft. Aircraft skins and major structural components typically have an epoxy-based primer coat applied to the surface followed by a highly cross-linked polyurethane top coat. Coatings are essentially made up of four ingredients: resin, pigment, fillers, and solvent. The resin acts as a matrix for the pigments and fillers. Primer coats may also have chromium pigments that act as excellent corrosion barriers. Fillers may include, among other ingredients, talc and calcium carbonate. The solvent liquefies the mixture facilitating delivery to a spray gun. The coalescing action of the solvent fuses the resin together as it evaporates. Depending on the chemistry, different properties can be modified as required by the coating system.

There are several coating characteristics that one can measure. One of the typical properties of the coating is hardness, which can be measured with a pencil hardness tester as per ASTM D-3363. Most aerospace coatings measure between 4H and 6H. Harder coatings are sometimes difficult to remove.

The adhesion of coatings may also be a factor affecting the efficacy of coating removal methods. The adhesion of a coating system can be measured with the crosshatch method as per ASTM 3359; typically an aerospace coating system should have at least 4B on the scale. A low number would suggest poor surface preparation before the application of the coating. However, a low number has the benefit of easy coating removal.

The thickness of a coating, which can be measured by eddy current, is typically 60–80 μm thick (2.2–3.2 mil). Older aircraft may have thicker coats. Thick coating systems may reduce the effectiveness of some coating removal methods. Finally, even the color of a coating system can also affect the performance of certain coating removal methods.

10.2.2 What lies beneath?

The surface beneath, to which aircraft coatings adhere, is typically referred to as the substrate, which in the case of aircraft can be a variety of different materials. High-strength aluminium alloys have been the main material used for fuselage and wing construction since the 1930s. They are lightweight and strong, and have relatively good corrosion resistance making them the material of choice. Typically, aluminium fuselage skins vary in thickness between 0.5 and 2.0 mm (0.020 to 0.080 in.) Flight control skins are typically thinner than fuselage skins. The aluminium alloy is, in general, protected by other systems in addition to the coating. Chemical conversion (anodizing) is often used to add additional protection, and a thin layer of pure aluminium cladding (thickness less than 0.05 mm) is sometimes applied by certain manufacturers to increase corrosion protection. Unprotected aluminium is susceptible to corrosion, particularly when exposed to salt water. It is very sensitive to alkaline (caustic) solutions, and sometimes even to high humidity air.[1] High-strength steel is used for critical applications in aircraft construction such as landing gear components, engine mounts, and critical wing fasteners. High-strength steel is resistant to caustic solutions but is very sensitive to acidic solutions. When a steel substrate is exposed to an acid, a condition known as hydrogen embrittlement may occur; this can significantly reduce the strength of the material creating a potential for sudden brittle fracture well below its design limits.[1]

Since the 1980s, more and more non-metallic materials, generally referred to as composite materials, have found their way into aircraft construction. Two of the earliest materials were fibreglass and aramid fibre in an epoxy matrix (Kevlar™). Because they are lightweight, these materials were used to replace metallic components in less critical secondary structures. Another advantage of these materials was that they do not corrode in the classical sense; however, they can degrade from environmental exposure. For example, when Kevlar was introduced, it was not known that this material was susceptible to damage caused by long-term ultraviolet (UV) radiation exposure or that they lost their performance characteristics because of moisture pickup when exposed to hot wet environments.

Higher performance carbon-fibre composites were introduced in the 1970s. The first major use of this material in the aerospace industry was on the F-18 aircraft where approximately 16% of the total weight of the aircraft consisted of carbon fibre composites. Carbon fibre is much stronger than fibreglass or aramid-based composites and may be used for primary structures, wings and empennage sections. Driven by rising energy costs, the use of carbon fibre composites permitted construction of lighter aircraft resulting in significant reductions in fuel cost compared with all-metal aircraft. Fibre placement and other advanced processing technologies are enabling complete fuselages and wings to now be made from carbon-fibre composite materials. In general, composite materials are sensitive to chemical solution attacks. On the down-side, depending on the chemical type, the time of exposure and temperature, composite materials can sustain an attack resulting in the dissolving or softening of the matrix of the material, thereby weakening the structure dramatically affecting its performance and long-term integrity.

10.3 When to remove the coating

Coatings are sometimes completely removed as part of standard inspection processes. After complete removal of the coating, small corrosion defects such as filliform or exfoliation can be detected visually, these defects are not always apparent without removing the coating system. If inspection requirements are not a factor, then the appearance of the aircraft may be a reason for coating removal and reapplication of a new coating system. As mentioned previously, the coating system will age and deteriorate after five to six years in service. On commercial aircraft, after this period of time, the coating will have lost its original gloss, suggesting to the observant passengers an old potentially deteriorating aircraft. In an effort to reduce cost and to maintain fresh appearances many organizations have adopted the scuff sanding technique.

Scuff sanding is a partial coating removal method whereby the surface is manually abraded with sandpaper to roughen the existing top coat. After washing the surface, a thin layer of primer (also know as the tie coat) is applied to the treated surface followed by an application of a new top coat. This method is by far the most rapid and economical way to enhance the appearance of an aircraft for at least another five years. However, the extra layers of coating do have long term negative implications. On an aircraft the size of a Boeing 767, an extra coating system layer was determined to add an extra 360 lb (163 kg) to the aircraft and cost 20000 $ US per year in extra fuel costs (2006).[4] In 2003, a B-1 aircraft (originally manufactured in 1986) was completely stripped for the first time after many scuff sanding/re-coat cycles. After stripping the aircraft was found to have shed 1800 lb (800 kg).[3]

Another potential problem with thick multiple coats is the tendency for them to crack. Depending on the location of these cracked coatings, time-consuming inspections may be required, involving nondestructive examination (NDE) and engineering support.

10.4 A brief history of coating removal

First isolated in 1840, by Henri Victor Regnault, dichloromethane (DCM), also known as methylene chloride, had been the primary solvent for most chemical coating removal methods since metallic aircraft were first introduced. Its volatility, powerful solvent characteristics and very low flammability made it a very effective paint stripper. Other applications for this solvent were in the foam-making industry, where it was used as a blowing agent. Other non-paint stripping uses included the manufacture of decaffeinated coffee, cosmetics and pharmaceutical applications.

Dichloromethane has been used in the aerospace industry for over 70 years. To increase its effectiveness for coating removal applications, a caustic or acidic component was often added to the solvent. In the 1980s, after many medical studies suggested that human exposure to dichloromethane had potential cancer causing and other severe health effects, several organizations, including the US Environmental Protection Agency (EPA) and the World Health Organization (WHO), initiated and implemented regulations to reduce human exposure to this chemical. DCM was also found to be bio-persistent in water and air, contaminating the environment.

The aerospace industry was one of the first to act to replace DCM. As early as 1984, dry stripping coating removal methods using plastic particles were being developed in the United States and were implemented by the USAF. In the meantime, the chemical industry also reacted by developing less toxic and more environmentally acceptable chemical strippers. By 1990, quite a few non-chemical alternatives were available as alternatives to dichloromethane coating removal for the aerospace industry.

10.5 How to remove coatings

The majority of newly developed coating removal methods use one of the following mechanisms:

- molecular disassociation,
- thermal, and
- impact.

Molecular disassociation categories, which include all chemical strippers, belong to three different groups, depending on the formulation of the strippers: acidic (below pH 7); neutral (pH 7); or alkaline (above pH 7).

Thermal methods for delicate aerospace components include lasers and heat lamps, both of which usually require some form of robotic application. Lasers are a non-kinetic form of coating removal; whereas heat lamps are usually accompanied by a kinetic CO_2 blast stream to loosen the heat-damaged coating debris.

Impact methods include:

- high-pressure water,
- sodium bicarbonate,
- agricultural by-products,
- sponge media,
- petroleum-based plastic media, and
- engineered bio-based media.

10.5.1 High-pressure water

Developed for aerospace applications in the early 1990s, use of high-pressure water for removal of aerospace coatings involved a delivery system (robotic) with a variety of different nozzles to concentrate very precise amount of hydraulic kinetic energy (up to 4000 bar (60 000 psi)).

10.5.2 Moh's scale

The remaining impact methods involve particles projected at the surfaces at high speeds (100–200 $m s^{-1}$ at the nozzle outlet) to remove the coating systems. When discussing dry stripping media, hardness characteristics are often mentioned. The Moh's scale is generally used to rate the hardness of different dry-stripping media types. Developed in the 19th century by Friedrich Moh, it is a scale from 1 to 10 that rates the scratch resistance of different materials. The principle is that the harder material will scratch the softer ones. Moh's scale ranges from diamond at 10 to talc at 1. It is not a linear scale; for example, 4 on Moh's scale is 10 times harder than 2.

10.5.3 Sodium bicarbonate

Adapted for the aerospace industry in the mid 1990s, bicarbonate of soda also known as sodium bicarbonate is a pure mineral with a relatively soft hardness of 2.0 on Moh's scale. Propelled at 2–6 bar (30–90 psi) it can be used in either dry or wet modes.

10.5.4 Agricultural by-products

This dry-stripping method used for decades in heavy industries and the aerospace industry is commonly used for removing coatings from non-delicate materials such as steel and titanium. This method utilizes a pressurized delivery system. Typical agricultural by-products (grits) include ground corn cob, walnut shells and cherry pits. These relatively inexpensive materials have an inherently wide hardness range which may vary from 2.0 to 4.0 on Moh's scale. They are most often propelled with pressurized dry air at 1.5–6.0 bar (20–90 psi).

10.5.5 Sponge media

This dry-stripping process, developed in the mid 1990s, involves relatively large particles manufactured of a pliant matrix material such as foam or fibre, and a percentage (by weight) of various other abrasives with hardness varying from 2.0 to 9.0 on Moh's scale. The abrasive particles are embedded in a pliant matrix, the nature of which allows a reduction in kinetic energy transfer to the substrate upon impact.

10.5.6 Plastic media

Plastic media blasting (PMB) is a dry-stripping method developed in the early 1980s. The original material used on aircraft was made of crushed plastic buttons (thermo-set urea formaldehyde). The process evolved with the US military creating a military specification in 1988 for plastic media (Mil-P-85891 A) to be used as a manufacturing quality assurance specification in the aerospace and non-critical aerospace applications. In the latest revision (1998), there were seven different types of plastic media. The plastic media hardness varies by type from 2.0 to 4.0 on Moh's scale and is usually propelled with dry air at 1.5–4.0 bar (20–60 psi).

10.5.7 Engineered bio-based media

This dry-stripping method, introduced in the early 1990s, involves a highly purified polysaccharide powder derived from plants (such as wheat starch), which, by means of a thermal hydro-mechanical transformation method (extrusion), is rendered plastic like. It was developed to produce a bio-based media with similar yet different properties from plastic media types. Other plant materials have since been introduced to this category. The media has an apparent hardness of 2.0 on Moh's scale. The major physical difference these media types have compared with other dry-stripping abrasives is their bound water content. The water content results in a reduction of kinetic energy transfer on impact. These media types are usually propelled with dry air at 1.0–3.0 bar (15–45 psi). Selective stripping (removing the top coat and leaving the primer coat intact) is sometimes feasible with these media types.

10.6 Chemical strippers

In response to health warnings and new legislation about dichloromethane, starting as early as the mid-1980s, new chemical formulations were developed by dozens of different companies to replace DCM. The principal objective was to produce a product that was as effective as DCM-based strippers, but with reduced toxicity and environmental impacts. After 20 years, the quest is still on, to-date no chemical strippers have been developed that match the effectiveness and rapidity of DCM-based strippers.

10.6.1 Chemical stripping mechanism

The three steps in a chemical removal operation are

- penetration,
- reaction and molecular disassociation, and
- vapour pressure and coating lifting (also called tenting).

Penetration

Typically the surface is wetted to allow full stripper contact with the coating surface; the product is brushed on, sprayed on or in the case of off-aircraft components, e.g. wheels, completely submerged. To produce a clinging effect, thickening ingredients, such as cellulose, are sometimes added to enhance adherence to non-horizontal surfaces. Penetration occurs when the solvent component of the stripper enters the coating by diffusion and capillary action. The preferred entry paths of the stripper include pre-existing coating defects, such as scratches, gouges, micro cracks, pin holes, and through the pigments in the coating.

Reaction and molecular disassociation

Once the coating is penetrated, a secondary chemical component of the stripper attacks and breaks the hydrogen and dipolar bonds of the resin to the pigments and fillers.

Vapour pressure and coating lifting (also called tenting)

The primary and secondary chemical components reach the uncoated substrate (usually metallic) and a reactive action takes place producing gases that help lift the paint from the substrate surface.

10.6.2 Non-dichloromethane chemical stripping

As a result of the large numbers of different chemical alternatives being introduced to the aerospace industry, several international engineering studies have focused on determining the differences between the DCM products and the alternatives. One of the most extensive depainting studies ever conducted between 1994 and 1999 was by the National Aeronautics and Space Administration (NASA).[2] The principal objective of the study was to determine the new chemical formulations' performance and effects on metallic aerospace test coupons and compare them with DCM-based strippers. This study was sponsored in part by the US EPA.

One of the most significant differences in the characteristics of the new chemicals that emerged from that study was in the performance. It appeared that the new chemical stripper types worked much slower than the DCM-based benchmark: hours versus minutes using the same type of coating systems. This confirmed previously published results.[5]

10.6.3 The pH of chemicals and its effect on aerospace materials

All chemical strippers belong to one of three categories: alkaline, neutral or acidic. As previously mentioned, the pH of a chemical may cause severe detrimental effects to the metallic substrate. The factors that can increase the detrimental effects are dwell time and temperature.

Aluminium alloys are resistant to mildly acidic solutions but are readily attacked by alkaline solution. On high-strength aluminium skins, the attack is usually preferential grain boundary damage. The alloying elements in the aluminium alloy such as copper may be dissolved from the surface. This etching action allows corrosion mechanism such as filliform or exfoliation to start. Alkaline strippers are most often used on ferrous parts. Some alkaline agents include sodium hydroxide and potassium hydroxide.

In general, a neutral pH (7) is by far the safest to use on ferrous and non-ferrous surfaces. Neutral strippers are generally less effective than acid or alkaline-based strippers. However, slight manufacturing deviations can result in non-neutral pH values with unexpected results. For example, during the NASA study, a chemical stripper that had been identified as neutral had in fact a pH of 5.7. During the study, this slightly acidic variant of a neutral stripper was enough to induce hydrogen to the high-strength steel test coupons, which failed by sudden brittle fracture.

Acid strippers are usually more effective than neutral or alkaline strippers. As previously mentioned, great care must be taken not to allow acid-based strippers to come in contact with high-strength, low alloy steels. When an acid comes in contact with the steel substrate, nascent hydrogen enters the microstructure and creates a hybrid formation at the grain boundaries. The only method that can release the trapped hydrogen is the baking of the affected part or structure at 325 °F for a minimum of 23 h immediately after exposure. Hydrogen embrittlement cannot be detected visually. Some acid agents include formic acid and chromic acid.

Advantages

The new chemical strippers developed are generally less toxic and less environmentally problematic than DCM-based strippers. The advantages of chemical stripping include ease of use, minimal operator training, and a pristine post stripping surface can sometimes be obtained if mechanical scrapers or blades are not used to remove residual hard-to-remove coatings. The economics of using the new chemical strippers can also be slightly better than that of other more advanced coating removal methods; however, water treatment costs and worker health liabilities can erode any speed of removal or financial benefits.

Disadvantages

The newly developed chemical strippers do not eliminate the toxic exposure risk to workers, but can reduce it in some cases. For example, a new environmentally acceptable solvent for chemical strippers introduced in the early 1990s and used by several manufacturers to make 'green' products in 2001, was added to the California Proposition 65 list as a potential reproductive toxin. Entry was via dermal exposure. Other solvents have recently come under closer scrutiny and concerns over related central nervous system damage have been expressed.

One method of reducing the toxicity of the new chemicals was to use ingredients that had low volatility. DCM produces large amounts of volatile organic compounds, which means it dries and evaporates quickly. This characteristic is actually beneficial from a coating removal aspect since the risk of unwanted effects of trapped chemical residues (ingress) in the structure will be minimized due to the rapid evaporation of DCM-based strippers. The new low-volatility chemicals may take up to 50 times longer to evaporate. These longer-lasting chemicals may pose long-term ingress problems because of the low evaporation rates, this may cause the chemical stripper to be in contact with the surfaces for a much longer period of time than it was originally tested and approved for.

10.6.4 Chemical stripping and composite materials

Composite materials are made by combining two or more different engineered materials. One component is the fibre that gives the composite its strength; carbon fibre and fibreglass being typical examples. The other main component is the matrix, i.e., the resin that holds the fibres together and transfers the applied loads to the fibres. The matrix can be made from a wide variety of resins; for example, epoxy or polyethylene is often used. When a chemical stripper is applied to a coated composite material, there is a risk that the surface of the composite will be damaged by matrix attack. Simple logic dictates that if a given chemical product is powerful enough to remove highly cross-linked, tough epoxy based aerospace coatings by molecular disassociation, then softening and dissolution of the resin in the first few exposed layers of the composite is possible. Once a composite surface has been compromised this way, water can enter composite parts during service. This water and moisture can cause considerable damage to parts as they are subject to a freeze and thaw process as experienced when aircraft reach cruising altitudes, where ambient temperature is typically in the –40 °C range.

Although generally not recommended, chemical stripping of composite materials has been conducted by several organizations using a minimum-time-of-exposure technique to reduce damage potential. Most of these components were secondary and tertiary parts such as radomes and fairings. It was not unusual for a small percentage of parts stripped to be scrapped in the process. In 1998, the US Federal Aviation Administration (FAA) issued advisory circular AC-43-205 warning aerospace maintenance organizations about the risks of using chemical strippers on composite materials and about using chemicals that were past the manufacturer's expiry date.

10.7 Thermal coating removal methods

10.7.1 Lasers and heat lamps

Both the laser and heat lamp methods were developed for aerospace coating removal applications in the early 1990s and require robotic manipulators for their application.

Laser devices emit a highly focused beam of energy that volatizes and incinerates the coating at extreme temperatures. Unlike the laser cutting method, which uses a continuous beam, lasers for coating removal are pulsed and moved very rapidly over the surface to minimize overheating and damaging the surface. The energy level and scanning frequency can be adjusted for different types of coating removal applications. Early industrial users of robotic laser stripping include several maintenance units of the USAF, with the principal applications being for removing coatings from composite structures.

Heat lamps emit pulsing, high-energy infra-red energy that ablates and incinerates the coating to an ash. To rapidly cool the surface to prevent thermal damage or change to the metal's heat treatment, and to remove the ash, a jet of pressurized CO_2 is used in tandem with the heat lamp.

Advantages

- With precise controls, both thermal methods can be used to remove coatings from very delicate composite and aluminium surfaces with little or no harmful effect.
- Lasers can be used to prepare and enhance new composite surfaces for subsequent bonding or coating application.

- Both methods produce very little additional waste, other than the removed coating debris.
- Selective stripping is sometimes feasible.
- These non-chemical processes are relatively safe for the operators, but they do require special eye protection.
- No ingress potential.

Disadvantages

- Equipment cost, operator skill requirements, and engineering support level are considered high in all cases.
- Complex geometries are difficult to process.
- Light-coloured coatings (e.g. white) will slow down the process due to light reflection.
- Minimal overexposure can result in heat damage to the substrate, changing the mechanical properties of the affected areas.
- Manual application is considered very risky or impractical.

10.8 High-pressure water

Developed in the early 1990s, high-pressure water is typically projected via special nozzles at pressures of 2000 to 4000 bar (30 000–60 000 psi). At these pressures even the most tenacious coatings can be removed. The operation is done robotically with such high precision that even very thin aluminium skins (0.4 mm thick) can be processed with little or no effect on the mechanical properties.[2] Large robotic systems for processing complete airframes were built in the mid-1990s. However, the process was never fully implemented because of a variety of concerns including potential damage to the substrate by overexposure. The process has had success in smaller cabinet-size units with a closed cycle water treatment incorporated. The process is used to remove coatings from parts that are usually not sensitive to impact damage (steel, titanium). This method is now used with much success on naval ships.

Note that there is a category called medium-pressure water stripping with pressures of 1000 to 1500 bar. However, this method cannot remove typical polyurethane coatings by itself. It is more often used in combination with a chemical coating softener or abrasive grits, or both.

Advantages

- With precise controls, high pressure water can remove coatings from very delicate metallic parts and structures with little or no effect.

- The process produces very little additional waste, other than coating debris.
- Selective stripping is sometimes feasible.
- This is a non-chemical process and is relatively safe for the operators in robotic applications.
- No consumables other than water.

Disadvantages

- Equipment costs, operator skills, and engineering support requirements are considered to be moderate to high.
- Coating removal from delicate composite substrates is often not possible, unless selective stripping is performed.
- Owing to the excessive force of high-pressure water at the nozzle outlet, manual application and control for delicate operations is not possible.
- Water must be treated to remove coating debris.

10.9 Dry stripping

The remaining impact methods all use the pressure blast compressed air delivery system. The other compressed air delivery system, known as suction blast, is not suitable for dry stripping. In the pressure blast process, angular grit shaped blast media is loaded in a pressure vessel with a steep conical bottom correctly called an abrasive blast generator, but often referred to as a pressure pot. At the time of delivery, compressed air pressurizes the pot, typically at 0.2 to 0.4 bar above the desired process pressure at the media delivery nozzle. The blast media flows by gravity through a variable orifice metering valve or a powered feeding device (auger valve or positive displacement feeder for example) where it enters the moving compressed air stream in the blast hose. Adjustments can be made to increase or decrease the amount of media being delivered, typically called the media flow rate and expressed in kilograms per minute. Blast media velocity exiting the nozzle is controlled by the air pressure in the blast generator. After delivery and impact, most dry stripping media can be reused by removing the very fine particles. Removing fine dust from the media is usually done by pneumatic cyclone and mechanical sieving. The pressure processes can be used in small hand cabinets up to multi-nozzle systems in specially equipped aircraft hangars. The important parameters for dry stripping are: nozzle pressure, nozzle to surface distance, nozzle to surface angle, media flow rate, media particle size and hardness. Used in the manual mode, operator skill is also an important factor.

10.10 Dry stripping media types

10.10.1 Sponge media

Sponge media is an engineered dry stripping media developed in the mid 1990s. It is propelled with compressed dry air against the surface to be treated. The media is a combination of a soft foam or fibre carrier, which acts as a matrix, and different types of abrasive grits (plastic, minerals, etc.) that are embedded in the matrix. The particle sizes are much larger than typical dry stripping media, ranging from 3 to 6 mm in diameter. The foam or fibre matrix component of the media essentially serves two purposes: first, it reduces the kinetic energy potential of the embedded abrasive particles (Moh's hardness 2.0 to 9.0) by compressing upon impact, causing a scrubbing action instead of the full impact; secondly, it reduces the production of airborne nuisance dust considerably. Although some aerospace applications have been developed, this method has been more successful in non-aerospace applications (bridges, de-contamination projects). Typical operating pressures are 1–6 bar.

Advantages

- Depending on the abrasive types used, coating removal on delicate parts can be conducted with minimal effects.
- Media can be reused 6–12 times.
- Significant reduction of airborne dust compared with other dry stripping methods.
- Selective stripping is sometimes feasible.
- This is a non-chemical process, relatively safe for the operators.
- Cost, operator skills and engineering support: low to moderate.

Disadvantages

- Precise cleaning controls, screening and new media replenishment are critical to prevent extreme performance variations due to abrasive loss from the matrix and loose abrasive in the working mix.
- Low dust production may lead to complacency, like all dry stripping methods airborne coating particles containing toxic heavy metals such as chromium may be generated. Proper breathing protection is still required.

10.10.2 Bicarbonate of soda

Bicarbonate of soda, also known as sodium bicarbonate, is a type of salt. It is generally manufactured using the Solvay process. As a coating removal method, it is considered a soft friable abrasive with hardness just above 2.0 on Moh's scale. On the other hand, with a specific density of 2.15 g cm^{-3} sodium bicarbonate is much heavier than other dry stripping media types (plastic, agricultural media). The media has a pH of 8.1 and rapidly dissolves in water. With a particle size of 50 μm, the media can be projected either dry or in combination with water. The latter method reduces the airborne dust cloud while blasting. To reduce water sensitivity, manufacturers are coating the particles to make them waterproof until after impact when the media once again becomes water soluble. The media can only be used once. At temperatures over 60 °C, sodium bicarbonate starts to decompose into sodium carbonate, which is very corrosive.

Despite its relative softness, the high specific gravity of bicarbonate of soda can be aggressive on delicate substrates. During the NASA study,[2] which evaluated bicarbonate plus water stripping, it was the only impact method studied out of four that failed the thin aluminium skin tests (0.4 mm) by deforming the test panels to a point that fatigue testing could not be conducted on the test panels. During the same study, it was also found that it was the only impact method that had a significant negative effect on crack detection by the eddy current method. Although some aerospace applications have been successfully developed for off-aircraft components that are not delicate, complete airframe stripping using sodium bicarbonate has never been generally adopted. Typical operating pressures are 2–6 bar, pressure with water can be much higher.

Advantages

- Low dust production when used in wet mode.
- Water dissolving characteristic reduces the potential impact of ingress on off-aircraft parts with complex interiors.
- Benign non-chemical process, relatively safe for operators.
- Equipment cost, operator skills and engineering support: low to moderate.

Disadvantages

- Can be aggressive on delicate substrates.
- Corrosion by sodium carbonate is a concern.
- Media can only be used once.
- When used dry, dust produced may contain toxic heavy metals, proper breathing protection is required.

10.10.3 Agricultural media

Crushed corn cobs were first use by the US Navy during World War II to clean aircraft engines. As it is a by-product, its mechanical properties can vary significantly from batch to batch. Corn cob media has a Moh's hardness that may vary from 2.5 to 4.0 and it has a specific density of 0.8 to 1.2 g cm^{-3}. Particle size for agricultural media can range from 1 to 0.2 mm.

Crushed walnut shell media is also a popular agricultural media. A by-product, it is usually made from English walnut, which has a Moh's hardness between 3.0 and 4.0, and a specific density of 1.2 to 1.4 g cm^{-3}. This media type is generally more aggressive than corn cob media and is preferred for aerospace applications as it meets higher quality standards (MIL-G-5634). A typical use for walnut shell media would be carbon deposit cleaning from non-delicate aerospace components.

Normal operating pressures for agricultural media is 2–6 bar.

Advantages

- Corn cob and walnut shells are biodegradable renewable products.
- Depending on parameters, media can be used more than once, 4 to 8 cycles.
- A non-chemical process, relatively safe for the operators.
- Cost, operator skills and engineering support: low to moderate.

Disadvantages

- Variations of inherent material properties make delicate operations risky.
- By-product nature makes quality control difficult.
- Certain agricultural media may leave oily residue on surfaces.
- Dust produced may contain toxic heavy metals; proper breathing protection is required.

10.10.4 Plastic media

The concept of using plastic as a dry stripping media to remove coatings was first patented by the Dupont Company in the 1940s. The concept was not considered viable until the health warnings about DCM-based chemical strippers started to emerge in the early 1980s. The first pioneers to evaluate and adapt this coating removal method for aerospace applications were the United States Air Force (USAF) and the United States Navy (USN).

Owing to the popularity of plastic media and the increasing number and variety of plastic media types and manufacturers, the USN created a military specification in 1988 to ensure that quality products were being sold to the government and their subcontractors. The angular shaped media is available in particle sizes (mesh size) which can vary from 2 to 0.5 mm in diameter. The typical operating pressures are 1.5 to 4.0 bar (20–60 psi).

Table 10.1 Various plastic media types listed in MIL-P-85891A, Amendment 2, 26 June 1998

Type#	Material	Hardness/scale	Moh	Specific gravity
I	Polyester	34–42 barcol	3.0	1.15–1.25
II	Urea formaldehyde	54–63 barcol	3.5	1.47–1.52
III	Melamine formaldehyde	64–72 barcol	4.0	1.47–1.52
IV	Phenol formaldehyde	54–62 barcol	3.5	1.38–1.42
V	Acrylic	46–54 barcol	3.5	1.10–1.20
VI	Allyl diglycol carbonate	30–40 barcol	3.0	1.28–1.33
VII	Starch-g-acrylic	72–79 shore	2.0	1.38–1.43

In the 1998 revision of Mil-P-85891A, there were seven different plastic types listed as shown in Table 10.1. In the early 1980s, the USN and the USAF evaluated different types of plastic media for use on airframes and components; the media selected was Type II. The early results were promising and indicated that a substitute for chemical stripping had been found. However, the initial hope of eliminating chemicals was dampened as a result of some early studies. Several organizations found that Type II (urea formaldehyde) caused undesirable effects on aluminium substrates after blasting such as reduced fatigue life, increased fatigue crack growth rates and significant erosion of protective metals. Alcad aluminium which protects the high-strength aluminium alloy skins is made of almost pure aluminium and is relatively soft (approximately 2.0 on Moh's scale). Cadmium plating is used to protect critical high-strength fasteners; cadmium also has a soft hardness of approximately 2.0 on Moh's scale.

The next blast media of significance reviewed by the USN and the USAF was designated Type V acrylic. Type V media was found to have less negative effects on aircraft substrates when compared with Type II. However, the thermoplastic nature of acrylic sometimes resulted in a plastic residue deposit on the surface after blasting. This residue could result in poor adhesion characteristics of new paint if not removed. The tenacious residue can only be removed by strong solvent such as methyl ethyl ketone (MEK). Note that the residue problem can be attenuated, if only high-quality virgin acrylic extruded stock is used to make the acrylic plastic media. It should be noted that Type II urea media does not leave a tenacious residue because of its thermoset nature that will not permit a return to a plastic state after heating.

In the late 1990s, a new plastic media type designation was added to the Mil P-85891A specification, Type VII. The Type VII designation provided a category for a starch grafted acrylic copolymer product. Type VII starch-g-acrylic was found to have fewer negative effects on aircraft substrates than did Type V. It was also found not to produce residue. The earliest aerospace pioneers to study and implement this new media type for airframe coating removal were the Swedish Air Force (1999) and the United States Coast Guard (2002).

10.11 Mechanical effects of coating removal by impact methods

All impact methods (including high-pressure water) will induce residual stresses in the metallic substrates being processed. In the impact blasting process, as the media reaches the metal substrate surface, the surface will be mechanically deformed to a certain depth. This deformation, which takes place upon impact, consists of compressing the surface beyond the tensile yield limit of the metal immediately after impact, the contracted region that was deformed and yielded plastically under tension recovers only partially, leaving behind a permanent compressive residual stressed region.

To measure the applied residual stress in thin aluminium skins, the USAF developed a method based on the shot peening industry. In the 1930s, an automotive engineer named John O. Almen developed a method to gauge the level of residual stress inducement that the shot peening process had on metallic substrates. The shot peening process involves spherical steel or glass shots propelled at the surface to induce surface compressive stresses of a work piece to increase its fatigue life by counteracting applied surface tensile stresses encountered in service. The Almen strips 3.0 in. by 0.75 in. in size come in three thicknesses, which vary from 0.031 to 0.0938 in. and are typically made from 1070 steel. The Almen strips are described in detail in Mil-S-13165. The strips are attached to a metal fixture and secured by four screws; they are then subjected to the impact blast stream for a certain predetermined exposure time. The strips are removed and a measuring gauge (Almen gauge) is used to determine the level of mid-plane distortion (curvature) of the strips. The higher arc height numbers indicated high compressive stresses were being generated by the given blast stream.

When plastic media blasting (PMB) was first evaluated in the early 1980s it was determined that aluminium fuselage skins with thicknesses over 0.060 in. (1.5 mm) were not significantly affected by impact stripping methods. However, for thin skins such as 0.032 in. (0.8 mm) and below, high residual stresses were of concern. Because the traditional steel Almen strips were not sensitive enough to measure the very low and yet potentially harmful residual stresses of impact coating removal, a new Almen strip material was needed. Coupons made from 2024-T3 aluminum were chosen because this alloy was found to be the most sensitive to the process and was one of the most frequently used for aerospace skin applications. This new Almen strip was designated Aero Almen strip. Over the past decades, several military and commercial organizations have processed tens of thousands of these strips to study various media types and blasting parameters. The standard thickness for Aero Almen strip is 0.032 inches (0.8 mm) and is not clad with pure aluminum. Using this bench mark material, early USAF researchers established a desired arc height value of 0.006 in. (0.15 mm) or less for coating removal impact processes.

Another negative mechanical effect of PMB is the potential that cracks may be masked by the processes. In fact, all impact methods have the potential to cause crack closure which may interfere with Liquid Penetrant Inspections (LPI). Off-aircraft non-ferrous components and parts are periodically inspected by the LPI method to determine the presence of defects such has fatigue cracks which are open to the surface. After coating removal, LPI uses the principle of capillary action of special oils to enter and exit fine defects, thus aiding in their detection when they are not visible to the naked eye. The induced compressive residual stresses of impact methods can pinch together the crack walls of a service-induced crack at the surface affecting the sensitivity of the inspection by reducing the ability of the penetrating oil to enter the crack. Secondary eddy current inspections (not affected by minor surface crack closure) and selecting impact methods that induce the least amount of residual stresses for a given task have been successful in minimizing this risk. It should be noted that all coating removal methods have the potential to have a negative effect on LPI.[7]

In 2002, the following Almen arc data were reported showing the difference between Type II, Type V and Type VII plastic media types.[6] Note that the productivity (strip rates) was similar for the three different media types. The values presented in Table 10.2 are the average of eight Almen strips and show residual stress levels after one blast cycle and after four blast cycles. The results suggest that the decision of the aerospace industry to move from Type II to Type V was a prudent one and that the current trend towards Type VII is well founded.

Coatings from composite materials can be removed using PMB, but aggressive media types and poor operator skill can result in surface erosion and fibre breakage. The soft media types are usually preferred as well as

Table 10.2 Residual stress levels after one and four blast cycles for three media types

	Residual stress level	
Plastic media	1 blast cycle	4 blast cycles
Type II	0.0150 in.	0.0192 in.
Type V	0.0079 in.	0.0094 in.
Type VII	0.0033 in.	0.0044 in.

selective stripping where the primer is left intact. Automation of PMB using flat nozzles is considered the most effective means of removing coatings from composite while minimizing the potential for damage. Typical operating pressure can range from 1.5–3.0 bar. A 2003 National Aerospace Laboratory (NRL) study on paint stripping techniques for composite aircraft components, indicated that PMB gave superior results to those of the thermal and high-pressure water techniques.[8]

Advantages

- PMB is a mature well proven process.
- Depending on the media type, it can be reused 3–20 times.
- Non-chemical process relatively safe for the operator.
- Cost, operator skill and engineering support, low to moderate.
- Selective stripping is feasible under certain circumstances.
- Disposal options of spent media includes recycling.

Disadvantages

- Aggressive media types can damage metallic and composite parts.
- Dust produced may contain toxic heavy metals, proper breathing protection is required.
- Ingress of media in structures is possible if incorrect masking procedures are used.

10.12 Engineered bio-based media

The concept of engineering a bio-based polymer media for aerospace coating removal applications was first realized in 1991. The first generation bio-based media was made via a hydrothermal extrusion process with pure native wheat starch. Several aerospace Original Equipment Manufacturers (OEM) found that this soft media type, with an apparent Moh's hardness of 2.0 and a specific gravity of 1.45 g cm^{-3}, had unique characteristics compared with other plastic media types available at the time. Several studies showed that this media was the gentlest dry-stripping media available. One of the first major users was a special USAF program, the B-2 Spirit aircraft fleet. After evaluating all of the alternative coating removal methods, wheat starch media was found to be unique in its ability to remove the thick elastomeric coatings found on this aircraft without compromising the all-composite skin of the B-2 aircraft. Wheat starch media has been used to maintain the B-2 fleet since 1994. During the 1990s, several OEMs evaluated and approved wheat starch for a wide variety of delicate aerospace applications such as coating removal from various different composites and metal structures and components. In the aerospace manufacturing sector, several OEMs found that wheat starch could selectively remove excess structural adhesive from bonded surfaces without removing the bond primer. This application replaced mechanical and chemical processes that damaged the corrosion protection scheme.

In 1996, during a new product research project, a blind comparative laboratory evaluation of a new corn hybrid starch polymer (CHP) media, along with other different starch polymer candidates, was conducted. During the performance evaluation, it was found that there were no significant differences between the candidate media in terms of mechanical effects, productivity or consumption, and the standard wheat starch product. However, it was discovered that this particular corn starch type was more resistant to water than the standard wheat starch. Hydration capacity testing by centrifuge confirmed that CHP had a higher water absorption resistance than wheat starch.

When introduced in 1998, the majority of OEMs that had approved wheat starch media found that there were no significant differences in mechanical effects between the two media types. Tested for years in various industrial settings, CHP was found to be more stable than the standard wheat starch in extremely high or low relative humidity conditions. Compared with wheat starch, CHP was also found to perform better when using less than ideal blast equipment.

CHP was also found to be very well suited for use as the backbone material to manufacture Type VII plastic media. The typical operating pressure for the bio-based media types is 1.0–3.0 bar. (15–45 psi).

Advantages

- Bio-based media is biodegradable and is made from a renewable resource.
- Bio-based media dry stripping is a mature and well proven process.
- Non-chemical process, relatively safe for the operators.
- Considered to be the most gentle of all impact processes.
- Cost, operator skill and engineering support low to moderate.
- Selective stripping is feasible under certain circumstances.
- Depending on media type and parameters they can be reused 8 to 15 times.
- Spent media disposal includes recycling options.

Disadvantages

- Dust produced can contain toxic heavy metals (from coating), proper protection is required.

- Ingress of media in structures is possible if incorrect masking procedures are used.
- Compared with aggressive plastic media, bio-based media may be slightly slower to remove certain tenacious coating systems.

10.13 Future trends

Large-scale chemical-stripping operations will continue to be closely monitored and heavily regulated. Around the world, new legislation is being enacted to increase workers' health and safety by decreasing their occupational exposure to potentially hazardous chemicals. For example, in June 2007, a new European chemical regulation came into effect, this EU regulation is called REACH (Regulation, Evaluation, Authorization, and restriction of CHemicals). This new regulation is a radical change from the past where the government authorities had most of the responsibilities to evaluate potential health hazards (cancer, infertility, genetic mutations or birth defects) of old and newly developed chemical substances, whereas under REACH the principal onus is now on the manufacturers. When users decide that they must use a particular hazardous product in their workplace, they may obtain authorization from REACH, but they will be required to progressively move towards safer alternatives where a suitable alternative exists. With over 30000 chemical substances to register, the process is expected to last until 2018.

Owing to the increasing number of reports that suggest that maternal and paternal occupational exposures to solvents can significantly increase the possibility of birth defects, chemical techniques are becoming even more regulated.

Growing pressure on all industries to move towards using green and sustainable resources will include aerospace coating removal activities. New bio-based solvents and coating removal formulas have been developed and will continue to evolve and improve. These products made from corn and soybeans are being processed into esters and ethyl lactates. Compared with their equivalents made from fossil fuel they have the added benefit that they are usually less toxic. The popularity of dry-stripping and bio-based techniques in particular will also continue to grow. For example, new engineered bio-based plastics (polyhydroxylalkanoate – PHA) could be engineered with a much higher Moh's hardness for bio-based alternatives to the aggressive plastic media types.

Automation of coating removal will continue to evolve and become more affordable as these advances will support all non-chemical processes. Multi-axis robots will permit much higher productivity and better results. In addition, the on-going trend to remove chromium and other toxic materials from all coating systems will make all dry-stripping processes safer.

Selective stripping of aircraft fuselages will be the preferred standard of the future, aided by new coating systems being developed, which will make this process easier. Applying chemical sacrificial coats or impact barrier coats with hardness higher than the primer coat will permit impact removal of the top coat with better success. The advantage of impact selective stripping versus a non-impact chemical or thermal method is that if corrosion or impact damage under the primer coat is present it will be revealed by the impact methods because of adhesion loss. This may not be the case with chemical removal.

10.14 Conclusions

Because there are so many different factors and requirements involved, there is no simple solution for removal of coatings in the aerospace industry. In the past, priorities for coating removal techniques were protection of the workpiece (aircraft) and economics. Current and future trends, however, will also include increased protection of the workpiece because of ageing aircraft issues (more coating removal cycles) and the introduction of new non-metallic aircraft materials. Economics will always be a critical priority. Finally, worker health and safety will be an ever increasing priority when selecting a technique.

10.15 References

1 FONTANA M. G., *Corrosion Engineering 3rd Edition*, McGraw Hill, ISBN 0-07-021463-8, 1986, 556 pp.
2 NASA (National Aeronautics and Space Administration), Joint EPA/NASA/USAF interagency depainting study. *Report Number*, NP-1999-12-152-MSFC, 1999, 100 pp.
3 GRIMES J., Process strips paint off B-1 s, USAF press release, 7/18/2003.
4 GATES D., New 787 goes on diet to keep that svelte look airlines love, Seattle Times, 11/21/2005.
5 PAULI R., 'Alternative processes to methylene chloride chemical strippers, a review of progress to date', *1995 Aerospace/Airline plating and metal finishing forum*, American Electroplaters and Surface Finishers Society.
6 MONETTE D., OESTRIECH J., 'Enhanced type VII plastic media e-strip (MIL-P-85891a) for military aerospace applications', *DoD Industry Advanced Coating Removal Conference*, 2002, 11 pp.
7 LARSON B., A study of the factors affecting the sensitivity of liquid penetrant inspections: a review of literature from 1970–1998, *FAA Technical Report*, DOT/FAA/AR 01-95, 2002, 51 pp.
8 HART W. G. J., Paint stripping techniques for composite aircraft components, *National Aerospace Laboratory Technical Publication*, NLR-TP-2003-357, 2003, 13 pp.

11
Novel corrosion schemes for the aerospace industry

F. G U I, CC Technologies – a DNV company, USA

Abstract: The aluminum alloys used on aircraft because of their light weight and high strength-to-weight ratio are susceptible to localized corrosion in chloride-containing environments. To combat corrosion, aircraft are conventionally protected by a multilayer chromate coating system and other protective measures at locations where conventional coating systems alone are not sufficient for corrosion protection. Regulations recently introduced by the Environmental Protection Agency (EPA) on the usage of chromate has led to efforts to either seek a chromate replacement or develop novel protection methods. In this chapter, both the non-conventional methods that are currently being used and the novel corrosion protection methods emerging as the result of recent research in this area are reviewed.

Key words: aluminum alloy, corrosion, coating, inhibitor, corrosion prevention, CPC, surface treatment.

11.1 Introduction

Aluminum alloys are widely used on aircraft because of their light weight and high strength-to-weight ratio. Often these alloys contain different concentrations of copper, zinc, magnesium and other elements in order to achieve the desired mechanical strength. The addition of these alloying elements, however, also causes the aluminum alloys to be susceptible to localized corrosion, especially in chloride-containing environments. Because of this, aircraft are protected conventionally by a multilayer chromate coating system from the corrosive environment. Along with the coating system, other protective measures are used at locations where conventional coating systems are not sufficient alone for corrosion protection. Regulations introduced recently by the Environmental Protection Agency (EPA), however, restrict the use of chromate as an inhibitor because of its carcinogenic effect. Consequently, much research has been conducted to either seek a chromate replacement or develop novel methods to provide protection for aircraft aluminum alloys. In this chapter, both the non-conventional methods that are currently being used and the novel corrosion protection methods emerging as the result of recent research in this area are reviewed.

11.2 Corrosion prevention compounds

Corrosion prevention compounds (CPCs) are materials that can prevent new corrosion sites from forming and, more importantly, suppress corrosion that has already initiated. CPCs are applied as a post-production treatment to provide cost-effective, temporary corrosion protection and to control existing corrosion.

CPCs have been used on aircraft for many years as a relatively inexpensive method of combating corrosion.[1] One of the main advantages of using CPCs is that little or no preparation of the affected site is required before application. Consequently, these CPCs can be used at the field maintenance level instead of requiring application at the depot. Thus, corrosion can be suppressed early on, before substantial structural damage can occur. CPCs are not meant to replace high-performance coating systems, but they can be effective for on-site repair of coated regions that may have been damaged or degraded, for extending the service life of a coating, and for protecting regions of aircraft that did not receive corrosion prevention treatments during original manufacture. CPCs can, in principle, serve a key function as a component of a corrosion management strategy.

The detailed compositions of most CPCs are not available as they are proprietary. Most of them, however, include two fundamental components: (a) organic solvent; (b) film former.[1] In some cases, inhibitors are also included in CPCs although they usually are not widely used because of the cost. The film formers are not usually polymers but are low-molecular-weight hydrocarbons. An example of a CPC that includes inhibitors is Amlguard, which has barium petroleum sulfonate (2.2%) and alkyl ammonium organic phosphate (3.5%) as corrosion inhibitors. CPCs can be divided into various categories based on the type of the film they can form or their ability to displace water. In addition, almost every CPC has some particular specifications to characterize its chemical or physical properties, such as viscosity, flash point or density.

The organic solvent (e.g. aliphatic hydrocarbon) usually functions as both a carrier and a water displacing agent, as well as helping CPCs to maintain the viscosity needed to allow them to spread evenly on the surfaces where protection is needed. Low viscosity is also critical for CPCs in order to penetrate into occluded regions such as aircraft lap structures and tight cracks. Once the solvent evaporates, the film former (e.g. an oil, grease or resin[1]) is left behind to form a barrier film on the applied surface, thus blocking contact with a corrosive electrolyte. As mentioned above, CPCs are aimed at a cost-effective way to provide short-term protection. Cost considerations play a role in what inhibitors are implemented for a CPC application and, consequently, most CPCs only contain some weak inhibitors.

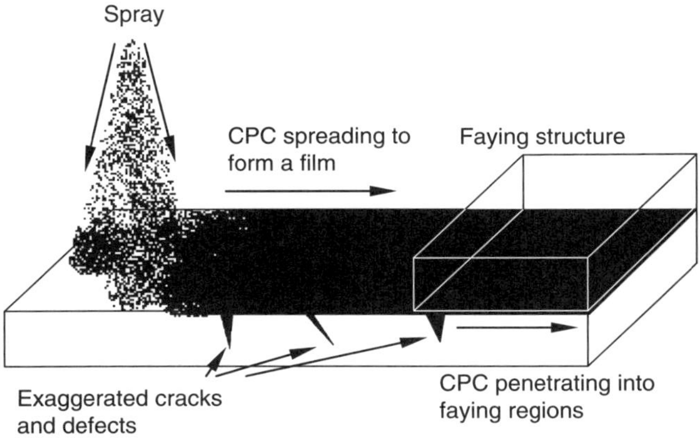

11.1 Schematic drawing illustrating the regions in which CPCs can provide protection.

A CPC effectively adheres to a target area through three mechanisms: (1) wicking into occluded regions quickly, independent of the surface condition and gap; (2) wetting of the surface while displacing water; and (3) forming and maintaining a protective film. Note that a CPC does not necessarily need to complete all of these mechanisms for certain applications. Figure 11.1 illustrates these processes.[2] The first process is not necessary when CPCs are applied to protect a boldly exposed surface. The second and third processes are more important in this case. CPCs need to cover the target surfaces and form and maintain a protective film. Water displacement may not be essential for a boldly exposed surface since its geometry does not lend to trapping water. Water displacement is essential for corrosion inhibition in occluded regions, however, because it is almost impossible to make the interior surface dry once water gets trapped inside owing to its extremely low egress rate.[3] When applied to a joint, the film formation is retarded because of slow evaporation of solvent in the occluded region.

Hinton *et al.*[1] assessed various commercially available CPCs and demonstrated the protection ability that CPCs can provide in both the laboratory and in the field. There are a considerable number of studies that apply CPCs to address particular corrosion problems. Salagaras *et al.*[4] describe the corrosion protection offered by CPCs regarding crevice, filiform corrosion, and stress corrosion cracks. Jacobs *et al.*[5,6] examined corrosion control of selected field problems with CPCs and Wilson *et al.*[7] evaluated the effect of water displacing CPCs on the corrosion behavior of two aluminum alloys.

Recent work by Gui[8,9] demonstrated the barrier properties offered by CPCs using electrochemical impedance spectroscopy (EIS). Through this technique, the impedance of a CPC-coated aluminum substrate was measured in a corrosive electrolyte. A higher impedance at low frequency

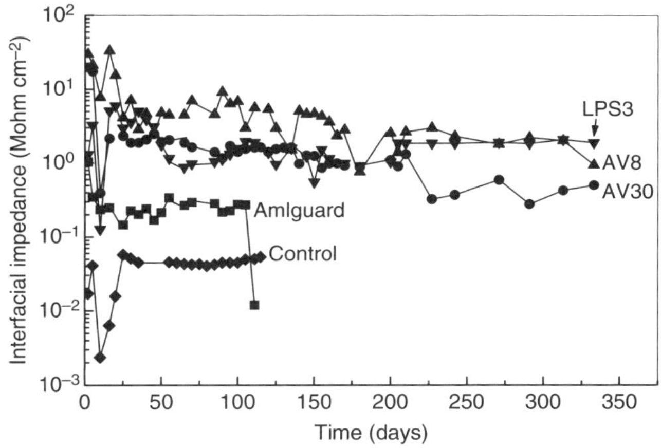

11.2 The evolution of the interfacial impedance of CPC-treated AA7075-T6 samples under constant immersion in LJSS.

usually reflects better barrier performance of the CPC. As shown in Fig. 11.2, specimens coated with a CPC exhibited much higher impedance than the bare control sample in a lap joint simulated solution (LJSS) that was developed to mimic the chemistry inside a joint structure on aircraft.[10] During these tests on various CPCs, a few specimens maintained a relatively high impedance for nearly one year of exposure, implying that some CPCs exhibit reasonable longevity. Gui[8] and Cooper *et al.*[11] also demonstrated that CPCs were capable of penetrating into lap joints and replacing the moisture present in the joint, fulfilling mechanisms (1) and (2) described above. In Fig. 11.3, the signal of a fiber optical sensor vs. time for a CPC when applied to a lap joint is shown. The detailed introduction on the mechanism of the sensor is described elsewhere.[8] Before the introduction of CPC, the aqueous LJSS was applied and detected. The subsequent sharp increase in the wavelength loss indicated the detection of CPC. Because the joint was wetted by LJSS, this suggested that CPC was able to displace water to reach the sensor.

Because of the wicking and water displacement properties that they possess, CPCs are ideal for providing protection to lap joint structures on aircraft. However, it should be noted that CPCs are designed to provide short-term protection and should be used in conjunction with conventional coating systems as complementary measures. The coating system should be used as the major barrier for isolating the structures from a corrosive environment. There are many reasons for this, one being that it is difficult for most CPCs to wick into joint structures or crevices when the interior surface is corroded and wet.[8,11] Thus, for joint structures, CPCs might not be able

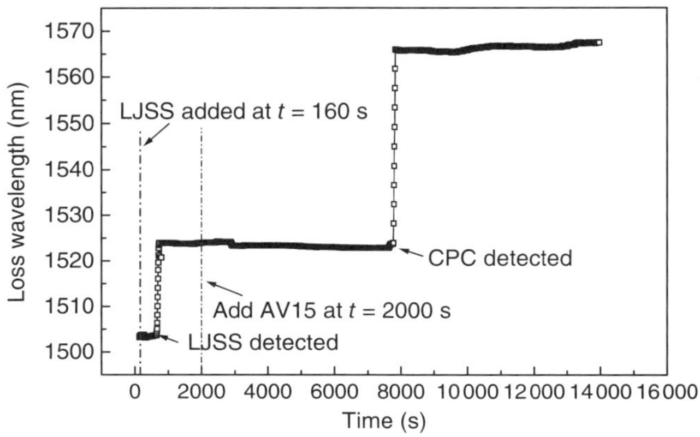

11.3 Wicking of LJSS into a dry, pristine AA7075-T6 riveted joint and subsequent wicking of AV15 against LJSS.

to provide sufficient protection once corrosion is initiated although they could wick into non-corroded joint structures and provide protection. This is because water can easily penetrate into these structures but the egression is extremely difficult and, thus, the CPC cannot wick and create a film in corroded cracks, crevices, or joints.

11.3 Novel inhibitors and coatings

Aircraft coatings typically consist of three layers: (1) conversion coating, (2) primer and (3) top coating. Conversion coating is a thin chemically grown layer on the surface of the aluminum alloy that could enhance the bonding of the organic coating and the metallic substrate. To optimize the coating performance and provide effective corrosion protection, inhibitors are usually incorporated into conversion coating or primer. Homogeneity and uniformity of the top coat (i.e. defects and 'holidays' are absent) is paramount for optimum corrosion protection. The most common inhibitor used in coating systems for aircraft is hexavalent chromium, which can effectively inhibit the oxygen reduction reaction – the main cathodic reaction on aluminum alloys. There have been many advances in inhibitor research during the last decades driven by the desire to replace hexavalent chromium to meet EPA regulations.

Although, to date, no inhibitors have exhibited superior performance to chromate with respect to inhibiting localized corrosion on aluminum alloys, several researchers have demonstrated the performance of a few promising inhibitors, such as the rare-earth compounds. Hinton *et al.*[12] first demonstrated a reduced corrosion rate of immersed AA7075 after the addition of trace amounts of cerium ions in a chloride containing environment

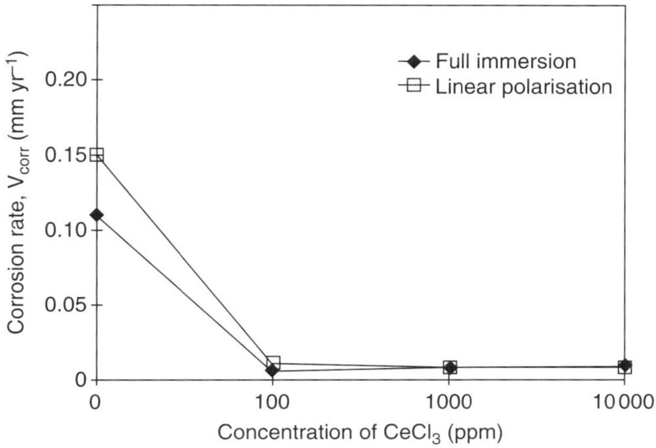

11.4 The corrosion rate of AA7075 in NaCl solutions as a function of $CeCl_3$ concentration.

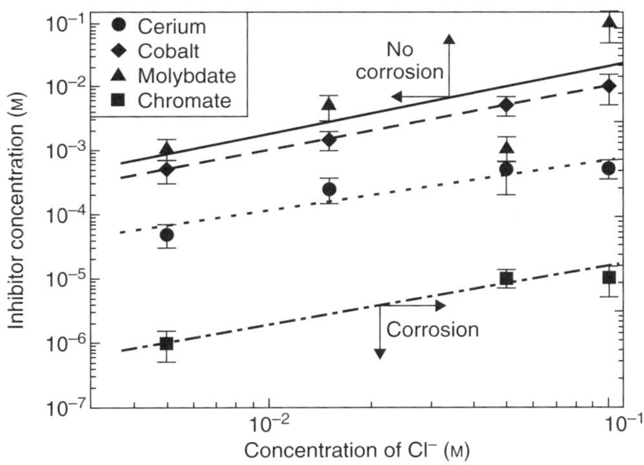

11.5 Critical (•) Ce(III), (♦) Co(II), (▲) Mo(IV), and (■) CrO_4^{2-} inhibitor concentrations as a function of chloride concentration determined by inhibition of corrosion leading to copper replating on AA2024-T3 over a range of NaCl concentrations. The error bars indicate standard deviation of the measurements.

(Fig. 11.4). Others also demonstrated the inhibiting effect of rare-earth metal compounds on corrosion of aluminum alloys.[13–18] Recently, Jakab *et al.*[19] showed that cobalt, cerium, and molybdenum ions released from an amorphous Al–Co–Ce (–Mo) alloy can provide sound corrosion inhibition on AA2024-T3. Figure 11.5 is a comparison of the critical inhibitor concentration as a function of chloride concentration between Ce(III), Co(II), Mo(IV) and CrO_4^{2-}. Compared with chromate, Ce(II) appeared to be the most potent inhibitor of the three investigated ions because it has the lowest critical concentration at any give chloride level. The other two ions also inhibit the corrosion of AA2024-T3 but they require much higher concentrations than Ce(II).

The inhibition of localized corrosion on AA2024-T3 may be attributed to the formation of hydroxides or oxides formed as a result of the reactions below:

$$Co^{2+} + 2OH^- \rightarrow Co(OH)_2, \; pK_{sp} = 14.23$$

$$Co^{3+} + 3OH^- \rightarrow Co(OH)_3, \; pK_{sp} = 44.30$$

$$Ce^{3+} + 3OH^- \rightarrow Ce(OH)_3, \; pK_{sp} = 26.15$$

$$MoO_4^{2-} + 4H^+ + 2e^- \rightarrow MoO_2 + 2H_2O$$

The formed hydroxides (or oxides) and adsorbed ions act as physical barriers to oxygen transport, not unlike the function of a coating layer.[19] Consequently, the cathodic reaction is inhibited.

The inhibiting ions Co(II), Ce(III) and Mo(IV) can be made available at the corroding site by being released from an amorphous Al–Co–Ce (–Mo) alloy layer. As shown in Fig. 11.6, the concentrations of the Ce^{3+} and Co^{2+} ions were observed to increase with the exposure time of the $Al_{87}Co_7Ce_6$ amorphous alloy in 0.01M HCl solution at pH 2. More importantly, the ion concentrations soon surpassed the critical concentration needed to inhibit localized corrosion of AA2024-T3 at this chloride level. The release kinetics

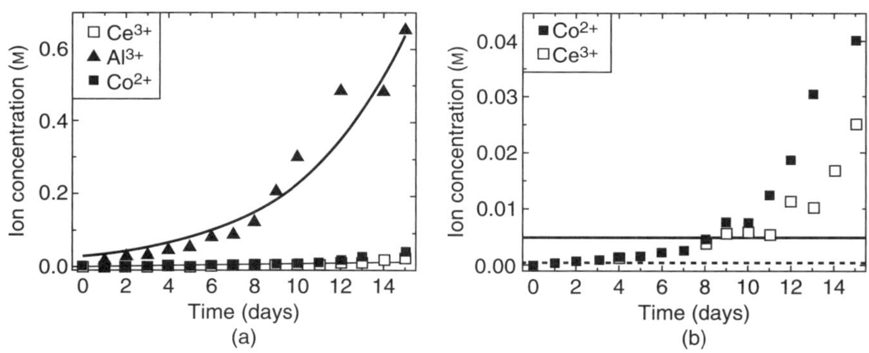

11.6 Measured concentrations of Al^{3+}, Ce^{3+}, and Co^{2+} ions in aliquots obtained after exposure of $Al_{87}Co_7Ce_6$ amorphous alloy to HCl solution (pH 2). The solid and dashed lines indicate the critical inhibitor concentrations for Co^{2+} and Ce^{3+}, respectively, in 0.05M NaCl. Ratio of the surface area of the alloy to the solution volume was 56 cm^{-1}.

of the inhibiting ions were confirmed in the same work. It was concluded that the inhibiting ions can be released from the amorphous alloy coating layer and delivered to the sites where corrosion was initiated, although the authors did not address the issue of throwing power, i.e., whether a threshold distance exhibits beyond which it would not be possible to deliver the inhibitors efficiently to provide corrosion inhibition.

Ho and co-workers investigated synergistic effects of several rare-earth metals when they were coupled with a multifunctional organic component.[17,20,21] The mixed rare-earth organophosphate, mischmetal diphenyl phosphate [$Mm(dpp)_3$], was demonstrated to significantly reduce the cathodic current density, as illustrated in Fig. 11.7, and it also showed stronger inhibition than $Ce(dpp)_3$. Since the main rare earth metals present in $Mm(dpp)_3$ are Ce (48–56%), La (20–27%), Nd (12–20%), and Pr (4–7%), the strong inhibiting effect observed with $Mm(dpp)_3$ may stem from the synergistic effect of several different rare-earth metals. This was confirmed by Markley and Forsyth in their recent work.[22] It was found that $Ce(dpp)_3$ and $Pr(dpp)_3$ appeared to work synergistically, resulting in a greater degree of corrosion inhibition of AA2024-T3 in 0.1M NaCl solution than that observed when either compound was used individually.[22] Among these two compounds, they found that $Ce(dpp)_3$ contributed more to reducing the cathodic reaction kinetics than $Pr(dpp)_3$.

Vanadium-based oxyanions, also referred to as vanadates, have also been investigated as inhibitors for aluminum alloys.[23–26] The speciation of

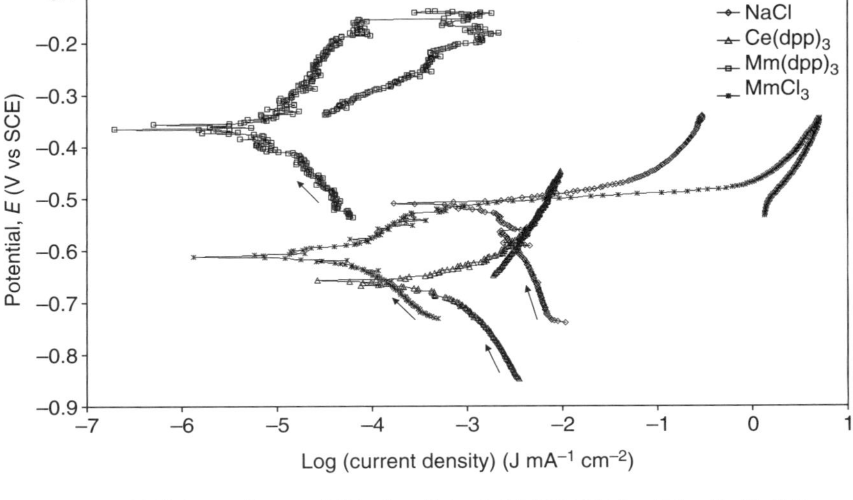

11.7 Thirty minutes CPP of polished AA2024-T3 in: 0.1M NaCl, 0.2mM $Ce(dpp)_3$ + 0.1M NaCl, 0.2mM $MmCl_3$ + 0.1M NaCl.

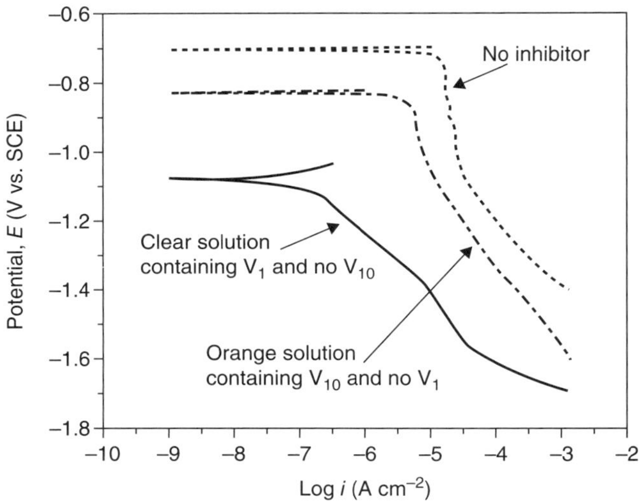

11.8 Inhibition of oxygen reduction reaction by vanadates.

vanadium is very complex and strongly depends on the pH of the solution. Because of this, an agreement on the vanadates species that are responsible for corrosion inhibition is lacking. Guan and Buchheit[26] attributed the inhibition of corrosion to the formation of a protective film involving the polymerization of pentavanadate (V_5). Iannuzzi *et al.* demonstrated that the inhibition from vanadates was mainly contributed by monovandates (V_1), which formed naturally in solutions containing vanadium salts.[27,28] As shown in Fig. 11.8, the oxygen reduction reaction was significantly inhibited in the solution containing monovanadates. The authors suggested that the inhibition from vanadates was due to the adsorption of the inhibitors on intermetallic compounds and thus the active sites on the intermetallic compounds were blocked. Overall, this resulted in the reduction in the oxygen reduction reaction kinetics.[29]

In parallel with efforts to find new inhibitors to replace chromate, some advances have also been made to improve the barrier properties of coating systems. One of the new advances is the use of sol–gel coatings. The sol–gel process is a method in which oxide films can be deposited on a substrate at much lower temperatures than the traditional ceramic process method.[30] Through sol–gel synthesis, coatings can be tailored to have functionally gradient properties that have strong adhesion to aluminum alloys due to the strong covalent bonding. The coating can also act as a potent barrier to reduce the permeation of corrosive electrolytes to the alloy surface.[31] To synthesize sol–gel coatings, a metal alkoxide usually is used as a precursor. The precursor goes though hydrolysis when deposited onto a metal surface

to form an intermediate product, which subsequently undergoes condensation on the metal surface to form a sol–gel coating. This process is demonstrated with the following reactions.[31]
Hydrolysis:

$$M(OR)_n + H_2O \rightarrow M(OR)_{n-1}(OH) + ROH$$

Condensation:

$$2M(OR)_{n-1}(OH) \rightarrow (OR)_{n-1}M - O - M(OR)_{n-1} + H_2O$$

or

$$M(OR)_{n-1}(OH) + M(OR)_n \rightarrow (OR)_{n-1}M - O - M(OR)_{n-1} + ROH$$

Sol–gel coatings, as well as sol–gel coatings doped with inhibitors, have been demonstrated to protect aluminum alloys to some extent.[32–36] Moutarlier and co-workers studied the corrosion protection of sol–gel coatings doped with inorganic inhibitors. The inhibitors investigated include Cr(VI), Cr(III), Ce(III), and Mo(VI). The inhibition offered by Cr(VI) was observed to be the strongest among the investigated inhibitors. Other inhibitors did not show very promising performance owing either to additive-induced changes (decrease the sol–gel network stability) in the sol–gel structure or to the high solubility of the additives (cause rapid leaching of the inhibitors). Hamdy also studied the anti-corrosion performance of a cerium-treated sol–gel coating on aluminum alloys.[32] It was found that direct cerium treatment did not improve corrosion protection for aluminum alloys in a NaCl solution. However, when combining with surface treatment by etching and oxide thickening before introducing the sol–gel process, the cerium treatment significantly improved the performance of the coating. The improvement was attributed to the formation of a more compact film on the substrate with surface modification.[32] It appeared that the etching process was beneficial to enhance the distribution of aluminum oxide film uniformly. Khramov *et al.* found that the sol–gel derived organosilicate hybrid coating exhibited pronounced corrosion protection on aluminum substrate when some organic inhibitors were incorporated in the coating.[33] The incorporation of the organic inhibitors was achieved such that the inhibitors were activated in a corrosive environment and released gradually to provide protection.[33,37] Because the release of the inhibitors was controlled, the coatings embedded with inhibitors exhibited self-healing properties and demonstrated superior performance over an extended period of time. As shown in Fig. 11.9, the coatings with an imbedded organic inhibitor maintained high impedance after four weeks whereas the coating without inhibitor showed decreased impedance after four weeks of exposure, indicating reduced coating performance.

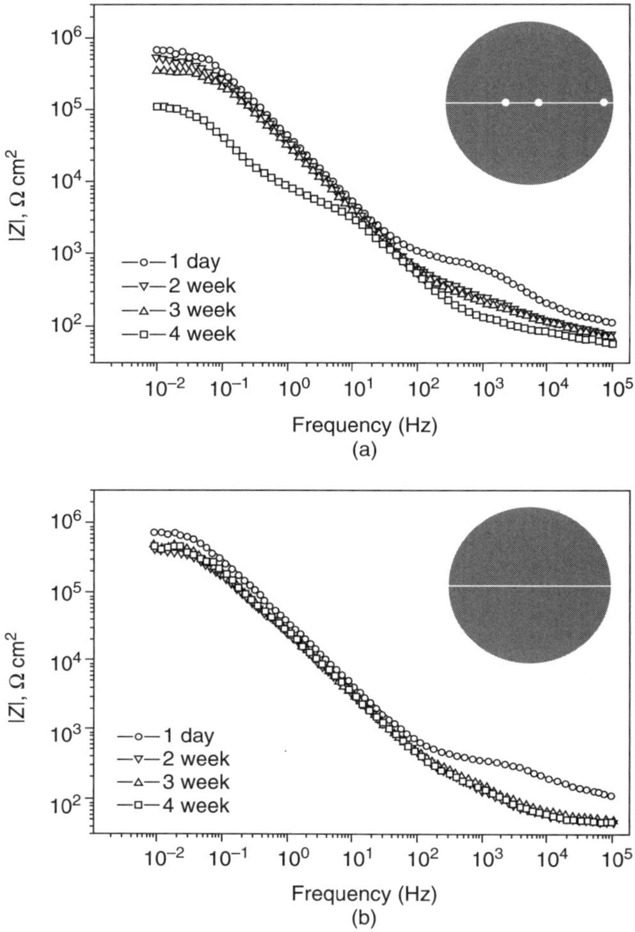

11.9 The electrochemical impedance spectra for scribed coatings at different immersion times in dilute Harrison's solution (a) without inhibitor and (b) with MBI/β-cyclodextrin complex.

As part of the research into nanomaterials, some work has been conducted to change the synthesis process of sol–gel coatings such that functionalized silica nanoparticles can be formed *in situ*.[38] This synthesis method is called a *self-assembled nanophase particle* (SNAP) process and consists of various stages. First, the sol–gel process is initiated using the conventional hydrolysis method, which also forms nano-sized, particle-like, silane macromolecules. Crosslinking agents and additives are then added to the sol–gel solution. When the solution is applied to a substrate, the crosslinker chemically connects nano-sized siloxane oligomers to form an organic coating.

The nanoparticles formed in the solution reduce the coating porosity that normally results from hydrolysis and condensation reactions taking place in the conventional sol–gel process. Because the SNAP process can form a thin and dense film on metal substrate, this process is a promising surface treatment method to form a 'conversion coating'. The SNAP film can provide inorganic functionality to form strong bonds to the metal substrate and tailorable organic functionality to interact with the subsequent organic coating layers.[38] Donley and co-workers demonstrated that the SNAP treatment on AA2024-T3 combined with a topcoat showed superior corrosion protection, although no comparison has been made with the film formed through a conventional sol–gel process. The protection offered by a SNAP film can be further enhanced by incorporating inhibitors, as observed by Voevodin *et al.* during a 2000-hour salt spray test.[39]

To improve the corrosion resistance of non-metallic composite materials and organic coatings, many researchers have also been attempting to incorporate nanomaterials by mixing nanoparticles or nanofibers into the bulk material.[40] There is evidence of the improvement of corrosion resistance of these materials after being modified with nanomaterials. However, whether such a method is economically practical or even technically sound for practical use still remains unclear.

In summary, tremendous efforts are ongoing in seeking novel inhibitors and coating systems to provide corrosion protection of aluminum alloys without using toxic hexavalent chromium as an inhibitor. However, because of the superior inhibition efficiency that chromate demonstrates, it is extremely difficult to find alternative inhibitors that are as effective as chromate. There is an additional challenge to make replacement inhibitors that are environmentally friendly. Nevertheless, the inhibitors discussed above are very promising, although they are still under development. Furthermore, even these newly identified inhibitors do not perform as well as chromate does when combined with other measures, such as the novel coating systems described above. In the future, these inhibitors may be able to provide potent protection for aluminum alloys.

11.4 Novel surface treatment

Another route that has been investigated to impart protection on aluminum alloys is to modify the alloy surfaces; surface modification methods include laser surface treatment and ion implantation amongst others. Laser technologies have been successfully used to modify the surface of steel in earlier years to improve the resistance to corrosion.[41–46] The use of laser technologies to enhance the corrosion resistance of aluminum alloys has been rising over the last several years. Generally, the laser technology used to modify alloy surfaces are either laser surface melting (LSM) or laser surface alloying (LSA).[47] In the LSM technique, the surface layer of the alloy is melted by laser then cooled under non-equilibrium cooling conditions established when the thin melted layer is allowed to solidify in contact with the substrate.[47] As a result of the non-equilibrium cooling process, a unique microstructure forms that is believed to improve the corrosion resistance of the alloy. In the LSA technique, a melt pre-/co-deposited layer of alloying elements – frequently in powder form – and the underlying substrate forms an alloyed zone through the assistance of the laser beam.[48] Similarly, novel microstructures can be formed because of the rapid solidification and the high concentration of beneficial alloy elements.

A number of researchers have demonstrated improvements in corrosion resistance after LSM of aluminum alloys, although the improvement seems to be dependent on alloy type.[49–52] Liu and co-workers studied the corrosion resistance of laser melted AA2014 and AA2024 alloys to 1M NaCl solution. It was found that the AA2014-T6 alloy showed improvement in the corrosion resistance after laser treatment by showing a 170 mV positive shift in the pitting potential. However, no improvement was observed on AA2024-T351 after LSM, and the pitting potential remained approximately the same or even more negative than that before treatment.[51] It was concluded that the effect of LSM on pitting corrosion behavior is dependent on the electrochemical characteristics of second-phase particles relative to the aluminum matrix. In the case of AA2014-T6, the laser melting process extended the solubility of copper in the aluminum matrix. Consequently, the difference in the potential between the aluminum matrix and the dominated second phase particle (θ-Al_2Cu) was reduced and thus the driving force for localized corrosion was decreased. Conversely, when the copper solubility was increased through LSM in AA2024-T351, the potential difference between the aluminum matrix and the dominated second phase particle (Al_2CuMg) increased. As a result, the likelihood of localized corrosion increased. Xu *et al.* demonstrated the improvement in the corrosion resistance of aluminum alloy 6013 after laser surface melting.[52] Specimens examined with EIS after laser melting (performed under both air and nitrogen) showed increased low-frequency impedance compared with untreated samples. The corrosion fatigue lives of the laser-treated samples under air and nitrogen both were much longer than the untreated samples. The improvement in the corrosion resistance and thus the corrosion fatigue life was attributed to the formation of a homogeneous microstructure on the aluminum surface without the presence of any constituent particles.

Little work has been reported with respect to the improvement of the corrosion resistance of aluminum alloys through LSA compared with the LSM technique. Man *et al.* produced a NiCrSiB alloy layer on an Al6061 aluminum alloy using the laser surface alloying technique.[53] To obtain an alloying layer on the aluminum substrate, NiCrSiB powder was mixed with

4% PVA to form a paste. The paste was then painted on the substrate that was roughened by sand blasting. The surface was subsequently alloyed with a continuous-wave Nd-YAG solid-state laser. After surface alloying, the cavitation erosion resistance of the aluminum alloy was improved significantly. The surface alloy also slightly increased the pitting potential. However, it is uncertain if it is beneficial from the perspective of localized corrosion because the treated sample demonstrated spontaneous pitting behavior whereas the untreated sample showed a fairly wide passive potential region.

Another surface treatment method that has attracted a significant amount of attention is the ion implantation technique. An example of the ion implantation process is shown in Fig. 11.10. The ion source is maintained at high voltage and the target is at earth potential. As such, a positively charged beam of ions can be produced. Upon being generated, the beam is accelerated to a high velocity and directed at the solid surface of the target.[54] After penetration, the speed of the ions is reduced by collision with the atoms in the solid. A doped layer is formed as a result of implantation into the surface. The ion implantation technique originated from the semiconductor industry, where this technique was used to dope the silicon surface to obtain a desired surface layer. Compared with conventional diffusion techniques to produce semiconductor devices, this technique offers controlled deposition and uniformity, because the ion beam can scan a relatively wide surface and the energy of the beam can be well controlled.[54]

The use of the ion implantation technique has been broadened to several other fields to achieve custom surfaces with desired properties. In the field of corrosion, there has been a number of studies conducted to implant ions

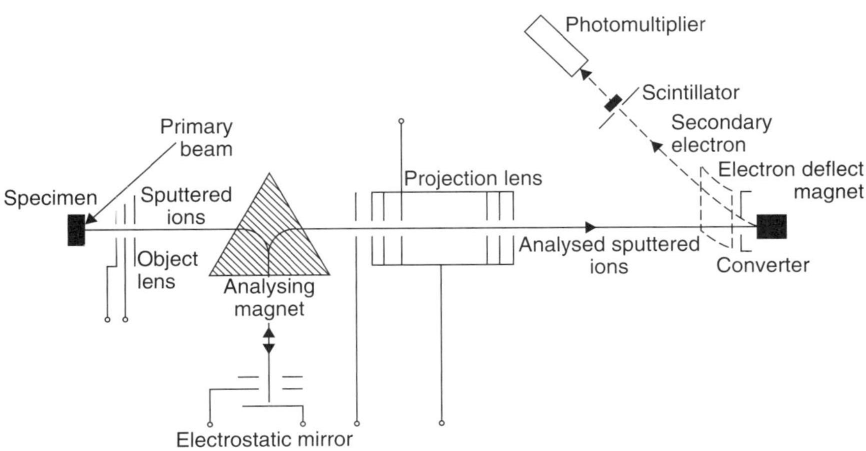

11.10 An example of ion implantation process.

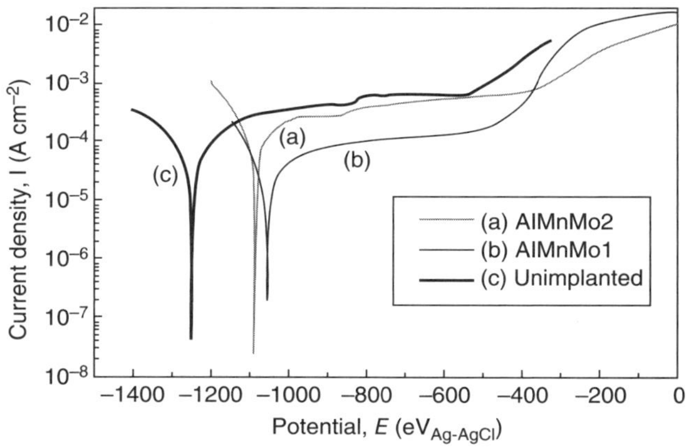

11.11 Anodic polarization curves of Al–Mn alloy samples in deaerated 0.1M NaCl solution buffered with bicarbonate/carbonate solution.

such as Mo, Cr, N, and Ni to improve the resistance of aluminum to localized corrosion.[55–61] Zhang *et al.*[61] demonstrated the reduction in the passive current of Al–Mn alloy after molybdenum implantation as shown in Fig. 11.11. This reduction was attributed to the replacement of a portion of Al—O bonds with Mo—O bonds. Consequently, the molybdenum oxides formed offered the possibility of inhibiting Cl^- adsorption.[61] However, it was observed that the pitting corrosion propagation rate increased after Mo implantation due to the ennoblement of the surface layer with respect to the relatively active Al–Mn substrate. Tian *et al.* implanted nickel ions into aluminum 1070 alloys, improving the corrosion resistance in a 1M NaCl solution as indicated by anodic polarization and impedance experiments.[60] The authors concluded that the improvement in the corrosion resistance resulted from the formation of a more uniform and compact NiO-containing film, as confirmed by x-ray photoelectron spectrometry (XPS) and atomic fluorescence microscopy (AFM) characterization work.

It is not the purpose of this chapter to cover all novel corrosion protection techniques. Rather, this review only reflects a few of the techniques that have been extensively investigated. There are many other non-conventional techniques that can enhance the corrosion resistance of aluminum alloys, including, but not limited to, the use of live biofilms,[62] magnetron sputtering,[63] and conductive polymers.[64,65] Some of these techniques are being used in the aircraft industry to extend the service life of the aluminum structural components. For instance, CPCs are constantly used during the maintenance period of aircraft and relatively large-scale experiments are ongoing with conversion coatings containing a few non-chromium inhibitors. However, many of them are still in development, and even if the knowledge has been well developed, some barriers still need to be overcome to allow large-scale manufacturing.

11.5 References

1 B. HINTON, P. TRATHEN, P. HABERECHT and P. BUSHELL, in *Proceedings of the Fourth International Aerospace Corrosion Control Symposium*, Jakarta, Indonesia (1996).

2 ZIP-CHEM, Document CD for Customers.

3 L. B. SIMON, J. L. ELSTER and R. G. KELLY, in *2000 USAF Aircraft Structural Integrity Program (ASIP) Conference*, San Antonio, TX (2001).

4 M. SALAGARAS, P. G. BUSHELL, P. N. TRATHEN and B. HINTON, in *5th Joint FAA/DoD/NASA Conference on Aging Aircraft* (2001).

5 H. JACOBS, in *47th AHS Annual Forum*, p. 1483, Phoenix, AZ (1991).

6 H. JACOBS and M. J. LLOYD, in *44th AHS Annual Forum*, p. 237, Washington, DC (1988).

7 L. WILSON and R. S. G. DEVEREUX, *Metal Forum*, **7**, 50 (1984).

8 F. GUI, 'Development of a performance test protocol for corrosion prevention compounds for aircraft', PhD Dissertation, University of Virginia, Charlottesville (2006).

9 F. GUI and R. G. KELLY, *Corrosion*, **61**, 119 (2005).

10 K. S. LEWIS, 'Determination of the corrosion conditions within aircraft lap splice joints', Masters Thesis, University of Virginia (1999).

11 K. R. COOPER, K. FURROW and R. G. KELLY, *Corrosion*, **61**, 155 (2005).

12 B. R. W. HINTON, D. R. ARNOTT and N. E. RYAN, *Metal Forum*, **7** (1984).

13 J. HU, X. H. ZHAO, S. W. TANG, W. C. REN and Z. Y. ZHANG, *Applied Surface Science*, **253**, 8879 (2007).

14 X. HUANG, N. LI, H. WANG, H. SUN, S. SUN and J. ZHENG, *Thin Solid Films*, **516**, 1073 (2008).

15 S. YOU, P. JONES, A. PADWAL, P. YU, M. O'KEEFE, W. FAHRENHOLTZ and T. O'KEEFE, *Materials Letters*, **61**, 3778 (2007).

16 A. K. MISHRA and R. BALASUBRAMANIAM, *Materials Chemistry and Physics*, **103**, 385 (2007).

17 T. A. MARKLEY, M. FORSYTH and A. E. HUGHES, *Electrochimica Acta*, **52**, 4024 (2007).

18 A. K. MISHRA and R. BALASUBRAMANIAM, *Corrosion Science*, **49**, 1027 (2007).

19 M. A. JAKAB, F. PRESUEL-MORENO and J. R. SCULLY, *Corrosion*, **61**, 246 (2005).

20 N. BIRBILIS, R. G. BUCHHEIT, D. HO and M. FORSYTH, *Electrochem. Solid-State Lett.*, **8**, C180 (2005).

21 D. HO, N. BRACK, J. R. SCULLY, T. MARKLEY, M. FORSYTH and B. R. W. HINTON, *Journal of the Electrochemical Society*, **153** (2006).

22 T. MARKLEY, A. E. HUGHES, T. C. ANG, G. B. DEACON, P. JUNK and M. FORSYTH, *Electrochem. Solid-State Lett.*, **10** (2007).

23 R. G. BUCHHEIT, H. GUAN, S. MAHAJANAM and F. WONG, *Progress in Organic Coatings*, **47**, 174 (2003).

24 B. D. CHAMBERS, S. R. TAYLOR and M. W. KENDIG, *Corrosion*, **61** (2003).

25 R. L. COOK and S. R. TAYLOR, *Corrosion*, **56**, 321 (2000).

26 H. GUAN and R. G. BUCHHEIT, *Corrosion*, **60**, 284 (2004).

27 M. IANNUZZI, J. KOVAC and G. S. FRANKEL, *Electrochimica Acta*, **52**, 4032 (2007).
28 M. IANNUZZI, T. YOUNG and G. S. FRANKEL, *Journal of the Electrochemical Society*, **153**, B533 (2006).
29 M. IANNUZZI and G. S. FRANKEL, *Corrosion Science*, **49**, 2371 (2007).
30 R. L. TWITE and G. P. BIERWAGEN, *Progress in Organic Coatings*, **33**, 91 (1998).
31 L. S. KASTEN, J. T. GRANT, N. GREBASCH, N. VOEVODIN, F. E. ARNOLD and M. S. DONLEY, *Surface and Coatings Technology*, **140**, 11 (2001).
32 A. S. HAMDY, *Materials Letters*, **60**, 2633 (2006).
33 A. N. KHRAMOV, N. N. VOEVODIN, V. N. BALBYSHEV and R. A. MANTZ, *Thin Solid Films*, **483**, 191 (2005).
34 V. MOUTARLIER, B. NEVEU and M. P. GIGANDET, *Surface and Coatings Technology*, **202**, 2052 (2008).
35 R. SUPPLIT and U. SCHUBERT, *Corrosion Science*, **49**, 3325 (2007).
36 N. N. VOEVODIN, J. W. KURDZIEL and R. MANTZ, *Surface and Coatings Technology*, **201**, 1080 (2006).
37 A. N. KHRAMOV, N. N. VOEVODIN, V. N. BALBYSHEV and M. S. DONLEY, *Thin Solid Films*, **447–448**, 549 (2004).
38 M. S. DONLEY, R. A. MANTZ, A. N. KHRAMOV, V. N. BALBYSHEV, L. S. KASTEN and D. J. GASPAR, *Progress in Organic Coatings*, **47**, 401 (2003).
39 N. N. VOEVODIN, V. N. BALBYSHEV, M. KHOBAIB and M. S. DONLEY, *Progress in Organic Coatings*, **47**, 416 (2003).
40 A. AGLAN, A. ALLIE, A. LUDWICK and L. KOONS, *Surface and Coatings Technology*, **202**, 370 (2007).
41 T. R. ANTHONY and H. E. CLINE, *Journal of Applied Physics*, **49** (1978).
42 E. MCCAFFERTY and P. G. MOORE, *Journal of the Electrochemical Society*, **133** (1986).
43 U. K. MUDALI, R. K. DAYAL, J. B. BNANAMOORTHY, S. M. KANETKAR and S. B. OGALE, *Materials*, **33** (1991).
44 Y. NAKAO, K. NISHIMOTO and W. P. ZHANG, *Transaction of the Japan Welding Society*, **24** (1993).
45 Q. Y. PAN, W. D. HUANG, R. G. SONG, Y. H. ZHOU and G. L. ZHANG, *Surface and Coatings Technology*, **102**, 245 (1998).
46 S. VIRTANEN, H. BOHNI, R. BUSIN, T. MARCHIONE, M. PIERANTONI and E. BLANK, *Corrosion Science*, **36**, 1625 (1994).
47 K. G. WATKINS, M. A. MCMAHON and W. M. STEEN, *Materials Science and Engineering A*, **231**, 55 (1997).
48 P. KADOLKAR and N. B. DAHOTRE, *Applied Surface Science*, **199**, 222 (2002).
49 P. H. CHONG, Z. LIU, P. SKELDON and G. E. THOMPSON, *Applied Surface Science*, **208–209**, 399 (2003).
50 R. LI, M. G. S. FERREIRA, A. ALMEIDA, R. VILAR, K. G. WATKINS, M. A. MCMAHON and W. M. STEEN, *Surface and Coatings Technology*, **81**, 290 (1996).
51 Z. LIU, P. H. CHONG, A. N. BUTT, P. SKELDON and G. E. THOMPSON, *Applied Surface Science*, **247**, 294 (2005).
52 W. L. XU, T. M. YUE, H. C. MAN and C. P. CHAN, *Surface and Coatings Technology*, **200**, 5077 (2006).
53 H. C. MAN, S. ZHANG, T. M. YUE and F. T. CHENG, *Surface and Coatings Technology*, **148**, 136 (2001).
54 V. ASHWORTH, W. A. GRANT and R. P. M. PROCTER, *Corrosion Science*, **16** (1976).

55 A. H. AL-SAFFAR, V. ASHWORTH, A. K. O. BAIRAMOV, D. J. CHIVERS, W. A. GRANT and R. P. M. PROCTER, *Corrosion Science*, **20**, 127 (1980).

56 P. M. NATISHAN, E. MCCAFFERTY and G. K. HUBLER, *Journal of The Electrochemical Society*, **135** (1988).

57 L. M. PRUDENCIO, R. C. DA SILVA, M. F. DA SILVA, J. C. SOARES, O. CONDE and R. VILAR, *Surface and Coatings Technology*, **128–129**, 166 (2000).

58 T. REIER, S. SIMSON and J. W. SCHULTZE, *Electrochimica Acta*, **43**, 149 (1998).

59 S. SIMSON, T. REIER, J. W. SCHULTZE and C. BUCHAL, *Surface and Coatings Technology*, **83**, 49 (1996).

60 L.-P. TIAN, Y. ZUO, X.-H. ZHAO, J.-M. ZHAO and J.-P. XIONG, *Surface and Coatings Technology*, **201**, 3246 (2006).

61 X. ZHANG, S. LO RUSSO, S. ZANDOLIN, A. MIOTELLO, E. CATTARUZZA, P. L. BONORA and L. BENEDETTI, *Corrosion Science*, **43**, 85 (2001).

62 R. ZUO, E. KUS, F. MANSFELD and T. K. WOOD, *Corrosion Science*, **47**, 279 (2005).

63 H. SCHAFER and H. R. STOCK, *Corrosion Science*, **47**, 953 (2005).

64 C. B. BRESLIN, A. M. FENELON and K. G. CONROY, *Materials and Design*, **26**, 233 (2005).

65 R. C. PATIL and S. RADHAKRISHNAN, *Progress in Organic Coatings*, **57**, 332 (2006).

12
Greases and their role in corrosion control in the aerospace industry

K. D. AKIN, Nye Lubricants, Inc., USA

Abstract: This chapter provides an overview of greases used in aerospace applications, with an emphasis on corrosion prevention. Specific chemistries of base oil, thickener, and additives are reviewed, with special focus on corrosion inhibitors. Grease manufacturing techniques and specific physical property testing are highlighted. Requirements for proper grease selection are included, along with future trends in lubrication technology and a listing of journals, symposia, and books specific to grease and corrosion.

Key words: grease, lubrication, corrosion inhibition.

12.1 Introduction

With the host of areas subject to corrosion in aerospace vehicles, many different mitigation techniques must be used. Materials (such as alloys and composites) and coatings are utilized for many structural items. When these components need to move relative to one another, some sort of lubrication technique is incorporated to ensure smooth operation and long life. These can be intricate load-carrying mechanisms and are ideal areas for corrosive attack. Greases can act as corrosion inhibitors as well as performing their role in reducing friction and minimizing wear.

What is grease? The classical definition of lubricating grease is 'a solid-to-semifluid product of a thickening agent in a liquid lubricant',[1] or in lay terminology, grease can be thought of as a 'sponge of oil'. All lubricating greases consist of three fundamental components: a lubricating base fluid, a thickener, and performance-enhancing additives.[2] The concentration of thickener determines the consistency of the finished product; however, it is the nature of the oil that generally determines the grease functionality. The myriad of oil, thickener, and additive combinations possible provide manufacturers with a great deal of flexibility in formulating products with many different physical and chemical attributes.[3] Although much of aerospace grease is petroleum-based, the more demanding applications are moving toward synthetic fluids for improved performance in temperature serviceability, wear reduction, corrosion prevention and material compatibility.

The basic building blocks of any lubricating oil come from nature. Animal, vegetable, and mineral oils, including petroleum, are harvested and refined. Synthetic oils undergo another step: they are manipulated at the molecular level to improve lubrication characteristics. For example, synthetic hydrocarbon oil starts with ethylene, a petroleum product. The ethylene is polymerized to create a pure oil with a narrow range of molecular weights. The result is a synthetic hydrocarbon that is much less volatile than any petroleum base stock or, in more practical terms, an oil that has a longer operating life and a broader operating temperature range. In short, synthetic oils rely on nature for their raw materials but the unique properties of synthetic oils are the products of rigidly controlled chemical processes.[4] In broad terms, the thermooxidative stability of synthetic oils contributes directly to improved performance and extended life of the lubricated component. The majority of synthetic oils also have lower cold temperature service limits than petroleum. For example, a number of synthetic hydrocarbon oils remain fluid at –60 °C, while other chemistries can withstand temperatures down to –90 °C, whereas petroleum fluids typically become intractable at –20 °C.

12.2 Grease composition

12.2.1 Base oil

In all lubricating greases, the base fluid represents the principal ingredient. All lubricating oils must have the ability to separate adjoining moving surfaces to prevent or at least minimize wear. Each oil has different advantages and limitations as described in Table 12.1. Of primary importance is the fluid's operating temperature range which is shown in Fig. 12.1. An understanding of these differences helps an engineer choose the best possible lubricant for the job at hand. A brief overview of seven oil families follows.

Petroleums

Petroleums (or mineral oils) comprise the largest share of grease made today. They are created from plant and animal life buried deep underground over millions of years. These chains, comprised of primarily hydrogen and carbon, are refined into various materials such as gasoline as well as the lubricating oils discussed here. Petroleum oil is the most economical fluid available, but also has the narrowest operating temperature range.

Table 12.1 Base oil comparison

Base oil	Advantages	Limitations
Petroleum	Very low cost	Limited temperature range Product variability
Synthetic hydrocarbons	Excellent thermal stability Good friction reduction and lubricity Wide range of viscosities Low-temperature serviceability Good plastic and elastomer compatibility Long and growing list of applications in many industries	Not suitable above 125 °C
Polyglycols/polyethers	Non-carbonizing, no residue Good lubricity and film strength Wide range of viscosities Unusually good elastomer compatibility Good load carrying Only synthetic oils that include water-soluble versions Good high-temperature stability with proper antioxidant Commonly used in arcing switches, and particularly effective in large worm and planetary gears	Not compatible with some plastics and elastomers Poor volatility above 100 °C
Synthetic esters	Excellent oxidative and thermal stability Low volatility Excellent anti-wear properties Outstanding lubricity Good low-temperature properties Minimal viscosity change with temperature Excellent load-carrying ability for bearing applications	Not compatible with some plastics and elastomers

Table 12.1 Continued

Base oils	Advantages	Limitations
Silicones	Excellent oxidative and thermal stability Low volatility Wide range of viscosities Minimal viscosity change with temperature Excellent plastic and elastomer compatibility Good wetting capability Commonly used with plastic and elastomer components, including gears, control cables, and seals Higher viscosities provide mechanical damping	Poor load carrying Tendency to migrate
Perfluoropolyethers (PFPE)	Excellent oxidative and thermal stability Low volatility and vapor pressure Nonflammable and chemically inert Excellent plastic and elastomer compatibility Resistant to aggressive chemicals and solvents Commonly used in extreme-temperature environments and applications that require chemical, fuel, or solvent resistance	High cost Reduced effectiveness under heavy loads
Polyphenylethers (PPE)	Highest thermal and oxidative stability of all oils Excellent radiation, chemical, and acid resistance Excellent lubricity Excellent high-temperature stability Non-spreading even in thin film Traditional lubricant for noble metal connector applications; also used for high-temperature, specialty bearings	Not suitable for temperatures below 10 °C Not compatible with some plastics and elastomers High cost
Multiply alkylated cyclopentanes (MAC)	Proprietary fluid that combines the low vapor pressure of a PFPE with the lubricity and film strength of a synthetic hydrocarbon	Not suitable above 125 °C High cost

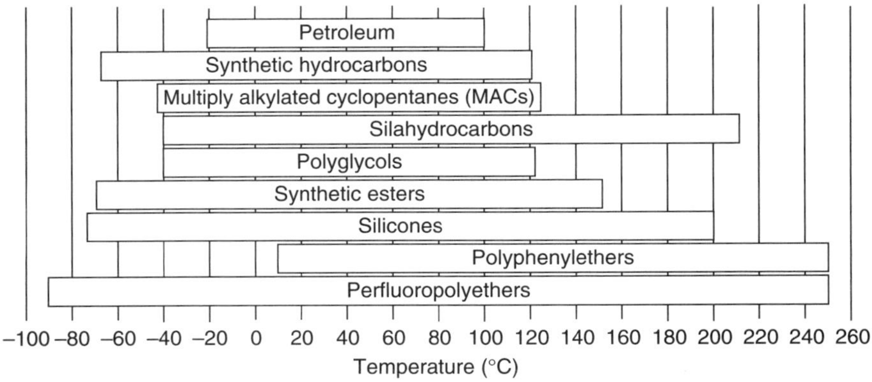

12.1 Base oil operating temperature ranges.

Polyalphaolefins

Polyalphaolefins (PAOs) are the most widely used synthetic base oil. (Polyalphaolefins are sometimes referred to as synthetic hydrocarbons, or abbreviated as SHCs). Because they are generally compatible with mineral oils, paints, plastics, and elastomers, switching from a natural to a synthetic hydrocarbon is relatively easy. They offer excellent cold-temperature performance and oxidative stability compared with petroleums. They are also relatively inexpensive compared with other synthetic oils.

Synthetic esters

Synthetic esters are similar to PAOs, but have an inherent polarity that makes them even less volatile and more lubricious. They are often blended with PAOs or other synthetic oils in lubricant formulations. Owing to their affinity for metal, especially steel and iron, esters provide significant wear protection. They are ideal for loaded bearings, potentiometers, and cut-metal and powdered-metal gearing, if proper seals are used. Since esters can withstand temperatures of 175 °C (or higher with proper formulation), they have become the choice for aircraft engine lubrication fluids and other severe duty applications. The disadvantages of esters are that they can compromise certain plastics and elastomers, and are susceptible to hydrolysis (molecular breakdown in contact with water).

Polyglycols

Polyglycols, like esters, have an affinity for specific metals, such as brass or phosphate bronze. They offer good lubricity and film strength. Because they offer good load-carrying ability, polyglycols are also particularly effective in large worm and planetary gears to reduce friction and improve efficiency. They are the only oil family that includes water-soluble versions. Like esters, however, they present compatibility problems with some plastics and elastomers, particularly polycarbonates, ABS resins, natural rubber, Buna S, and butyl.

Perfluoropolyethers and silicones

Perfluoropolyethers (PFPEs) and silicones are compatible with nearly all plastics. Both are suitable for broad temperature applications, and have shown exceptional, low-temperature torque characteristics. PFPEs are also resistant to chemically aggressive environments and are unaffected by sulfuric acid, hydrochloric acid, alkalis, halogens, and petroleum solvents. They do not react with oxygen – even at 300 °C under 500 psi of pure oxygen. In addition, some PFPEs have very low vapor pressure, which is essential for vacuum chamber and aerospace applications where outgassing can be problematic. The major advantages of silicones are that they show little change in viscosity with temperature and resist evaporation at elevated temperatures. However, silicones have limited load-carrying capacity. The lack of effective boundary lubrication additives limits their use to relatively low load applications. Neither fluid forms good hydrodynamic films, which can lead to high wear. Both fluids have high surface energies (low surface tension), which can be both beneficial (providing a complete and self replenishing oil film within the contact area) and detrimental (increased oil migration and potential contamination of nearby components).

Polyphenyl ethers

Polyphenyl ethers (PPEs) are the traditional lubricants for noble metal connectors. With the highest thermal and oxidative stability of all oils, they are also used for high-temperature, specialty bearings. PPEs are radiation resistant, which make them candidates when the potential for exposure to radiation is high. PPEs are not suitable for temperatures below 10 °C as they become solid, nor are they compatible with some plastics and elastomers.

Multiply alkylated cyclopentane

Multiply alkylated cyclopentane (MAC), a type of synthetic hydrocarbon, is one of the newest synthetic lubricants. Its uniqueness lies in the fact that its low vapor pressure, approximately 3.5×10^{-11} torr, rivals the vapor pressure of PFPEs, but its sturdier hydrocarbon backbone makes it able to handle heavier loads better than a PFPE. It also allows more traditional additives to be combined into the grease matrix. This fluid is used in several space-specific applications where load capacity is of utmost concern.

12.2.2 Thickening systems

Although it is the oil that characterizes much of the functionality of the grease, the thickener type completes its identity. For example, a lubricating grease prepared from a polyalphaolefin oil thickened with lithium 12-hydroxystearate would be referred to as a lithium-soap thickened synthetic hydrocarbon. An ester thickened with organomodified clay would be described as a clay-based synthetic ester.

Greases are prepared from both organic and inorganic thickeners as listed in Table 12.2. Organic thickeners are prepared from the reaction of a suitable alkali metal with either high-molecular-weight carboxylic acids or fats. When the chemical reaction takes place in the oil used in formulating the grease, it is referred to as in situ neutralization or in situ saponifica-

Table 12.2 Thickener comparison

Thickener	Advantages	Disadvantages
Paraffin wax	Lower cost	Low melting point: low load/low friction only
Alkali soap	Lower cost Water resistance Pumpability	Reacts with some oils and metals
Organoclay	High loads Melting temperature >250 °C	Limited oil content/oil separation
Alkali complex soap	Water resistant Pumpability Low oil separation Melting temperature >250 °C	Reacts with some oils and metals
Polyurea	Water resistant Pumpability Low oil separation Melting temperature >250 °C	Stability at low shear Storage hardening
Silica	Water resistant Low oil separation Very high melting temperature	Mechanical instability with some base oils
PTFE	Lubricity Inertness Melting temperature >300 °C	Moderate loads only
Metal oxide	Thermal conductivity Inertness Very high melting temperature	Limited oil content, oil separation

tion, depending upon whether an acid or fat is the co-reactant. Table 12.2 also identifies some of the commonly employed alkali metals used to make greases. The alkali metals are usually treated with stearic acid, myristic acid, 12-hydroxystearic acid, or hydrogenated castor oil, a triglyceride that liberates 12-hydroxystearic acid during saponification.

Inorganic thickeners, such as chemically modified clay, amorphous silica, and polytetrafluoroethylene (PTFE), can also be used to form grease, but without the need for a chemical reaction for grease formation to occur. The efficacy of a particular thickener to convert synthetic oil into grease is dependent on the ultimate surface area of the thickener, its ability to hydrogen bond, and its tendency to associate with the fluid on a molecular level. The thickener must have an affinity for the base fluid that is intermediate between the forces that lead to greater solubility and those forces tending to induce phase separation.

12.2.3 Additives (non-corrosion prevention)

The third component of grease is the performance enhancing additives, which are chemicals that allow base fluids to function in harsh environments.[5] In many cases, the oil does not possess the properties necessary to perform effectively in today's demanding lubricating environment. Additives provide grease manufacturers with the ability to formulate greases with specific physical and chemical characteristics. The additives used in a lubricant provide even greater design flexibility. Additives are mixed in small concentrations with the oil and thickener – usually less than 5% by weight – to enhance critical performance properties of grease, such as low-temperature torque, fluid oxidation resistance, and wear or extreme pressure demands. As an example, the high-temperature stability of polyol ester base oils can be dramatically improved by using different antioxidants.[6]

Greases have the luxury of not requiring additives to be soluble. Oils must have their additives fully soluble or separation will occur, and in both application and operation, the additives may not be in the locations required. Grease matrices allow additives to be dispersed homogeneously throughout the product.

Additives can be either liquid or solid. Two typical solid lubricant additives are polytetrafluoroethylene (PTFE) and molybdenum disulfide (MoS_2). PTFE imparts a very low coefficient of friction to a grease, which can reduce start torque, eliminate stiction, and reduce wear on some plastic components. 'Moly' is a load-carrying additive that is attracted to metal surfaces (due to the sulfur) and acts as a sacrificial barrier between sliding surfaces. Owing to their platelet morphology, both additives can also reduce friction. One downside to the use of solid lubricants is once depleted or forced out of the contact area, replenishment is difficult.

Nearly all hydrocarbon-based greases have an anti-oxidant package which protects the base oil from oxidative degradation. The chemistries are aminic or phenolic and are soluble in hydrocarbon fluids. Anti-wear additives are typically phosphate or sulfuric and create organo-films to protect the surface of bearings, gears, lead and ball screws, and other mechanical parts.

Other additives may provide additional attributes to the grease in which they are included. Oxides, metallic or carbon additives can enhance the thermal or electrical conductivity of the lubricant for specific applications. Viscosity Index (VI) or pour point improvers help with the temperature serviceability at the low end of the range (but can have negative effects at higher temperatures). Color or ultraviolet (UV) dyes aid in inspection or help differentiate a product for its intended use.

12.3 Corrosion-inhibiting additives

All metals (other than noble metals) react under atmospheric conditions and oxidize. As this corrosion can occur in the presence or absence of electrolytes, grease manufacturers are concerned with protection of metals from both electrochemical and chemical reactions.[7] The contributing factor of both the material and the environment must be considered. The lubricant formulation itself may contribute to corrosion, such as anti-wear additives (MoS_2) with their reactive products. Aircraft are subject to many different environments which can provide the electrolytes (fresh or salt water, cleaning agents, etc) for electrochemical corrosion to occur. Depending on the type of metals the corrosive process may be different, but the result is always a compromise of the component, either structurally or in its operation.

The two basic methodologies to resist corrosion are via acid neutralization and protective film formation. Acids can be controlled by reacting with basic materials to neutralize their destructive potential. Film passivation is similar to anti-wear protection, except in this case the goal is to 'seal' the metallic surface from its local environment. These long-chain organic molecules have polar groups attached which interact with the surface to form a protective layer via physical or chemical absorption as shown in Fig. 12.2. The chemistries used in corrosion protection are wide and varied (there are several companies who develop additives and complete additive packages), but include phenols, sulphonates, triazoles, phosphates, and amines.

12.4 Lubricant selection and the design cycle

While the performance of a lubricant depends on many variables, early evaluation of key lubricant selection criteria can help avoid design pitfalls and shorten product development time.

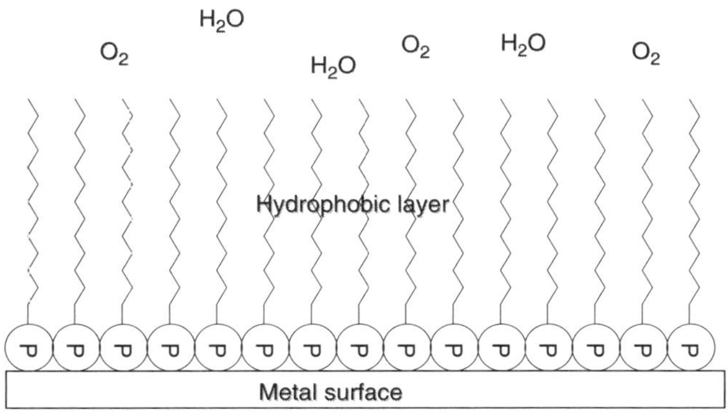

12.2 Corrosion additive structure representation. Courtesy of King Industries.

12.4.1 Operating temperature

The most important design variable is the operating temperature range of the device. At the high-temperature limit, the lubricant must be chemically stable, exhibit low volatility and have sufficient film strength to adequately prevent wear. At the lowest expected temperature, it must remain sufficiently fluid to allow the component to operate.

12.4.2 Material compatibility

Some lubricants can 'attack' certain plastics and elastomers. The base oil can infiltrate the material or cause the polymeric components to leach into the lubricant. Good design tests the compatibility of specific plastics and elastomers by evaluating physical properties such as tensile strength, dimensional stability, and gravimetric stability after immersion in the lubricant. Higher temperatures (more energy) and lower base oil viscosities (smaller molecules) usually exacerbate chemical incompatibility. Certain metals that come in contact with the lubricant may exhibit accelerated corrosion or lead to undesirable polymerization or 'varnishing' and failure of the lubricant base oil by acting as catalysts for oil oxidation. Esters (with their high polarity) will compromise many polymeric materials. These problems can be avoided by identifying early in the design process the materials used in the device and testing their compatibility with candidate lubricants.

12.4.3 Load and wear

For most applications, the prevention of wear is the primary reason for the use of a lubricant. Thus, the load at the interface is an obvious concern. In general, higher viscosity base oils support heavier loads, within a specific family of base fluids, because of the hydrodynamic film created by the fluid under pressure in the contact zone. If the load in the contact zone is too great or the speed is too slow for film formation, asperities on the rubbing surfaces can collide, causing excessive wear. In this situation, which is referred to as boundary lubrication, extreme pressure (EP) additives may be necessary. There is also a mixed regime, called elastohydrodynamic lubrication falling between direct surface contact and continuous film separation.[8] Synthetic ester greases are particularly suited for preventing heavily loaded metal-on-metal wear. Under relatively light loading, the outstanding viscosity–temperature properties of a silicone grease may be useful.

12.5 Lubricant testing

12.5.1 General industry standard test methods

The American Society of Testing and Materials (ASTM), along with its European and Japanese counterparts, have standardized a vast number of tests used to measure a specific chemical or physical property of grease. Some of these tests may be classified as those that measure characteristics particular to the composition, while other tests are more suitable for assessing batch-to-batch variability. Volatility per D-972, water washout per D-1264, and the four-ball wear test per D-2266 are examples of tests that measure properties of the grease inherent to the formulation. Some lot variation in these properties is expected, owing to either the product or the test method, but the magnitude of the variation should be minor from one batch to the next. However, it is necessary and prudent to conduct tests that are sensitive to the grease composition and performance. The specific testing can be jointly determined by the lubricant manufacturer and the end user.

Table 12.3 lists tests that can be conducted on a batch-to-batch basis to monitor the consistency of material manufactured. The tests and criteria listed in this section represent only a small number of methods and procedures that are routinely used to characterize lubricating greases. Other tests determine the water resistance, electrical properties, and extreme pressure properties of lubricating grease. These test methods and others are available through ASTM in the US, DIN, NF or IP in Europe, and JIS in Japan. Although similar in nature, there can be subtle differences in methods, so care must be taken when comparing data. There are also military and governmental test methods which are required to meet military product demands. Grease manufacturers may also develop their own internal test methods if standard tests are unavailable to appropriately measure parameters important for specific end-use applications. Some of the more typical tests are listed below.

Table 12.3 Grease properties

Property	Method	Purpose
Unworked penetration	ASTM D-217	Measures the consistency of the grease before the input of mechanical energy. This is what the customer's pump is required to transfer.
Worked penetration, 60 strokes	ASTM D-217	Measures the mechanical stability of the grease. Excessive change may signal problems in applications imparting high shear to the grease.
Oil separation	FTM 791B Method 321.2	This test assesses the amount of oil released from the grease structure after 30 h at 100 °C.
Evaporation	ASTM D-972	Measures evaporation of any volatile ingredients in the grease or residual by-products generated during manufacture.
Dropping point	ASTM D-2265	Determines the high temperature attainable before a drop of oil separates from the grease.
Oxidation stability	ASTM D-942	Measures the thermo-oxidative stability of the lubricant at elevated temperature under pure oxygen pressure. The reduction in pressure indicates oxidation.
Water washout	ASTM D-1264	This test is used to asses the resistance of a grease formulation to resist displacement from a rolling element bearing by fresh water under dynamic conditions. Theoretically, one would not expect a particular grease formulation to show deviation beyond the limits inherent in the tests. However, minor changes in grease consistency, within the NLGI grade, could affect results.

Penetration

The consistency of a grease measures its resistance to deformation under an applied force. Consistency attempts to quantify plastic behavior as viscosity tries to delineate fluidity. The National Lubrication Grease Institute (NLGI) has developed a numerical scale to classify the consistency of greases by measurement of a depth (in tenths of millimeters) to which a metal cone penetrates a sample of the grease in free fall under defined test

Table 12.4 Grease stiffness

NLGI grade	ASTM worked penetration, 60X	Analogous consistency for household food item (unworked penetration)
000	445–475	Ketchup
00	400–430	Apple sauce
0	355–385	Brown mustard
1	310–340	Tomato paste
2	265–295	Peanut butter
3	220–250	Vegetable shortening
4	175–205	Frozen yogurt
5	130–160	Smooth pâté
6	85–115	Cheese spread

conditions. The NLGI defines nine distinct grades of grease, ranging from NLGI Grade 000 to 6, based on the sixty stroke worked penetration P_{60} as shown in Table 12.4. Each range is thirty penetration units wide and fifteen units separate the grades. For instance, the worked penetration of an NLGI Grade 2 Grease, a common grade for many bearing applications, is 265 to 295, while a Grade 3 Grease is 220 to 250. Semifluid greases have a triple-zero rating while the hardest greases would receive a rating six. Although the worked, 60X, penetration determines the grade, the unworked penetration, P_0, of grease is a very useful parameter since it identifies the consistency of the lubricant that dispensing equipment may deal with. Changes in P_0 and P_{60} with time may or may not be a manifestation of lubricant deterioration due to oxidation. These measurements are taken shortly after grease is manufactured and represent the state of the thickener dispersion at that time.

Oil separation

The oil separating tendency of a lubricating grease under thermal stress is determined by the cone and beaker technique. The method consists of placing approximately 10 g of the test specimen in a metal cone made from a 60-mesh screen and suspending the fixture in a covered beaker. The grease sample is placed in a constant temperature oven for some specified time and temperature. The amount of oil that has been separated from the grease and captured in the beaker is determined gravimetrically at the completion of the test. Twenty-four hours at 100 °C is a typical duration and temperature to rate oil separation and make comparisons between different greases. Temperatures exceeding 150 °C should be avoided to prevent false positives due to the likelihood of oil evaporation from the covered beaker following separation. The results of oil separation should not be used initially to judge the appropriateness of a lubricating grease for an application. Low oil separation is good for some applications while high oil separation is appropriate for others. An example where high separation is beneficial is an enclosed gear box, where oil migration is not a concern but the free oil helps to replenish the contact zone. After a grease has been selected based on performance merit, oil separation is a useful quality control check. Oil separation is determined in accordance with FTM 791 Method 321.2 or ASTM D-6184.

Evaporation

The evaporation of volatile components in a lubricating grease is measured by ASTM Methods D-972 and D-2595. In this test, the grease sample is placed in a metal sample holder and the entire test fixture is immersed in an oil bath (or equivalent heating block) for the required time and temperature. Preheated air is blown across the surface of the grease to facilitate the removal from the test apparatus of volatile grease constituents. Sample weight loss is used to determine volatility, although at higher temperatures the weight loss could include oxidative degradation.

Dropping point

The dropping point of a lubricating grease is a useful parameter but less so today than years ago. Early thickening systems softened appreciably at temperatures approaching 100 °C and melted before 150 °C. As a result, the dropping point of the grease, which is defined as the lowest temperature at which a drop of oil separates from grease, was used to determine the upper temperature performance limit. However, with the advent of high-melting thickener systems such as organically modified clay, complex soaps, PTFE, and others, the dropping point of grease no longer serves as a reliable indicator of high-temperature usefulness. Shortcomings include poor results in the test from thermal separation by relatively soft greases made from heat-resistant thickening systems. Today, the suitability of a lubricating grease at elevated temperatures is predicated on the thermooxidative stability of the base oil, thickener, and additives rather than the grease's dropping point. With modern grease, the dropping point should not be considered the operating temperature limit.

12.5.2 Corrosion-specific test methods

In order to compare the corrosion prevention properties of a grease, there are specific tests developed to simulate a variety of conditions in which the lubricant and component may be exposed.[9] In general, metallic components are coated with a grease and then exposed (in either a static or dynamic state) to an environment for a predetermined time. The component is then examined for signs of corrosion and rated. Although the tests listed below are specific to grease, some oil tests can be modified to utilize greases, such as the Salt Spray/Salt Fog (ASTM B 117).

Copper corrosion

Probably the most typical corrosion test is ASTM D 4048, which determines not only the preventive properties of a grease to corrosion on copper, but also if the chemistry of the grease itself promotes corrosion of yellow metals. These materials are commonly used in bushings and bearings in aircraft. A cleaned and polished copper coupon is completely immersed in a container that holds the grease sample. The time and temperature for the test is not specified, but 24 h at 100 °C is commonly used. The coupon is then removed and lightly cleaned, and then compared with a reference standard, which is rated in 4 classifications (slight, moderate, and dark tarnish, along with corrosion) and sub classifications (Fig. 12.3 gives an indication of the differences). This chart and other procedural information is referenced in ASTM D 130, which is for oil rather than grease, but virtually identical in process.

'Emcor' test

The 'Emcor' is a dynamic bearing test which simulates wet conditions and can include a variety of aggressive chemicals. Lubricated ball bearings are cycled off and on (at 80 rpm) for one week. Once the bearings are removed and cleaned, they are inspected and rated on a scale of 0 to 5 based on the

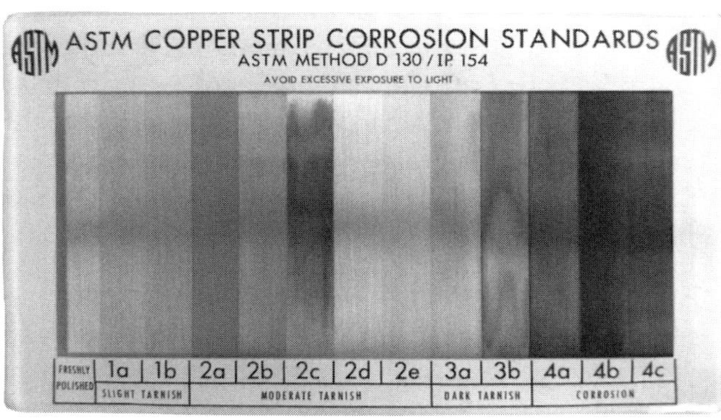

12.3 Copper strip corrosion color reference.

percentage of corrosion observed. A rating of '0' would have no corrosion and a rating of '5' has more than 10% of the bearing race covered in corrosion. The ASTM rating system is shown in Fig. 12.4. As the testing is done in duplicate, a dual number such as '1,2' is provided to rate each bearing separately. This procedure was originally developed in Europe as IP 220 and later adopted by ASTM (D 6138).

Corrosion preventative properties

Two related tests are ASTM D 1743 and D 5969, which utilize tapered roller bearings in the procedure. These tests fill and distribute the lubricating grease in a bearing then have the bearing placed in a covered jar with a

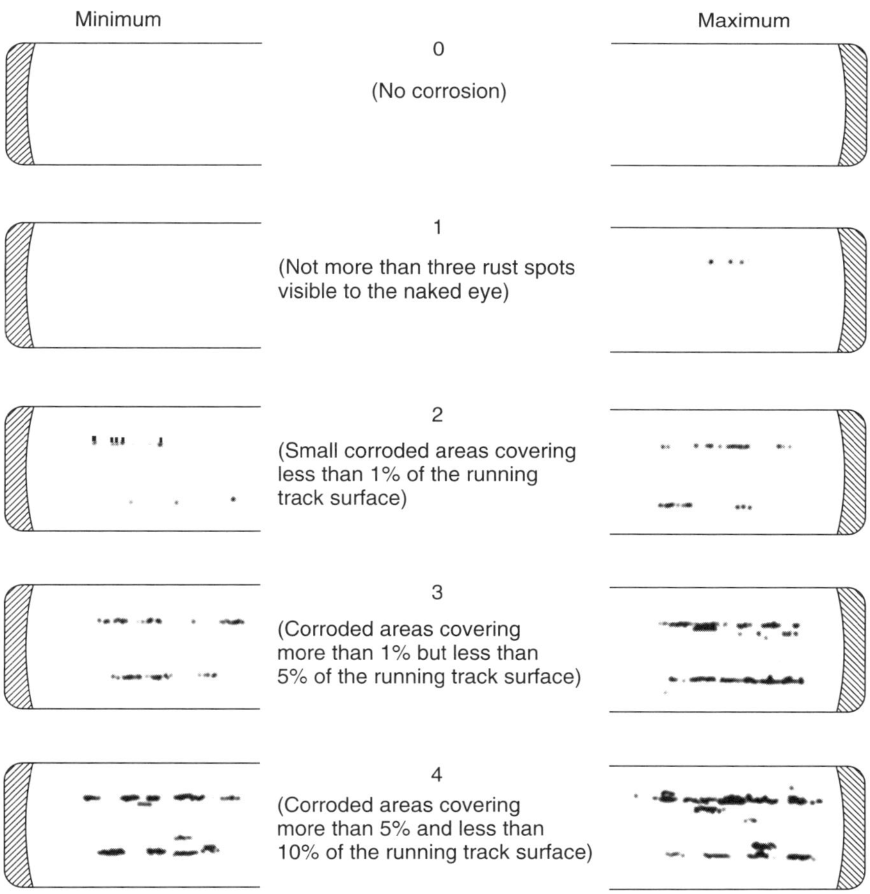

12.4 EMCOR bearing corrosion value ratings.

water solution. The jar is then placed in an oven which provides effectively a 100% humidity environment. The 1743 uses only distilled water and lasts 48 hours. The 5969 adds a percentage of synthetic seawater (not specified, but typically between 1 and 10%) but reduces the exposure time to 24 h. A fail rating is based on any non-transparent corrosion spot greater than 1 mm in any direction.

Corrosion rate evaluation procedure

The US Air Force has developed an oil or grease corrosion 'screening' test called 'Corrosion Rate Evaluation Procedure' or CREP.[10] This procedure places suspended grease or coated metallic coupons in a glass reaction kettle placed on a hot plate. An uncoated or non-additivized sample coupon is used as a reference. Either distilled water or an acetic acid/sodium acetate solution is added to the kettle and heated until the liquid boils, along with a positive air flow. The kettle includes a condenser to recover gases. The test set up is shown in Fig. 12.5. Generally, visual differences are seen within 1–2 h, and more telling than weight loss to determine corrosion or protection potential.

12.5.3 Analytical and application-specific test methods

Modern analytical laboratory equipment with integrated computer software can provide key insights into the makeup and function of synthetic lubricant formulations. Spectroscopic methods such as Fourier Transform Infrared (FTIR) spectroscopy can identify molecular signatures for various thickeners, base oil species and additives. A thermogravimetric analyzer (TGA) and pressure differential scanning calorimeter (PDSC) are used for detailed quantitative analysis of melting points, phase changes, and thermo-oxidative stability of lubricant formulations.

Functional testing is often performed by lubricant formulators to pre-screen product candidates for component manufacturers before full-scale qualification tests in the actual OEM device. Prescreening can include measurement of special functional properties (electrical or thermal conductivity, viscosity vs. shear rate, etc.), material compatibility (lubricant vs. plastic, elastomer, or solvent), and the effects of low and high temperature extremes on the lubricant.

12.6 Grease manufacturing

Organic greases are usually manufactured in kettles. The size of the vessels ranges from laboratory units capable of manufacturing only 2–5 kg per batch, to very large units capable of manufacturing 18000 kg (or greater)

12.5 CREP test equipment. Courtesy of US Air Force Materials Research Laboratory.

of grease in a single operation as shown in Fig. 12.6. Since heat is required to initiate the reaction of the ingredients used to manufacture soap-based grease, or, in general, to promote the solvency of lubricating fluids and additives, grease kettles are heated. Most are jacketed to accommodate either steam or hot oil. Steam is an advantageous thermal medium because cold water can be circulated through the same jacket to cool the batch on completion of the chemical reaction. The primary disadvantage of heating the grease mixture with steam is that high pressure is required to attain temperatures above 230 °C. Along with steam, oil jackets and electric elements are also used to heat grease kettles. Generally, a soap-thickened grease is manufactured by adding a small portion of base oil to the kettle along with all of the fatty acid. At this stage, only enough heat is applied to melt the acid. Once the fatty acid has melted into the base oil, an aqueous solution of the alkali metal is added to the kettle incrementally.

12.6 Grease manufacturing kettle.

The kettle contents are continuously stirred, usually with counter rotating blades, to facilitate the dehydration of the soap mass as the reaction proceeds. After dehydration, additional base oil is gradually added to the kettle. The addition of oil must proceed slowly in order to maximize proper mixing of the intractable soap mass and the oil being added. After the addition of the required quantity of oil, the kettle contents are heated to some predetermined temperature and maintained at that temperature for several hours. After the heating cycle, the grease is rapidly cooled to optimize the dispersion of the thickener. The rate at which the kettle contents are cooled has a pronounced effect on the finished consistency of the grease. Additives are usually added to the grease after the temperature of the batch has fallen below 100 °C. When the kettle contents have reached ambient temperature, the grease may be either milled or homogenized, which adds additional shear to insure uniform consistency. As a final step, the grease may be filtered to remove contaminates. Quality control testing usually occurs before discharging the grease from the kettle for packaging.

12.7 Future trends

The trend in aerospace lubrication (as in many other industries) tends to focus on extremes. Increasing the maximum (or minimum) operating temperatures of a grease continues to push synthesizers of fluids and additives to improve the thermal stability of the finished formulated grease. Unfortunately, trade-offs are a reality in the lubrication world. PFPE-based greases provide the most thermally stable lubricants, but their non-reactivity and inertness limit the additives available to improve their thermal properties, as well as corrosion and wear prevention. Jet engine exhaust gases can reach temperatures above 300 °C, which requires frequent relubrication for components such as thrust actuators, as present lubricant technology cannot handle long-term exposure to this heat. Much development work is going into fluorinated additives to improve properties of PFPE-based lubricants.[11]

Ionic liquids are a new category of function fluids which may hold promise for high-temperature application. These unique salts (with low melting points) are liquid at room temperature and are reported to withstand extreme elevated temperatures. Although their use is presently limited to battery electrolytes and solvents, the potential for demanding lubrication applications is under investigation.[12] Their corrosion potential is unknown at this point, but warrants further study.

A study states that 25% of commercial aircraft are over 20 years old[13] and military aircraft certainly exceeds that percentage. This issue is expanded upon in other chapters. These aircraft are extremely expensive to maintain and repair. This offers the opportunity to develop improved greases which can reduce the frequency of maintenance (which equates to more flight time) and minimize the potential for component failure. One such lubricant is being considered by both the US Air Force and Navy as an improved general purpose aviation grease.[14] It has improved corrosion protection and increased load capacity compared with the previously used material, and is in use in transport landing gear and under test in wing flaps and other high-load gear applications. The Navy estimates it could save over $500 000 per year on a single aircraft alone.

12.8 Sources of further information and advice

The best starting point for grease information is the National Lubrication and Grease Institute – NLGI – in the US (www.nlgi.org) and its counterpart the European Lubricating Grease Institute – ELGI – (www.elgi.nl). Both organizations hold annual symposia which offer an opportunity to attend technical presentations specifically related to grease, such as production, research developments, and environmental concerns. NLGI publishes the ‘Lubricating Grease Guide’, which is an excellent primer.[1]

Focused on greases used in the aviation industry is the Society of Automotive Engineers (SAE) AMS-M committee (http://aerospace.sae.org). This group is involved with the creation of grease specifications primarily for the commercial aviation industry as well as a forum to discuss grease related issues amongst airframe manufacturers, grease manufacturers, and military entities.

Not specific to grease is the Society of Tribologists and Lubrication Engineers, STLE (www.stle.org), which also provides resources for grease and lubrication study. Other societies relating to the study of lubrication are the American Society of Mechanical Engineers, ASME (www.asme.org), which hosts a Joint Tribology Conference with STLE. The primary technical group associated with corrosion is NACE International (www.nace.org), which has conferences, committees, and training courses related to corrosion.

Many journals are available with peer-reviewed technical papers. These include: *ASME Journal of Tribology*, *Tribology Transactions*, *Tribology Letters*, *Corrosion* and *Journal of Synthetic Lubrication*.

A handbook on various grease base oils is *Synthetics, mineral oils, and bio-based lubricants* edited by Leslie Rudnick. Much of this chapter was originally published in this book written by Dr. Joseph Braza. Copyright 2006 *Synthetics, mineral oils, and bio-based lubricants* by Rudnick. Reproduced by permission of Taylor & Francis, a division of Informa plc.

12.9 References

1 (1996), 'Lubricating Grease Guide', National Lubricating Grease Institute, Kansas City, MO, 4th edition, 1.01.

2 VENDURA T M, BRUNETTE G, and SHAH R (2003), 'Lubricating Greases', in Totten G, *Fuels and Lubricants Handbook: Technology, Properties, Performance, and Testing*, ASTM International, 557.

3 BESSETTE P and STONE D (1999), 'Synthetic Grease', in Rudnick L and Shubkin R, *Synthetic Lubricants and High-Performance Functional Fluids*, 2nd edition, Marcel Dekker, 519–538.

4 LAY J and WEIKEL J (2001), 'Gaining A Competitive Advantage with Synthetic Lubricants', *Appliance*, April 2001, 56–60.

5 RIZVI S Q A (2003), 'Additives and Additive Chemistry' in Totten G, *Fuels and Lubricants Handbook: Technology, Properties, Performance, and Testing*, ASTM International, 199.

6 HOLLEY B (2001), 'Next-Generation Ester Grease Survives Higher Temps', *Lubricants World*, **11**(1), 13–17.

7 RIZVI S Q A (2003), 'Additives and Additive Chemistry' in Totten G, *Fuels and Lubricants Handbook: Technology, Properties, Performance, and Testing*, ASTM International, 200–205.

8 LUDEMA K (1996), *Friction, Wear, Lubrication – A Textbook in Tribology*, CRC Press, 114–120.

9 HUNTER M and BAKER R (2003), 'Corrosion' in Totten G, *Fuels and Lubricants Handbook: Technology, Properties, Performance, and Testing*, ASTM International, 831–832.

10 ROBERTS M, GSCHWENDER L, SNYDER C and FULTZ G (2004), 'Corrosion Rate Evaluation Procedure (CREP): A Convenient Reliable Method for Determining Corrosion Inhibition Ability of Lubricants', AFRL-ML-WP-TR-2006-602.

11 MACCONE P, BOCCALETTI G, PETRICCI S and RADICE S (2005), 'A Very Effective Stabilizer for Perfluoropolyether Lubricants', *World Tribology Conference*, Washington DC.

12 SHORT P, 'Out of the Ivory Tower', *Chemical & Engineering News*, **84**(17) 2006, 15–21.

13 NACE INTERNATIONAL WHITE PAPER, 'Aircraft Corrosion', www.nace.org/nace/content/publicaffairs/media/air.asp.

14 AKIN K, GSCHWENDER L, SNYDER C and FRANCO G (2005), 'Comparative Study of Military Specified General Purpose Synthetic Lubricants in regards to Wear Performance, Corrosion Prevention, and Salt Water Resistancy', Tri Services Corrosion Conference 2005, Orlando, FL.

13
Business strategies for corrosion control in fleet maintenance

M. W. SCHLEIDER, Mercer University, USA

Abstract: This chapter provides guidance into the business strategy for corrosion concerns in fleet management, starting with acquisition and continuing through sustainment. It discusses the importance of training of personnel, as well as corrosion and costs tracking as important metrics for monitoring a fleet's corrosion management program. Means of communicating ideas and exchanging information on corrosion with other organizations are also presented.

Key words: corrosion training, corrosion costs tracking, business strategies for corrosion management.

13.1 Introduction

As one of the primary causes of life limitation of an airframe, corrosion is more than a technical parameter of material behavior that must be reckoned with during maintenance. The cost of corrosion for the United States Department of Defense (DoD) is estimated to be $20 billion per year and affects fleet availability and readiness. With this magnitude of cost impact to the DoD's budget, organizations within the DoD must have a sound corrosion business strategy for their systems from cradle to grave. Corrosion can become a cost driver in any phase of the life cycle from design and development, through acquisition, operational deployment and sustainment, to retirement and decommissioning. As such, it must be considered appropriately in the business strategy for life cycle management of an aerospace vehicle.

The fundamental technology advances in materials are normally driven by the need for greater strength and durability, lower weight to strength ratios, lower cost production and machining, and lower total ownership costs. Corrosion susceptibility, normally a secondary property of material, is more often an unknown or unintended consequence than a basic goal of the design process in creating new materials. The corrosion issue then is normally relegated to one that design and management teams must build into trade studies during the vehicle design and construction phases. Since corrosion and its impact is important to life cycle management, but not necessarily a primary driver in fundamental material selection, or top line design goals, it becomes essential that an integrated and effective set of precautions, evaluations, and provisions be included as appropriate in each step of the aerospace vehicle life cycle management process to manage the unavoidable impacts of corrosion.

This chapter provides guidance into the business strategy for corrosion concerns in fleet management, starting with acquisition and continuing through sustainment. The training of personnel as an important part of the strategy is discussed, with this training also providing a solid basis for corrosion prevention and control decision-making. Corrosion tracking and costs tracking are presented as important metrics for monitoring a fleet's corrosion management program. Means of communicating ideas and exchanging information on corrosion with other organizations are also discussed. Lastly, other sources of information are provided to better aid the reader in acquiring additional data.

13.2 Acquisition requirements

The best opportunity to avoid the mechanisms that cause corrosion is at the initial design of an aerospace system. Though corrosion management and treatment should be a life-long consideration, once designs have been formulated, materials have been selected, and assembly methods have been chosen, the remaining flexibility is in how to treat and minimize the problems that will eventually occur. Such is the importance of the acquisition phase in managing corrosion of airframes. One would think that this is obvious, but even now the DoD procures aircraft that are minimally improved versions of older aircraft when it comes to resistance to corrosion. There are improved corrosion-resistant materials and assembly methods in existence which have been widely applied in aviation. The acquisition requirements should include the submission of a technical basis for the materials, coatings, and assembly methods selected in terms of corrosion avoidance and the proposal evaluation should include criteria which address anticipated cost of corrosion in sustainment. Personnel who have knowledge of corrosion maintenance activities should be brought into the selection process so that total life cycle cost considerations can be realistically included in the acquisition process.

Given the stringent requirements of performance for aerospace design, the use of light materials with high strength-to-weight ratios is essential. In addition, the use of dissimilar materials in close proximity can be a recurring requirement. Therefore, the corrosion avoidance gains possible from a well-executed acquisition program are focused not on totally avoiding all potential for corrosion, but on making corrosion management a visible and properly valued element in the design trade processes. In places where the necessary performance can be achieved and corrosion avoided, that should be the goal. There have been significant improvements in more corrosion resistant materials in the last 50 years, yet often these materials are not used in new acquisitions, possibly due to the initial expense. Total life cycle costs are rarely considered in the decision process for material selection. Even with the selection of materials and assembly of materials designed to reduce corrosion, a Corrosion Prevention and Control plan should still be provided during the acquisition process.

13.3 Sustainment requirements

Fleet structural management efforts can yield major returns on investment during the sustainment phase. A formal structural management plan should be prepared with specific inclusions intended to flow necessary corrosion management actions derived during design and acquisition into the sustainment processes. Only with the precise knowledge and information captured in the early phases of the program can adequate and detailed management and maintenance practices be fully included in a robust sustainment program.

Materials selection happens well before sustainment begins, but the materials selection process normally forms the basis for unique considerations for sustainment activities and techniques such as those for coatings, barrier layers, and assembly processes, which are necessary to control the natural phenomenon of corrosion. Likewise, special material treatment, coatings, and assembly practices important in the original design are also important to the maintenance, repair, and overhaul processes required to improve or maintain the corrosion resistance of the original design.

Sustainment activities, though normally envisioned as a logical extension of the acquisition phase, are most often carried out by a different part of the organization or even a separate agency. The significance is that well-executed design tradeoffs performed during the acquisition phase are not always well-documented and fully integrated into the sustainment activities. That, coupled with the fact that corrosion by its very nature is a delayed phenomenon, can result in significant loss of useful life and major costs for correction if the proper management, treatment and correction of corrosion are not included from the beginning of the sustainment period.

Corrosion is a significant problem for a large number of aerospace systems. The details of the maintenance program have a huge effect on the cost of corrosion and availability of the system. The sustainment process should include a systematic assessment of maintenance activities in the form of a Reliability-Centered-Maintenance program (RCM). The RCM program should determine when corrosion prevention and control should take place, and point the way to potential changes in materials, coatings, and processes. For example, RCM can be used to determine the optimum conditions or intervals for aircraft washings and/or corrosion preventive compound application.

13.4 Training

An important part of the business strategy for corrosion management is the training of personnel. In fact, the DoD Corrosion Prevention and Mitigation Strategic Plan, published in June 2007, lists training of technical and management personnel as one of the key metrics to judging the success of a corrosion prevention and mitigation plan. Optimization of corrosion prevention and mitigation can only be accomplished through training of personnel. A comprehensive training program is necessary for both acquisition and sustainment of aircraft systems. Training is essential for maintainers, program managers, systems engineers, logisticians, and contracting personnel.

As part of an acquisition, training personnel to cite specifications and requirements for reduced corrosion is essential to minimize the total cost of ownership. More can be accomplished in cost reduction in the acquisition process than in the entire sustainment process. In the materials and design of aircraft systems, the requirement for corrosion resistance is often overlooked or minimized, even in more recent acquisitions. Budgets for the acquisition of new systems must include corrosion resistance as a high priority requirement. It must be recognized that corrosion is a 'pay me now or pay me more later' phenomenon and what is paid is not only dollars, but fleet availability. The acquisition team, including contracting personnel, must be trained in the requirements and specifications to include in the Request for Proposal which address a corrosion-resistant design and the testing required to prove the design's corrosion-resistance. The CPC Planning Guidebook provides guidance for these requirements and specifications and determination of compliance with the specifications. This Guidebook is available on the DoD Corrosion Exchange website.[1]

In the sustainment process, a comprehensive training program is also needed. The corrosion prevention and mitigation training should be tailored to the many technical and management personnel responsible for aircraft system sustainment. For example, one of the DoD Corrosion Strategy metrics is for 100% of all the maintenance personnel working on aviation equipment to have training on the proper applications and techniques of corrosion compounds, sealants, and coatings. The training should be presented at the appropriate level. For example, the US Army has some excellent training material for maintainers on their website, presented in a graphical and interactive manner.

Aircraft systems require the formation of a Corrosion Prevention and Advisory Team (CPAT) as part of the corrosion management structure. The CPAT is considered an expert body for providing corrosion management advice. As such, the CPAT members must be trained in technical and programmatic topics pertaining to corrosion. Too often an engineer is designated as the aircraft corrosion 'expert', but has not had any additional corrosion training. The corrosion engineer's knowledge is often reactive, that is, gained by the experience of significant corrosion problems, rather than by formal training before his/her designation as the expert.

Technical training for the various skills of personnel should be established to understand the causes and types of corrosion, corrosion prevention and control methods, and the cost of corrosion. The DoD Office of Corrosion Policy and Oversight has established a website (www.Corrosionexchange.org) which includes training documents and listing of courses offered by professional societies. The established training program and associated budget should include courses and certification programs offered by National Association of Corrosion Engineers (NACE) International and The Society for Protective Coatings (SSPC). Both organizations offer technical training and certification programs that serve a variety of functions.

13.5 Corrosion prevention and control

Corrosion prevention and control (CPC) entails the characteristics of a system design to preclude or reduce corrosion, materials selection, nondestructive inspections for corrosion detection, coatings, finishes, cleaning materials and washings, repairs, and other maintenance activities. There are numerous choices to make from the acquisition through retirement of a system. At times, there seems to be too much information, but at other times there is not enough data to guide those for whom the responsibility of corrosion prevention and control has been given.

A Corrosion Prevention and Control (CPC) Plan should be prepared as early in a program as possible and updated through the life of the system. The Plan should define the CPC requirements and include goals and metrics of the program. For example, metrics may be in terms of dollars, nonmission capable rates, or maintenance manhours. It is important to be able to track the effectiveness of various initiatives throughout the program. The metrics for effectiveness must be quantitative. If the Plan is written in the acquisition phase, the corrosion-related design needs should be included in the Plan. Goals of the program may include a specified amount of fielded time that the system is corrosion-free, or the maximum amount of dollars spent on corrosion maintenance for the first five or fifteen years of the life of a system.

Whether in the acquisition or sustainment phase, adequate materials and/or protective coatings to minimize the deterioration of the system are essential. Another important attribute is the ability to detect corrosion, ideally without extensive amounts of disassembly of the system. Also, as systems age, they tend to lose some of the ability to prevent moisture induction or accumulation. What might have been a good design for preventing moisture intrusion initially, can deteriorate with time.

After a system has been fielded for many years and significant corrosion exists throughout the fleet, as is the case with many military aircraft, it becomes a matter of first removing the corrosion, then making changes in design features or maintenance practices to reduce future corrosion. Washing an aircraft to remove corrosive agents can extend the life of an aircraft's structure, but it must be performed in a manner that only the corrosive materials are removed, not the protective paints and other materials. When designing a repair for a component which has experienced significant cracking across a fleet, the emphasis is often on extending the fatigue life or providing adequate strength. Often, not enough attention is paid to preventing corrosion in the repair design. It is important to select appropriate materials and the means of joining the repair to the basic structure which will not include dissimilar metals nor create an environment for moisture entrapment.

There are many available documents referenced in the Corrosion Prevention and Control Planning Guidebook which should be used by those designing new systems or repairing old ones. Too often the same mistakes occur over and over due to the lack of training and knowledge of basic material and design considerations for corrosion prevention and control. The Guidebook also contains an example of a Corrosion Prevention and Control Plan to guide the CPAT and Contractor Corrosion Teams (CCT). Specific materials are listed, as well as finishes and coatings which should be part of a corrosion-resistance design. There are also guidelines pertaining to the joints and faying surfaces to avoid moisture intrusion and entrapment.

13.6 Corrosion tracking

There is a clear requirement to track corrosion in aerospace vehicles based on the same fundamental need to monitor and characterize structural crack damage. Though there are parallels to tracking structural cracks and corrosion, implementation of corrosion tracking has generally not kept pace with crack tracking. The release of MIL-STD-1530C, United States Air Force

(USAF) Aircraft Structural Integrity Program (ASIP), includes a requirement for a corrosion history database, thus placing corrosion in the same height of importance as fatigue cracks.

Ideally, a corrosion database should be generated for each aircraft from its initial acquisition, although it is never too late to begin assessing the state of corrosion. To be most useful, a corrosion database should capture, collate, and characterize information adequately to fully describe individual corrosion events, patterns of corrosion across the platform, as well as patterns that result from fleet usage and deployment. Capturing the data at the point of maintenance is the most efficient and reliable method and should be part of a complete maintenance data tracking system for the specific tail number, and become part of the Individual Aircraft Tracking Program (IATP), which includes base assignments and deployments and flight usage.

A corrosion tracking system should capture observations that indicate lack of damage in suspect or critical areas, as well as events of observed damage to provide as much fidelity as possible for timing and rates of damage, in addition to location and scope of damage. Specific database fields should be utilized for part names and locations on the part, utilizing graphics and populated listings wherever possible. The environment in which the aircraft has flown should already be part of the IATP. The US Coast Guard, Air Force, and Navy, have all begun efforts to electronically capture the corrosion state of some of their fleet. They have begun building databases for individual aircraft, resulting in fleet availability effects and trends. All of this feeds into corrosion prevention plans for the fleet, including decisions on material changes or coatings for parts found to have a high incidence of corrosion.

In addition to the capturing of corrosion on the aerospace system itself, an indirect indicator of corrosion is a set of environmental corrosion sensors. These sacrificial devices are intentionally placed in observable areas to measure the amount and type of corrosion that occurs during the use of an operational platform. If sufficient information can be derived to correlate actual airframe corrosion to that experienced by the sensors, the sensors can be used to determine the amount of damage and the timing of corrective actions. In addition, long term analysis could actually provide strategies of treatment and protection that would prolong the ability of basic structure to survive corrosive environments with acceptable damage for operation.

Unfortunately, there have been many efforts where corrosion data has been captured for aircraft only to have the data reside on a compact disk (CD), on an individual desk top computer, or other inaccessible sources. The use of a common website, such as the DoD Corrosion Exchange Website or other designated location for the database is extremely important for the communication of information and prevention of lost data.

13.7 Costs tracking

A recent study commissioned by the US Congress reported the cost of corrosion to the US market to be $276 billon per year. Costs isolated to the aircraft industry were estimated to be in excess of $2.2 billion per year. These excessive costs make corrosion a subject that requires increased study, analysis, emphasis and, in the long run, increased effort at cost avoidance. In addition to the cost of corrosion, the relationship between costs and age in aircraft is particularly significant. The USAF costs alone were documented to be in excess of $600 million for corrosion in 1997, with over $400 million going to the maintenance of the aging fleets of the KC-135, C-141, and C-5, which represent less than 20% of the fleet.

Despite these costs and the need to control the budgets they unintentionally absorb, additional corrosion control efforts are warranted. A 2003 GAO report serves to surmise the situation: 'DOD and the military services do not have an effective approach to prevent and mitigate corrosion'.[6] While representing an assessment of only a part of the larger segment of the economy suffering these effects, the conclusions probably parallel that of the larger economy. Corrosion is often assumed to be an unavoidable part of the aging process and therefore cannot be avoided. The gaps in effective corrosion management and prevention appear to stem from a lack of knowledge rather than any neglect or carelessness. Corrosion is an insidious process and its effects, including costs, are equally insidious.

While our current database for corrosion costs may be less than what is ultimately needed, it is certainly sufficient to begin the front end analysis of major investments and to quantify the projected costs of corrosion with and without a preplanned corrosion prevention and control effort. As with all cost projections, the computed return on investment (ROI) of an effective corrosion control program will be limited to an estimated range, which in most cases, will be sufficient to justify high payoff efforts while precluding high cost initiatives that may have doubtful benefits. In addition, given the state of activity-based cost accounting, the inclusion of data elements and reports to quantify costs of corrosion is warranted where cost tracking is an ongoing phenomenon. While no problem as intractable as corrosion is easily solved, there are ad hoc examples that show the value of corrosion management and prevention. With effective cost tracking, the state of the art as practiced in sensitive technology areas can be improved to the point where major losses in investment can be avoided.

The cost of corrosion tracking should include direct costs, such as labor and materials used for corrosion control. It should also include indirect costs of training and research and development. Costs should be captured as they occur, if possible, and tagged as corrosion-related. The general types of costs to include and retain as separate types of cost in the analysis, each contributing to 100% of the costs are: 1) depot versus field maintenance costs, 2) preventative maintenance versus corrective maintenance costs, and 3) structure-related versus parts-related costs. These costs should be determined and calculated as percentages of other costs. For example, the cost arising from maintenance, including repairs or replacement of parts, at the depot should be compared with the total cost of depot maintenance.

The most effective means of tracking corrosion cost is by capturing the cost at the source. The costs may be in terms of man hours or material expense. Several of the military services are in the process of developing a means for easily tracking maintenance expenditures related to corrosion. Some organizations are evaluating hand-held portable aids for their maintenance personnel to use to enter all maintenance actions, and to code them as corrective or preventative maintenance actions. The cost of preparing an aircraft for corrosion-preventative treatment, as well as the application of paints, coatings, or other compounds should be captured. In addition to the cost of maintenance actions, it is just as important to capture the cost of parts replaced due to corrosion damage. Compiling the data in a database and relating the activities to costs provides an effective means of tracking corrosion costs.

Tracking costs arising from corrosion is important not only for the purpose of capturing current costs, but for estimating future costs. This is essential in evaluating the effectiveness of corrosion prevention and control programs. It is also significant in retirement decisions and identifying where material or design changes may be required.

13.8 Communications

Communication and collaboration is essential to succeed in corrosion management. Communication must be a two-way street with other organizations. This provides synergy and substantially lessens the chances of repeating mistakes that have been made in the past. Communication and collaboration also stretches the budget, allowing much more to be accomplished with the same investment than if corrosion programs are managed in isolation.

The DoD Corrosion Exchange Website (www.dodcorrsionexchange.org) is designed to be a one-stop shop for communication with other organizations in government, industry, and academia. As the knowledge about this website grows, its usefulness will continue to grow. Information on the website includes basic information on DoD policy, technical definitions and explanations of corrosion, training documentation, guidance on designing corrosion-resistant systems, spreadsheets for calculating the cost of corrosion, and much more. The website provides forums for communication of issues and an exchange of questions and answers.

Organizations should look outside themselves to improve their corrosion programs. Partnerships among government, industry, and academia members serve to communicate successful processes and products, as well as provide a means for technology insertion. To learn what organizations are doing and as a means to identify potential partners, numerous corrosion conferences as well as the Corrosion Exchange Website are available as means to make contact with other organizations. Best practices should be shared with others in the field.

Government agencies are limited in their budget for corrosion studies, control, and management. Where practical, government agencies should consider pooling their resources, including dollars, manpower, and other assets, to fund joint efforts, especially on common aerospace systems. These efforts provide direct feedback to the funding organizations and multiple organizations immediately benefit.

13.9 Sources of further information and advice

Other sources of information regarding business strategies in corrosion management are found in this section.

1. www.dodcorrosionexchange.org lists many sources of information
2. NACE International – The Corrosion Society. NACE is a professional technical society that offers technical training and certification programs, sponsors conferences, and produces industry standards and reports, publications, and software.
3. The Society for Protective Coatings (SSPC; formerly the Steel Structures Painting Council) is a professional technical society/trade organization that offers technical training and certification programs, sponsors conferences, and produces industry standards and reports, publications, and software.
4. The Advanced Materials, Manufacturing and Testing Information Analysis Center (AMMTIAC), which was established in 1996 and receives management and technical oversight from OSD(DDR&E), is sponsored by the Defense Technical Information Center (DTIC). AMMTIAC (formerly AMPTIAC) provides a wide range of corrosion-related functions, including inquiry services, newsletter, data gathering and analysis, and product development (state-of-the-art reviews, technology assessments, and databases).

13.10 References

1 DOD CORROSION EXCHANGE WEBSITE, www.dodcorrosionexchange.org.
2 WYNNE MICHAEL W (March–April 2004), 'Corrosion Prevention and Control: Status and Update', *Defense AT&L*, 1 Mar 2004.

3 CORROSION POLICY AND OVERSIGHT OFFICE (1 June 2007), *Corrosion Prevention and Mitigation Strategic Plan*, US DoD.
4 FEDERAL HIGHWAY ADMINISTRATION Research Report FHWA-RD-01-156, *Corrosion Cost and Preventitive strategies in the United States*, September 2001.
5 G. COOKE *et al.*, 'A Study to Determine the Annual Direct Cost of Corrosion Maintenance for Weapon Systems on Equipment in the United States Air Force,' Final Report, CDRL No A001, February 1998.
6 US GOVERNMENT ACCOUNTING OFFICE, 'Defense Management: Opportunities to Reduce Corrosion Costs and Increase Readiness', July 2003, Report GAO-03-753.
7 MIL-STD-1530C, Department of Defense Standard Practice Aircraft Structural Integrity Program, 1 November 2005.

Index

'A' rating, 204
AA7178-T6
  results from foil penetration studies, 137
AC-43-205, 235
acoustic emission
  crack and corrosion detection, 155–6
  definition, 153
  due to corrosion growth detected by R15I sensor, 161
  instrumentation, 154–5
  origin, 154–5
    macroscopic, 154–5
    microscopic, 155
  possible sources on a metal structure during environmental exposure and stress, 156
  schematic testing and basic instrumentation, 154
Acoustic Emission Working Group, 153
acquisition, 289–90
active corrosion detection
  on a magnesium gearbox housing, 161–2
    acoustic emission due to corrosion growth detected by R15I sensor, 161–2
  on storage tanks floors, 162
additives, 273–4 *see also* corrosion-inhibiting additives
  molybdenum disulfide, 273
  polytetrafluoroethylene, 273
Advanced Materials, Manufacturing, and Testing Information Analysis Center, 118
Aero Almen strip, 242, 243
Aero Commander 680, 41
aerospace industry
  corrosion, 1–13
    cost assessment, 17–33
  novel corrosion schemes, 248–63
AFGROW, 5, 133, 145
aging aircraft, 37, 38
Aging Orbiter Working Group, 210, 212
Aging Vehicle Assessment program
  major products, 211
aircraft *see also* specific aircraft
  corroded skin, 3
  corroded structure, 4
aircraft coating
  components, 226
  conversion coating, 252
  description, 226
  primer, 252
  properties, 226
  surface
    aluminum alloy, 227
    carbon-fiber composites, 227
    fiberglass and aramid fiber in epoxy matrix, 227
  time of removal, 228
  top coating, 252

aircraft coating removal techniques, 225–247
chemical strippers, 231–5
and composite materials, 234–5
mechanism, 232
non-dichloromethane, 232–3
pH of chemicals and its effect on aerospace materials, 233
dry stripping, 237
media types, 238–42
engineered bio-based media, 244–6
future trends, 246–7
high-pressure water, 236–7
history, 228–9
chemical strippers, 229
dichloromethane, 228–9
impact methods, mechanical effects, 242–4
methods, 229–31
impact, 229
molecular disassociation, 229
thermal, 229
thermal removal method, 235–6
Aircraft Structural Integrity Program, 50, 293–4
aircraft structure
corrosion pitting and fatigue modeling, 175–80
damage tolerance approach, 178–9
full-scale fatigue testing and corrosion finding, 176–7
fundamentals of modeling, 175–6
holistic structural integrity process, 174–5, 179–80
life determination, 173
safe-life and durability analysis, 173, 177–8
exfoliation corrosion and modeling, 180–3
fuselage, 120
residual fatigue life analysis, 184
risk assessment, 184–9, 187–9
corrosion maintenance actions, 187–9
methods and tools, 186–7
wing, 120
7075-T651aluminum extrusions, 42
aliphatic hydrocarbon, 249
alloys *see also* specific alloys
corrosion performance in Space Shuttle launch environment, 217–18
austenitic nickel-base super-alloy, 218
duplex, 218
low-carbon, 218
molybdenum-containing austenitic, 218
nickel-molybdenum-chromium-iron-tungsten, 218
nickel–chromium–molybdenum, 218
super austenitic, 218
superferritic stainless steel, 218
Almen strips, 242
residual stress levels, 243
Aloha Airlines flight 243, 2, 37, 125
aluminum alloy, 41, 227 *see also* specific alloy
corrosion behavior, 68–70
heat-treatable, 68–70
hydrogen embrittlement, 70
effect of prior corrosion, 70–1, 76–7, 79, 80–8, 94–7
damage tolerance behavior, 86–8, 94–7
fatigue behavior, 80, 82–6
tensile behavior, 71, 76–7, 79–80
aluminum-lithium alloys, 70
AL6XN tubing alloy, 218
American Society for Nondestructive Testing, 111
American Society for Testing and Materials, 45, 276
Amlguard, 249
anodization, 134
'anticipate and manage' approach, 172
APC 850 PZT, 164–5
ASTM 3359, 226
ASTM D-3363, 226
ASTM E647, 181

back-scattered electron (BSE)-SEM, 206
image of Enterprise Koropon showing chromium distribution, 208
image of the surface of Koropon sample, 206
base oil, 267–72
comparison, 268–9
multiply alkylated cyclopentane, 271–2
operating temperature ranges, 270
perfluoropolyethers and silicones, 271
petroleums, 267
polyalphaolefins, 270
polyglycols, 270–1
polyphenyl ethers, 271
synthetic esters, 270
beachside atmospheric exposure test site, 214
Boeing, 6
airline annual cost of corrosion, 7
fasteners from upper wing skin
exfoliation damage and grinding marks from repairs, 120
metallographic section around fastener hole
multiple layers of intergranular attack from pits in countersink, 120
section of wing plank
exfoliation damage and grinding marks from repair, 121
Bombardier Aerospace, 179–80
borescope, 113, 209
bottom-up analysis, 25
possible flaws, 29
Bragg grating *see also* fiber Bragg grating
demodulation, novel fiberoptic, 153
strain sensor, 159–60
Britain's de Havilland Comet, 47–8
burst emission, 153
business strategies
corrosion control in fleet maintenance, 288–97
acquisition of requirements, 289–90
corrosion tracking, 293–4
costs tracking, 295–6
prevention and control, 292–3
sustainment requirements, 290–1
training, 291–2

C-130 Hercules, 60
CF-18 wing test, full-scale, 176, 179–80
crack initiation program, CI89, 177
fatigue crack on right upper outboard longeron, 177
CH-46 helicopters, 6
chemical stripping
and composite materials, 234–5
mechanism
penetration, 232
reaction and molecular disassociation, 232
vapor pressure and coating lifting (tenting), 232
pH of, and its effect on aerospace materials, 233–4
chromate conversion coating, 134
coating *see* aircraft coating
combined top-down/bottom-up analysis
implications and advantages, 30
possible flaws, 29
communications, 296–7
Condition Based Maintenance, 6
contact stress, 46
continuous emission, 153
copper strip corrosion color reference, 280
corn hybrid starch polymer, 245
corrective cost, 21–2
and corrective cost curves, 21–2
and preventive cost curves, 22
corrosion, 7–13
3-phase corrosion model, 9
in aerospace industry, 1–13
aircraft availability, 4, 18
aircraft structure pitting and fatigue modeling, 175–80

damage tolerance analysis, 178–9
full-scale fatigue testing and corrosion finding, 176–7
fundamentals of modeling, 175–6
holistic structural integrity process analysis, 179–80
safe-life and durability analysis, 177–8
airline annual cost, 7
behavior of aluminum alloys, 68–70
carbon steel calibrating specimens at various locations, 196
civilian transport aircraft, 6–7
corrosion prevention compounds, 249–52
CPC regions providing protection, 250
fiber optical sensor signal vs time, 252
film former, 249
interfacial impedance of CPC-treated samples, 251
organic solvent, 249
corrosion rate and salt content of atmosphere as function of distance from Atlantic Ocean, 196
cost, 4
design paradigms and structural effects, 47–51
damage tolerance, 51–5
fail-safe design, 50–1
holistic structural integrity, 55–61
safe life, 47–50
economic impact, 36–7
effect on load-carrying capability, 3–4
crack formation, 3–4
cross-sectional area of structure reduction, 3
repair of corroded structure, 4
effect on mechanical behavior of aluminum alloys, 67–105
exfoliation corrosion and modeling, 180–2
exfoliation and fatigue tests, 180–2
modeling, 182
forms
crevice corrosion, 145
general thinning, 145
intergranular/exfoliation corrosion, 145
insidious synergisms, 43–7
corrosion fatigue, 43–4
fretting fatigue, 46
role of surface integrity, 46–7
stress cracking, 44–6
inspection and repair
Dessault Falcon jet, 2
integrated health and monitoring system, 151–69
acoustic emission, 153–6
active corrosion detection, 161–2
strain monitoring, 159–60
ultrasonic guided waves (Lamb waves), 156–9
Lamb wave signal change due to corrosion, 169
living with, 35–6
maintenance action
corrosion prevention compound application, 188
grindout and corrosion prevention compound application, 188
no repair, 188
replacement, 188
mechanisms in Orbiter, 219
military aircraft, 4–6
modeling of effect on damage tolerance characteristics, 97–104
basic considerations of the LTSM-F model, 104
fatigue crack growth prediction, 101–4
fracture-mechanical model for residual strength prediction, 97–101
location-dependent degree of embrittlement of crack tip, 99
material element at crack tip under mode I monotonic loading, 99
typical aircraft stress spectrum, 102

monitoring, 168–9
nondestructive testing (*see* nondestructive testing)
novel inhibitors and coatings, 252–9
AA7075 corrosion rate in NaCl, 253
cathodic current density reduction, 255
cerium and cobalt ion concentration, 254
critical inhibitor concentration comparison, 253
electrochemical impedance spectra for scribed coatings, 258
oxygen reduction reaction, 256
novel schemes for aerospace industry, 248–63
anodic polarization, 262
ion implantation process, 261
other forms of attack, 42–3
parent group, 38
risk assessment, 184–9
corrosion maintenance actions, 187–9
methods and tools, 186–7
safety, 4, 18, 37–8
and structural integrity, 3
structurally significant and unique forms, 39–42
general attack, 39
intergranular and exfoliation corrosion, 41–2
pitting corrosion, 39–41
survival analysis, 8–13
censored corrosion data, 9–10
censored corrosion maintenance data, 11–12
Kaplan-Meier probability plots, 12
probability plots, 12–13
right-censored data, 10
survival function, 10–11
US Coast Guard HH-65 corrosion mapping program, instances of corrosion, 11
US Coast Guard HH-65 depot corrosion mapping program, man hours, 12
survival analysis application, 9
threat to aircraft structural integrity, 35–63
traditional modeling, 7–8
types and their impact on aircraft structure, 38
typical activities, 20, 34
variation of Orbiter corrosion environment during typical flow, 198
corrosion control
business strategies in fleet maintenance, 288–97
acquisition requirements, 289–90
communications, 296–7
corrosion tracking, 293–4
costs tracking, 295–6
prevention and control, 292–3
sustainment requirements, 290–1
training, 291–2
Corrosion Control Review Board, 210–11
corrosion cost, 18
assessment, 17–33
characterization, 21–3
classification into preventive or corrective natures, 22
corrective and preventive costs, 21–2
preventive and corrective cost curves, 22
structure and parts cost, 22
corroded helicopter, 19
data structure and methods of analysis, 31
database and data structure, 30
definition, 18–19
elements
definition, 33–4
identification, 19–20
prioritization, 20
general case studies, 31–3
Government Accountability Office recommendations, 33
Navy ships work team, 32
new Navy ship example, 31

measurement methodology, 23–30, 24
bottom-up analysis, 25
combined top-down/bottom-up analysis, 25–7
consolidation of budget categories, 24
final calculation, 27
household budget example using top-down analysis, 24
initial calculation, 26
potential spending expansion, 25
top-down analysis, 24–5
use of information, 20–1
corrosion costs
general types, 295–6
depot vs field maintenance, 296
preventive vs corrective maintenance, 296
structure-related vs parts-related, 296
Corrosion Effects on Structural Integrity (CESI) program, 135–40, 142–5
analysis of real aircraft inspection and repair data, 135, 139–40, 142–5
Battelle exfoliation depth data with lognormal distributions, 145
combined mild–severe distribution, 142
cumulative probability fits with 95% confidence, 141
exfoliation damage on KC-135 exterior skin, 143
grindout vs Ramey distribution data, 144
KC-135 grindout depth data (mild-moderate), 141
lateral corrosion data vs lateral grindout data, 144
laboratory accelerated testing, 135, 136–8
controlled humidity method, 136
foil penetration method, 136
results from foil penetration studies, 137
outdoor exposures, 135, 138–9, 142–5
exfoliation depth data from coupons exposed at various locations for one year, 139
wing coupon exposed to seacoast environment, 138
corrosion fatigue, 43–4
corrosion induced fatigue via load transfer, 44
corrosion-nucleated fatigue, 43
prior-corrosion fatigue, 44
corrosion growth rate, 185
corrosion-inhibiting additives, 274
basic methodologies
acid neutralization, 274
protective film formation, 274
structure representation, 275
corrosion prediction modeling, 131–49
alternative approaches, 132–3
data generation and correlation, 135–40, 142–5
laboratory evaluations, 136–8
outdoor exposure experiments, 138–9
real aircraft data, 139–40, 142–5
of growth, for aircraft maintenance planning, 133
material and environmental influences, 133–5
alloy/temper, 133–5
local vs global environment, 135
structural configuration and manufacturing processes, 134
surface finishes and coatings, 134–5
model development and implementation, 145–8
assumptions, 146–8
future trends, 148–9
Corrosion Prediction Module (CPM), 145, 148
Corrosion Prevention and Advisory Team, 292
Corrosion Prevention and Mitigation Strategic Plan, 291

corrosion prevention compound, 186
  effect on single flight probability failure, 189
cost *see* corrective cost; corrosion cost; preventive cost
cost tree, 24
  United States Department of Defense corrosion study, 28
crack detection
  acoustic emission, 155–6
  ultrasonic guided waves, 164–8
crack growth lives, 43
'crack initiation' program, 177
CREP test equipment, 283
crevice corrosion, 43
cube root power law, 180
Cyantek Al-12S, 168

2D finite element analysis, 184
3D finite element analysis, 184
damage tolerance, 51–5, 53–5, 173, 178–9
  arguments for augmenting of, 53–5
  assessment of corrosion fatigue influence, 52
  definition, 51
  fundamentals of concept, 51–2
  potential pitfalls regarding corrosion, 52–5
data fusion, 115–17
  definition, 116
  general categories of level, 116
Department of Defense (DoD), 4
Dessault Falcon jet
  corrosion inspection and repair, 2
dichloromethane, 228–9
digital speckle correlation, 113
distributed feedback laser, 162, 163, 164
DoD Corrosion Exchange Web site, 296
drain system, 209
dry stripping media types, 238–42
  agricultural media, 239–40
  bicarbonate of soda, 238–9
  plastic media, 240–2
    various types listed in MIL-P-85891A, 241
  sponge media, 238
durability analyses *see* reliability assessments
duralumin, 49

eddy current nondestructive testing, 114, 226, 243
  giant magnetoresistive sensors, 114
  multi-channel portable instruments, 114
8xxx series alloys *see* aluminum-lithium alloys
electrochemical impedance spectroscopy, 250–1
electromagnetic testing *see* eddy current nondestructive testing
electronic speckle pattern interference, 113
'Emcor' test, 280–1
  corrosion value ratings, 281
engineered bio-based media, 231, 244–6
  advantages, 245
  corn hybrid starch polymer, 245
  disadvantages, 245–6
  wheat starch media, 244
enhanced visual/optical nondestructive testing, 113
  digital speckle correlation, 113
  electronic speckle pattern interference, 113
  holography, 113
  laser interferometry, 113
  triangulation-based methods, 113
Enterprise *see* Space Shuttle Orbiter
environmental corrosion sensors, 294
environmental severity index (ESI), 140
EXCO test, 45
exfoliation corrosion, 42
  description, 180
  and modeling, 180–2
    boundary conditions for finite element analysis, 183

three-dimensional finite element mesh with soft inclusion, 183
residual fatigue life analysis, 184
probabilistic modeling based on grindout database, 185
results from fatigue tests carried on coupons with various levels of exfoliation, 181
Expanded Ships Work Breakdown, 32

F-4 Phantom II, 50
factor-of-safety, 3
fail-safe fatigue design, 50–1
fatigue behavior
effect of prior corrosion, 80
coexistence of pitting and intergranular corrosion, 83
crack growth curves and fatigue crack growth rates at different stress ratios, 93
crack growth curves and fatigue crack growth rates at $R = 0.1$, 90
crack growth curves and fatigue crack growth rates at R = 0.01, 89
crack growth curves and fatigue crack growth rates at R = 0.5, 91
crack growth curves and fatigue crack growth rates at R = 0.7, 92
crack growth data of corroded and uncorroded specimens, 88
critical crack length, 87
endurance limit after exfoliation corrosion solution for 36h, 85
performed fatigue crack growth tests, 87
S–N curves for 2024 T351 aluminum alloy (as-received specimen), 84
tests performed on 2024 T351 to obtain S–N curves, 83
fatigue crack growth prediction
calculation, 102–4
stress spectrum simulation, 101–2
fatigue crack nucleation, 42
fatigue failure, 46
Federal Aviation Administration, 37, 50, 51, 55, 118, 235
fiber Bragg grating, 160
ambient temperature signals monitors as function of time over 17 h, 163
detected Lamb wave signals, 165
integrated strain, temperature. and stress wave monitoring sensor, 162–4
locked-in based, interrogation prototype system, 166
and PZTs bonded on aluminum plate surface, 166
sensor response to Lamb waves, 167
fiber optic interferometric sensors, 160
fiber optic sensor, 160
filiform corrosion, 43
'find and fix' approach, 112, 172, 173
corrosion maintenance, 54, 57
flash thermography techniques, 115
focused ion beam/scanning electron microscopy (FIB/SEM), 206
image of region of control Koropon surface, 207
fractographic analyses, 94
fracture toughness tests, 95–7
results and *R*-curve behavior of corroded and uncorroded bare panels, 96
fretting fatigue, 46

galvanic corrosion, 43
general corrosion attack
net-section material loss, 39
giant magnetoresistive sensors, 114
glossy polyurethane, 134
Government Accountability Office, 5
recommendations for corrosion program, 33
grease
in aerospace industry corrosion control, 266–85
corrosion-inhibiting additives, 274
future trends, 284–5
grease composition, 267–74

lubricant selection and design cycle, 274–6
lubricant testing, 276– 82
manufacturing, 282–4
grease manufacturing, 282–4
grease manufacturing kettle, 284
ground support equipment
corrosion control, 212–18
coating evaluation and development at Kennedy Space Center, 214–17
corrosion performance of alloys in Space Shuttle launch environment, 217–18
inorganic zinc coatings, 213–14
NASA beachside atmospheric exposure test site, 214
materials selection, 200–3

Hall effect sensor, 122
HCl/alumina slurry spray, 218
heat-treatable aluminum alloys
corrosion behavior, 68–70
2xxx series alloys, 68–9
6xxx series alloys, 69
7xxx series alloys, 69–70
hexavalent chromium, 252
high-pressure water, 230, 236–7
advantages, 236–7
disadvantages, 237
hit-based (qualitative) analysis, 153
holistic structural integrity, 55–61
augmenting management of structure, 57–8
benefits, 57
elements, 59–61
criticality assessment, 59–60
data management and fusion, 59
life and residual strength analyses, 59, 60, 61
nondestructive inspection, 59
process documentation, 59
usage and environment characterization, 59, 60–1
notional trends of residual life vs discontinuity size, 58
primary elements of holistic structural analysis including effects of corrosion, 62
spectrum flight hours of two identical part geometries with identical amounts of corrosion, 61
Holistic Structural Integrity Process (HOLSIP), 173, 174–5, 179–80
accounting for the progression of discontinuity states, 175
corrosion fatigue analysis procedure, 176
corrosion pits and cracking scenario, 179
physics-based models, 174
holography, 113

impact coating removal technique
agricultural by-products, 230
engineered bio-based media, 231
high-pressure water, 230
plastic media, 231
sodium bicarbonate, 230
sponge media, 230–1
Individual Aircraft Tracking Program, 294
infrared testing *see* thermographic nondestructive testing
initial discontinuity states, 174–5, 179–80
as-manufactured, 174–5
as-produced, 174–5
inorganic zinc coatings, 213–14

KC-135, 5, 52
exfoliation damage on exterior skin, 143
grindout depth data (mild-moderate), 141
Kennedy Space Center, 195
beachside atmospheric exposure test site, 214
coating evaluation and development, 214–17
ceramic-filled epoxy, 216

high-gloss polysiloxane, 216
inorganic zinc-rich primers, 215–16
polyanilines, 216
sprayable silicone ablative coatings, 216
volatile organic compound, 216
corrosion control, 203–19
and treatment program, 218–19
Kevlar, 227

LabView program, 164
Lamb waves *see* ultrasonic guided waves
lap joint simulated solution, 251
laser and heat lamps, 235–6
advantages, 235–6
disadvantages, 236
laser surface melting, 259–60
launch pad
corrosion control, 212–18
coating evaluation and development at Kennedy Space Center, 214–17
corrosion performance of alloys in Space Shuttle launch environment, 217–18
inorganic zinc coatings, 213–14
NASA beachside atmospheric exposure test site, 214
materials selection, 200–3
Liquid Penetrant Inspections, 243
Los Gatos Research, 152, 164, 166
lubricant
selection and design cycle, 274–6
load and wear, 275–6
material compatibility, 275
operating temperature, 275
lubricant testing, 276–82
analytical and application-specific test methods
Fourier Transform Infrared, 282
pressure differential scanning calorimeter, 282
thermogravimetric analyzer, 282
corrosion-specific test methods, 279–82
copper corrosion, 280
corrosion preventive properties, 281–2
corrosion rate evaluation procedure, 282
'Emcor' test, 280–1
general industry standard test methods, 276–9
dropping point, 279
evaporation, 279
grease stiffness, 278
oil separation, 278–9
penetration, 277–8
grease properties, 277

magneto-optic imager, 115
magnetoresistive sensor, 122
Materials International Space Station Experiment, 199
methyl ethyl ketone, 241
microelectrical–mechanical systems, 159
microwave nondestructive testing, 115
MIL-P-85891A
various plastic media types list, 241
Mini 30 sensor, 161
Moh's scale, 230
moisture entrapment, 293
moisture intrusion, 293
MSFC-SPEC-250, 204, 205
class I, 204
class II, 204
class III, 204
MSFC-SPEC-522, 204
multi-layer thin structures, 120–3
corrosion pitting, 123
lap joint pillowing, 122–3
multiple-element damage, 51–2, 54, 55
multiple site damage, 51–2, 54, 55, 86
multiply alkylated cyclopentane, 271–2

NASA Corrosion Technology Laboratory, 195–6
NASA STD-5008 QPL, 213
NASA's Launch Operation Center, 195
see also Kennedy Space Center

National Association of Corrosion Engineers, 292
National Defense Authorization Act of Fiscal Year 2006, 5
National Lubrication Grease Institute, 277–8
National Transportation Safety Board, 37
Navy ship, 31–2
  highest contributors to corrosion cost by ESWBS, 32
  work team, 32
nickel–chromium–molybdenum–tungsten substitute alloy, 217
nondestructive testing, 35, 111–28
  classification, 112
    additional testing methods, 115
    eddy current, 114
    enhanced visual/optical, 113
    radiographic, 115
    thermographic, 114–5
    ultrasonic, 114
    visual, 113
  data fusion, 115–17
  effect on single flight probability failure, 190
  estimates of performance, 122
  reliability for corrosion, 117–19
  requirements, 112–13
  summary of techniques for aging aircraft, 127
  typical applications to corrosion in aerospace systems, 119–26
    coatings, 125–6
    engine components, 125
    muti-layer thin structures, 120–3
    non-structural systems, 125
    thick section structure, 123–5

Orbiter Enterprise, 205–6
  BSE-SEM image of the surface of Koropon sample, 206
Original Equipment Manufacturers, 244

PAC ARB-1410-150 pulser system, 164, 166
PAC S-9208 PZT actuator, 164
Paris law, 94, 103
PATRAN Command Language, 182
PCI-6111 National Instrument data acquisition board, 164
perfluoropolyethers and silicones, 271
petroleums, 267
Physical Acoustic Corporation (PAC), 152, 161
piezoelectric transducer (PZT), 153
  and FBG bonded on aluminum plate surface, 166
  response to Lamb waves after crack simulating damage, 168
pitting corrosion, 39–41
  and fatigue modeling, 175–80
plastic media blasting, 231, 242
Poisson effect, 160
polyalphaolefins, 270
polyanilines, 216
polyglycols, 270–1
polyphenyl ethers, 271
preventive cost, 21–2
primer coating, 134
probability of detection, 54, 117–18
  as function of both area and thickness of damage, 119
  as function of single metric of damage, 117
  nondestructive testing error on the level of damage and subsequent actions, 119
probability of failure, single flight, 184, 188
  effect of corrosion prevention compounds, 187
  effect of corrosion severity, 187
  effect of NDI uncertainty, 190
ProDTA (Probabilistic Damage Tolerance Analysis), 184
  analysis procedures, 186
  major elements of NRC in-house program, 185
PROF, 187, 188

programmed depot maintenance (PDM), 143
pulsed eddy current nondestructive inspection, 181
pulsed-eddy current (P-ET) technique, 122
purge system, 209

radiographic nondestructive testing, 115
  digital radiography, 115
Regulation, Evaluation, Authorization, and restriction of CHemicals (REACH), 246
reliability assessments, 53–4
Reliability-Centered-Maintenance program, 290–1
R15I sensor, 161
  acoustic emission due to corrosion growth detection, 161

safe-life fatigue design, 47–50
  analysis, 48
safe-life paradigm, 173, 177–8
scanning electron microscopy (SEM), 206
scuff sanding technique, 228
search shot peening, 180
self-assembled nanophase particle, 258–9
SEM-x-ray energy dispersive spectrometry, 206
  chromium region of control Koropon, 207
6013 T6
  chemical composition, 76
  tensile tests performed, 77
  thermal desorption spectrum, 79
6xxx series alloys
  heat-treatable aluminum alloys, 69
shot peening process, 242
7178-T6
  wing coupon exposed to seacoast environment, 138
7xxx series alloy, 41, 45
  heat-treatable aluminum alloys, 69–70
    copper-containing, high-strength alloys, 69
    low-copper alloys, 69
'soft inclusion' technique, 182
  boundary conditions for finite element analysis, 183
  3D finite element mesh with, 183
sol-gel coatings, 256–7
space launch vehicles
  corrosion control, 195–221
    future trends, 220–1
  environment, 196, 198–200
    corrosivity at Kennedy Space Center, 196, 198
    corrosivity at space shuttle launch pads, 198
    orbiter flight and ground environment, 198–200
Space Shuttle Orbiter, 200
  corrosion control, 204–12
    actions to consider for reduction of probability of stress corrosion, 204–5
    maintenance, 209–12
    water intrusion/entrapment design features, 209
  description, 200
  history of corrosion, 196, 219
  launch pad and ground support equipment
    fixed service structure, 201
    flame trench and deflector system, 202
    liquid oxygen and liquid hydrogen storage, 202
    major features of pads 39 A and 39 B, 201
    rotating service structure, 201–2
    shuttle era pad modifications, 202–3
  materials selection
    launch pad and ground support equipment, 200–3

Standard Design Manual, 205
strain monitoring
  fiber optic sensor, 160
    Bragg grating strain sensor, 159–60
    Fabry Perot, 160
    interferometric, 160
    microbend, 160
    polarimetric, 160
  microelectrical–mechanical systems, 159
stress corrosion cracking, 44–6, 124, 204
  corrosive environment, 44, 45
  susceptible material, 44, 45
  sustained tensile strength, 44, 45
Structural Damage Management Tool (SDMT), 145, 146
  estimated intergranular/exfoliation growth values, 147
Super Koropon, 205
surface finished and coatings, 134–5
sustainment, 290–1
synthetic esters, 270

TANKPAC, 162
tensile behavior
  effect of hydrogen, 79–80
    after thermal treatment for removal of trapped hydrogen, 81
    energy density reduction, 82
    thermal desorption spectra, 79
  effect of prior corrosion, 71, 76–7
    after mechanical removal of corroded specimen layers, 78
    degradation in EXCO environment for 2024 and 6013 aluminum alloys, 74
    degradation in EXCO environment for 8090 and 2091 aluminum alloys, 75
    degradation of 2024 and 6013 aluminum alloys, 72
    degradation of 8090 and 2091 aluminum alloys, 73
    gradual degradation with increasing exposure time in EXCO environment, 76
    tests performed on Al 2024 T351 aluminum alloy, 77
    tests performed on Al 6013 T6 aluminum alloy, 77
terahertz imaging, 115
thermographic nondestructive testing, 114–5
  thermal imaging cameras, 114–5
thermography nondestructive testing, 181
thick section structure, 123–5
  intergranular and exfoliation attack, 124–5
    corrosion pitting, 124
    depth location of intergranular corrosion, 125
    extent of intergranular corrosion, 125
thickening systems, 272–3
  comparison, 272
  inorganic, 273
  organic, 272–3
3-phase corrosion model, 9
time division multiplexing, 160
time-to-event phenomena, 10–11
TM-584C Revision C, 218–19
top-down analysis, 24–5
  possible flaws, 29
training, 291–2
2024-T3, 44, 133, 242
2024 T351 *see also* aluminum alloys; fatigue behavior
  chemical composition, 76
  embrittled fractured zones, 95
  fatigue tests performed to obtain S–N curves, 83
  fracture toughness test, 95
  pre-corroded, damage tolerance behavior, 86–8, 94–7
  tensile tests performed, 77
  thermal desorption spectrum, 79
2xxx series alloys
  heat-treatable aluminum alloys, 68–9

ultrasonic guided waves (Lamb waves), 156–9
  basic pulse echo and pitch catch techniques, 158
  crack detection, 164–8
  generation, 157
    active probing, 157
    cracked-induced acoustic emission in thin plates, 157
  sensors, 159
  signal change due to corrosion, 158, 169
ultrasonic nondestructive testing, 114, 181
upper outboard longeron, 176–7
  life predictions using various methods, 178
  safe-life analysis, 177–8
US Coast Guard HH-65 corrosion mapping program
  instances of corrosion, 11
  man hours to repair corrosion, 12
US Federal Aviation Administration, 7

vanadates, 255–6
vanadium-based oxyanions *see* vanadates
visual nondestructive testing, 113
water intrusion/entrapment design, 209
  drain system, 209
  purge system, 209
waveform-based (quantitative) analysis, 153
wavelength division multiplexing, 160
wheat starch media, 244
widespread fatigue damage, 51–2, 53, 54, 55

x-ray nondestructive inspection, 181

zinc-rich primer, inorganic, 215–16